Foundations of Regenerative Biology and Medicine

Foundations of Regenerative Biology and Medicine

David L Stocum

Indiana University-Purdue University, Indianapolis, USA

IOP Publishing, Bristol, UK

ISBN   978-0-7503-1626-2 (ebook)
ISBN   978-0-7503-1624-8 (print)
ISBN   978-0-7503-1625-5 (mobi)

DOI   10.1088/978-0-7503-1626-2

Version: 20181001

IOP Expanding Physics
ISSN 2053-2563 (online)
ISSN 2054-7315 (print)

British Library Cataloguing-in-Publication Data: A catalogue record for this book is available from the British Library.

Published by IOP Publishing, wholly owned by The Institute of Physics, London

IOP Publishing, Temple Circus, Temple Way, Bristol, BS1 6HG, UK

US Office: IOP Publishing, Inc., 190 North Independence Mall West, Suite 601, Philadelphia, PA 19106, USA

Contents

Part II Regenerative medicine

Preface

The purpose of this book is to give advanced undergraduate and beginning graduate students, as well as individuals in areas ancillary to biomedical science, citizen scientists, and lay persons an introduction to the field of regenerative biology and medicine that provides a fundamental framework for organizing and building a panoramic view of this science.

The book is organized into ten chapters divided into two parts. Part I, 'Regenerative biology' (chapters 1–5), begins with a broad overview of regenerative biology, followed by wound repair by fibrosis, stem cell and non-stem cell mechanisms of tissue regeneration, and vertebrate appendage regeneration. Part II, 'Regenerative medicine' (chapters 6–10), begins with an overview of the strategies of regenerative medicine, followed by specific research and potential clinical applications of these strategies: the pharmaceutical induction of regeneration, stem cell transplantation, and biomimetic tissue and organ constructs. The final chapter of the book looks to the future directions of regenerative medicine, and includes a section on the bioethics of research approaches and clinical applications.

The writing for this book ends with the literature of 2017, with a few references to papers published in early 2018. Where we stand today, and our vision of the future, is the result of gains in our understanding of the natural world over many centuries, slowly at first, then becoming exponential in the 20th and 21st centuries CE. The vast numbers of papers published monthly on regenerative biology and medicine means that I have been able to sample only a small fraction of the published literature. Two previous volumes that provide more detailed accounts are *Regenerative Biology and Medicine*, 2nd edition (Stocum, 2012, Elsevier/Academic Press) and *Principles of Regenerative Medicine*, 3rd edition (2018, Elsevier/Academic Press) edited by Atala, Lanza, Thomson, and Nerem. A more comprehensive review of the basic biology of regeneration may be found in Carlson's *Principles of Regenerative Biology* (2007, Elsevier/Academic Press). Over the next decade, we can expect to see many advances in our understanding of regenerative biology and medicine, the evolution of which will be fascinating to watch.

I thank the following colleagues who provided comments and critiques of earlier versions of the manuscript: Dr Jo Ann Cameron of the University of Illinois at Urbana-Champaign; Dr Michael Levin of Tufts University, Boston; and Dr Ashley Seifert of the University of Kentucky, Lexington. I am especially grateful to Dr Matt Allen, Indiana University School of Medicine at IUPUI, for allowing me to use his histological images of a number of tissues that are prime targets for the development of regenerative therapies.

Last, but not least, I thank my editors, Jessica Fricchione and Daniel Heatley, and my production editor Chris Benson, for their support during the publication process.

David L Stocum
Indiana University-Purdue University Indianapolis
Indianapolis, IN
July, 2018

Author biography

David L Stocum

David L Stocum is Emeritus Professor of Biology and Emeritus Dean of the School of Science at Indiana University-Purdue University Indianapolis. He holds an undergraduate degree in Biology and Psychology from Susquehanna University, a MS in Zoology from the University of Illinois at Urbana-Champaign and a PhD in Cell and Developmental Biology from the University of Pennsylvania. He pursued an active research and teaching career in the School of Life Sciences at the University of Illinois Urbana-Champaign for 21 years, and also served administrative terms as Director of the Honors Biology Program and Acting Head of Anatomical Sciences. Subsequently he served as Dean of the School of Science at Indiana University-Purdue University Indianapolis for 15 years, and then another ten years as Professor in the Department of Biology, during which he also held an appointment as Affiliate and Research Scholar in the Regeneration theme of the Institute for Genomic Biology at the University of Illinois at Urbana-Champaign. Dr Stocum is known for his research on the regeneration of amphibian limbs, and for his writings on the general subject of regenerative biology and medicine.

Part I

Regenerative biology

IOP Publishing

Foundations of Regenerative Biology and Medicine

David L Stocum

Chapter 1

Regeneration

Summary

Regenerative biology has its roots in studies carried out in the 18th century on hydra, planaria, crustaceans and amphibians, gathered momentum as a field in the late 19th and early 20th century, and has flourished ever since. Regeneration is a process essential for the persistence of life, and occurs on every level of biological organization. On the tissue level, cells with regenerative capability exist in niches that supply them with the appropriate signals to restore tissue lost to injury or disease. Depending on the tissue type, regeneration is achieved by one or more mechanisms: Cellular regrowth, as in transected peripheral nerve axons; lineage-specific regeneration from pre-existing differentiated cells; the transdifferentiation of one cell type into another cell type; and the differentiation of undifferentiated stem and progenitor cells into new tissue. A major feature of regeneration research has been to determine the origin of regenerated tissues and the signals that regulate the regenerative process. Several comparative models are useful in analyzing the differences in regenerative ability between regeneration-competent and regeneration-deficient species and tissues.

1.1 Introduction

The tissues of multicellular organisms, both plant and animal, are constantly threatened with breakdown and dysfunction caused by the stresses and strains of function, injuries inflicted by mechanical, thermal and chemical insults, and disruption by disease. Evolution has provided multicellular organisms with two mechanisms to maintain the structural and functional integrity of tissues and the organism. The first is regeneration, a homeostatic mechanism that maintains tissue architecture and function in the face of normal cell turnover, as well as reconstitution of the original tissue architecture and function after a loss of tissue due to injury or disease. The cells of many tissues, such as epithelium and blood, normally have a limited lifetime and so must be continually replaced to maintain tissue structural and functional integrity. Replacement due to cell turnover is termed *physiological* or

doi:10.1088/978-0-7503-1626-2ch1

maintenance regeneration. Injury that results in tissue loss induces a more intense response termed *injury-induced regeneration.* The second mechanism is scar formation, or fibrosis, which does not restore the original tissue architecture, but patches tissue gaps with fibrous connective tissue. Fibrosis is the default repair mechanism of tissues that are unable to regenerate, maintaining the local structural integrity of the tissue, but at the expense of reducing its functional capacity to a degree proportional to the extent of the scarring.

Modern medicine is able to replace some tissues and organs through autogeneic or allogenic transplants. Autogeneic transplants are limited to small tissue areas and volumes, while allogenic transplants are limited by donor availability, the need to take immunosuppressive drugs, and the specter of graft versus host disease, so we would prefer to enable tissue and organ restoration by regeneration. Achieving this goal, however, requires that we obtain a much deeper understanding of the mechanisms of regeneration and fibrosis than we currently possess. The mammalian, including human, body has limited powers of regeneration, but numerous other species of organisms have remarkable regenerative capacities. The study of regenerative mechanisms in the different tissues of regeneration-competent species, and how it differs from the fibrosis of regeneration-deficient species, constitutes the field of *regenerative biology.* The translation of knowledge gained from regenerative biology into therapies that promote regeneration in regenerative-deficient tissues, organs and appendages is the field of *regenerative medicine.* Regenerative biology and medicine is a 'convergent' science that draws on a wide variety of disciplines from biology, chemistry, physics, engineering, and computer and information science. This convergence has the potential to drive a revolutionary transformation of medicine.

1.2 Who and what regenerates?

Regeneration takes place to different degrees in all species, and on all levels of biological organization from the cellular to the whole body. All cells are continually adjusting their balance of protein synthesis and degradation in response to biochemical or mechanical load. Organelles are demolished and replaced. Cardiomyocytes, for example, replace most of their molecules over the course of two weeks. Under conditions of sustained hypertension, they elevate their rate of protein synthesis and become hypertrophied, which can lead to heart failure (Gevers 1984). Regeneration at the tissue level requires cells capable of division, an injury environment that provides stimulatory factors for proliferation and differentiation, and an absence or neutralization of factors inhibitory to regeneration.

On a phylogenetic level, one might think that the simpler the organism, the greater its ability to regenerate. Such a general correlation has not been found and every phylum has organisms that span the regenerative spectrum, although the ability to regenerate the whole body from a fragment is confined to worms and coelenterates such as planaria and hydra (Goss 1969, Carlson 2007). The free-living unicellular protozoans can regenerate complete cells after removal of large fragments of the cell, as long as nuclear material is present in the remaining part. Some vertebrates such as fish and amphibians can regenerate fins and limbs, while others cannot. Mammals of various sorts can regenerate the tips of their digits, antlers, and ear tissue (Stocum 2012, for a review). Nerve axons can regenerate after crush injury provided that the external guide tubes encasing their axons remain intact and in register at the site of injury (Yannas 2001, for a review). The best regenerators, however, are not animals, but plants, which regenerate leaves each year and can reconstitute whole plants from cuttings and in some cases, such as the carrot, even from single cells (Birnbaum and Sanchez-Alvarado 2008, Ikeuchi *et al* 2016, for reviews).

Regeneration is intimately related to reproduction. This relationship is direct in invertebrates such as planaria that undergo fission and subsequent regeneration as a means of asexual reproduction. Fission can be mimicked in planaria by cutting it into fragments, each of which can regenerate a whole worm in clonal fashion. In vertebrates, which reproduce sexually, the relationship is less direct, but just as important. Sexual reproduction requires that individuals live to reproductive age and, in the case of humans and some other large animals, well beyond. Regeneration clearly is essential for individuals to reach reproductive age, but regenerative powers decline thereafter and so, in accord with the second law of thermodynamics, individuals ultimately succumb to increasing entropy and die. Species as a whole, however, resist entropy by reproduction of individuals, as long as they remain adapted to an environment that can furnish them with an energy source. The major energy source for life on the surface of the earth is the sun, the light and heat energy of which is captured in the photosynthetic mechanisms of plants and some bacteria, and transferred up the food chain. Another energy source, however, is the internal heat of the planet. Deep in the oceans (and also reaching the surface in places such as Yellowstone), fluids welling up from hydrothermal vents support mats of chemosynthetic bacteria and archaebacteria, which constitute the base of a food chain that supports a diverse ecology of multicellular organisms. In fact, it has been proposed that life first evolved around these vents and expanded from there to the planet's surface (Wachtershauser 1990).

The relationship between reproduction, regeneration and individual and species survival has been summed up in two eloquent quotes by the great cell biologist Edmund B Wilson and the maven of regenerative biology Richard J Goss, writing in the first and second halves of the 20th century, respectively: '...life is a continuum, a never-ending stream of protoplasm in the form of cells, maintained by assimilation, growth and division. The individual is but a passing eddy in the flow who vanishes and leaves no trace, while the general stream of life goes forward' (Wilson 1925); 'If there were no regeneration, there could be no life. If everything regenerated, there would be no death. All organisms exist between these two extremes. Other things being equal, they tend toward the latter end of the spectrum, never quite achieving immortality because this would be incompatible with reproduction (Goss 1969).

1.3 A brief history of regenerative biology

1.3.1 Regeneration in mythology

The history of regenerative biology is as old as humanity itself, beginning with observations on how wounds heal, the cyclical regeneration of leaves and the production of seeds, and the regeneration of appendages in food animals such as the legs of crustaceans and the antlers of deer (Dinsmore 1991, 1996, 1998). Regeneration is a prominent theme in the ancient Greek myths of the Phoenix, the Hydra, and Prometheus. The Phoenix was a long-lived bird that died by fire and was regenerated from its ashes. The many-headed, snake-like monster Hydra could regenerate two heads for every one cut off. The Prometheus myth is a classic horror tale of regeneration. Prometheus was a Titan god who created humans out of clay. Having empathy and compassion for his creations, he stole fire from Zeus to keep them warm. As eternal punishment for this act, Zeus chained Prometheus to a rock where an eagle pecked out his liver by day, only to have it regenerate each night. Hercules killed the Hydra in the second of his twelve labors, and rescued Prometheus from his fate. Other myths from medieval times speak of the replacement of limbs lost in battle, for example the transplantation of the leg of a Moor to a Christian knight (Goss 1991).

1.3.2 The rise of modern biology

Modern biology has its roots in the intellectual awakening begun in the Renaissance of the 14th to 16th centuries, which focused on the study of the secular and worldly. DaVinci, Vesalius and others produced detailed descriptions of human anatomy. This intellectual momentum continued during the Enlightenment of the 17th and 18th centuries, which emphasized the acquisition of knowledge by reason and experiment, exemplified in the works of Kepler and Galileo in astronomy, Newton in physics and Descartes in philosophy.

Descartes introduced the philosophy of mechanism, which postulated that everything in the material universe, including organisms, had machine-like qualities that could be described according to mathematical and physical laws (Van Doren, 1991). This in turn gave rise to the materialistic biology of the 19th and 20th centuries (Allan 1977, Coleman 1977) that had by the mid-20th century resulted in five key ideas: (1) life depends on chemical reactions, as shown by Lavoisier in the 1700s; (2) the basic units of life that carry out those reactions are cells, as elaborated by Schleiden and Schwann, Remak, Virchow, Flemming, Strasburger, van Beneden, O Hertwig, and Kolliker in the 1800s; (3) the diversity of life is created by evolutionary change driven by natural selection of favorable traits, as argued by Darwin in his 1859 book, *The Origin of Species*; (4) physical traits are transmitted as particulate units (now known as genes), as deduced by Mendel in 1866 and Morgan from 1910–40 in his experiments on *Drosophila*; (5) embryogenesis is driven by nuclear-cytoplasmic and intercellular interactions as shown by the embryologists Weissman, Roux, Driesch, Boveri, and Spemann and Mangold from the late 1800s to 1921. These ideas would be unified at the molecular level by the discovery that DNA is the hereditary material (Avery *et al* 1944), and consists of a double helix of two deoxyribose sugar backbones joined by sequences of paired purine and pyrimidine bases (Watson and Crick 1953). The enormous power of this structure to explain the replication of DNA, heredity, cell function, development, and evolution today constitutes much of the framework within which studies of regeneration are conducted. In the 21st century, the role of ion channels and bioelectricity in patterning the distribution of biochemical mediators of developmental morphogenesis and regeneration specified by DNA is an emerging avenue of research (see Herrera-Rincon *et al* 2017, Moskvitz 2018).

1.3.3 Regenerative biology begins

Much of the history of regenerative biology has revolved around the ability of some organisms to regenerate complex body parts such as appendages, or complete individuals from body fragments (Birnbaum and Sanchez-Alvarado 2008). Crustaceans and fish can regenerate legs and fins, and salamanders can regenerate limbs, tails and jaws (Brockes and Kumar 2008). Hydra and the flatworm, planaria, share the ability to regenerate clonal copies of whole animals from fragments (Galliot 2012, Sanchez-Alvarado 2000). Plants replace lost branches by activating the undifferentiated cells of dormant meristems located along the axis of the branch. These meristems sprout new branches that more than compensate for the loss (Sugimoto *et al* 2011, Elliot and Sanchez-Alvarado 2013). Plants can also regenerate a whole plant from a cutting, tissue fragment or a single cell (Steward *et al* 1964).

Although Aristotle mentioned that lizards can regenerate their tails, a systematic study of appendage and whole-body regeneration did not begin until the 18th century, when Abraham Trembly (1719–84) performed detailed experiments on the regeneration of hydra (Lenhoff and Lenhoff 1991). In 1774, P S Pallas described the regeneration of planaria by fissioning (cited in Elliot and Sanchez-Alvarado 2013). Rene Reaumer (1683–1757) and Lazzaro Spallanzani (1729–99) made detailed observations on the regeneration of limbs in crustaceans and newts, respectively (Skinner and Cook 1991, Dinsmore 1991). In 1823, the English physician Tweedy John Todd (1789–1840) severed the sciatic nerve of amputated newt hind limbs and showed that limb regeneration is nerve-dependent (Singer and Geraudie 1991). All of this research took place prior to the cell theory, and so the physical basis for regeneration was still unknown.

Prior to the 20th century, limb regeneration in amphibians and crustaceans was explained in the context of preformation. The limbs of these animals were presumed to contain nested copies of preformed appendages that could be stimulated to develop by repeated amputations. Later investigators of the early 20th century recognized that regeneration was the manifestation of an inherent plasticity of mature limb tissues that restored the whole from the remaining part. A major research goal of Thomas Hunt Morgan, before he turned to the genetics of fruit flies in 1908, was to explain the regeneration of body parts in terms of internal cellular activities and external influences affecting the cells. In 1901, Morgan published his comprehensive and now classic book, *Regeneration*, that summarized the progress that had been made toward this goal. Over a century later, we are still in the process of formulating this explanation.

Many tissues harbor populations of stem cells that renew the tissue in the face of turnover or injury. The term stem cell was first used in the second half of the 19th century to refer to the fertilized egg in animals and to cells involved in the growth and regeneration of plants. In 1909, Alexander Maximow (1874–1928) presented evidence that a common multipotent stem cell of the bone marrow continually renews the different types of blood cells. As the 20th century progressed, it became clear that stem cells reside in many adult tissues and are a source of new cells for regeneration. Two major classes of stem cells were identified, those that reside in epithelial tissues such as the epidermis, airways and digestive tract, and those that reside in mesenchymal tissues such as muscle and bone. Major advances have since been made in identifying the molecular signatures that characterize the stem cells of different tissues. There are, however, several mechanisms of regeneration in addition to stem cells (see ahead).

1.4 The regenerative niche

The ability of a tissue to regenerate depends on the microenvironment in which its cells reside (called the niche) and the ability of these cells to respond to niche signals during physiological turnover or after injury (regeneration competence). Cell niches are complex and consist of extracellular matrix (ECM), nerves and blood vessels permeating the niche, immune cells, and a variety of soluble molecules. Some of these molecules are made locally by cells of the tissue themselves. Others, such as hormones, are made by distant tissues and circulate in the blood. The ECM provides structural support and signaling functions for tissues. The ECM, cell contacts and soluble molecules all act as signals (ligands) that bind to receptors embedded in cell plasma membranes or residing in the nucleus of the cells. These components act as a communication network to monitor niche conditions throughout the body and maintain or activate the patterns of gene activity and protein synthesis that result in physiological or injury-induced regeneration.

Anatomically, the niches of stem cells can be widely distributed throughout a tissue (satellite cells of muscle), confined to restricted spaces (bulge of the hair follicle), or scattered in facultative spaces (hematopoietic stem cells of the bone marrow). Niche composition has been particularly well studied in repair of skin wounds (Barrientos *et al* 2008) and regeneration by stem cells (Morrison and Spradling 2008, Voog and Jones 2010). The spatial position of cells within a niche is an important determinant of their activity and fate, as shown for both stem cells of the hair follicle (Rompalas *et al* 2013) and the hematopoietic system (Boulais and Frenette 2015). Much like families, parent stem or progenitor cells can provide niche signals essential for the differentiation of their daughter cells, for example airway stem cells in mice (Pardo-Saganta *et al* 2015).

1.4.1 Types of niche signals

The signaling molecules of the niche are classified as paracrine, autocrine, juxtacrine and endocrine (figure 1.1). *Paracrine* signals are secreted by cells, are soluble, and diffuse over short ranges to bind with receptors on neighboring cell surfaces. Some paracrine signals, such as retinoic acid (RA), bind to nuclear receptors (Deuster 2008). An *autocrine* signal is one that binds to receptors on the surface of the cell that produces it. *Juxtacrine* signaling is achieved by contact between two cells, in which a ligand on one cell surface binds to a receptor on the other. *Endocrine* signals (hormones) circulate in the blood and bind to nuclear receptors. Some cells can signal one another at long range by juxtacrine or paracrine means. Such signaling is

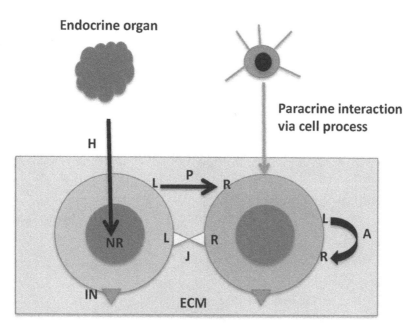

Figure 1.1. Types of intercellular niche signals regulating cell behavior. Two cells are shown next to one another embedded in surrounding extracellular matrix (ECM, light blue). Integrin receptors (IN, red triangles) on the surface of the cells respond to peptide signaling sequences from the ECM. L = ligand; R = receptor. P represents a paracrine interaction via a soluble signal between the cells; A represents an autocrine interaction of the cell with itself; J represents a juxtacrine interaction between the cell surfaces (yellow triangles). Signaling at a distance can take place by hormones (H) secreted from endocrine organs into the circulation that bind nuclear receptors (NR), or via a cell extension (such as an axon).

achieved by structural extensions of the signaling cell to the target cell, for example the extension of motor axons from spinal neuron cell bodies to muscles and the extension of filopodia from ingressing mesenchymal cells to the ectoderm in gastrulating sea urchin embryos (Rorth 2003).

Short amino acid sequences of ECM components act as signals by binding to cell surface receptors called integrins (Hynes 2002). For example, maintenance of connective tissue fibroblasts in a quiescent state relies on their adherence to the matrix they produce. Mechanical properties of the ECM, such as porosity, stiffness, elasticity, tension and compression, play key roles in cell behavior (Ingber 2006, Engler *et al* 2006, Guilak *et al* 2009). The behavior of cells is also influenced by whether or not they are interacting with niche factors in two dimensions (as in tissue culture monolayers) or three dimensions (as in a three-dimensional matrix). Three-dimensional culture, which more accurately reflects *in vivo* conditions, promotes more normal cell differentiation and tissue architecture.

1.4.2 Intracellular signaling pathways

Signaling ligands bound to cell surface receptors are transduced by intracellular signaling pathways that alter the cytoskeleton and/or patterns of gene activity of the cell (Gilbert and Barresi 2016). Six intracellular signaling pathways have been identified that fall into two groups based on receptor structure (figure 1.2). In the first group, ligands (often called agonists) bind to monomeric receptors. This group includes the Notch, Wnt, and Hedgehog pathways. Notch is a juxtacrine pathway, whereas Wnt and Hedgehog are paracrine pathways. In the second group, ligands bind to dimeric receptors. This group includes the receptor tyrosine kinase (RTK),

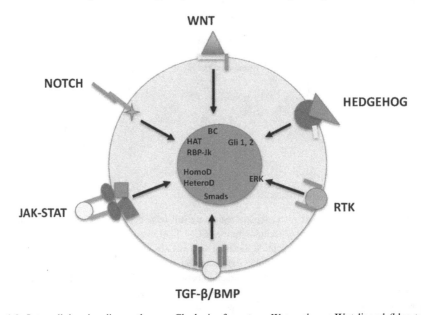

Figure 1.2. Intracellular signaling pathways. Clockwise from top. *Wnt pathway*: Wnt ligand (blue triangle) binds to its receptor complex consisting of co-receptors Frizzled (yellow bar) and LRP5/6 (red bar). This leads to phosphorylation of the Disheveled protein, blocking the destruction of beta-catenin (BC). BC can now accumulate in the nucleus where it synergizes with Lef/Tcf transcription factors to activate transcription. *Hedgehog pathway*: Unlike most ligands, the hedgehog signaling proteins (dark red triangle) repress the action of their receptor, Patched (purple), allowing the activation of Smoothened (Smo, yellow bar). Smo activates Gli transcription factors in the nucleus. *RTK pathway*: Receptor tyrosine kinases are transmembrane proteins dimerized (red bars) by their ligands (orange circle), which include a large variety of cytokines and growth factors. The dimerized receptors undergo conformational change that leads to the activation of G-proteins such as Ras. Ras activates a cascade of cytoplasmic phosphorylation reactions ending in the extracellular signal-regulated kinase (ERK), which enters the nucleus where it phosphorylates and activates transcription factors. *Transforming Growth Factor Beta (TGF-β) pathway*: This is a complex pathway activated by the two subfamilies of TGF-β ligands, TGF-β/Activin, and Bone Morphogenetic Protein (BMP). The two subfamilies act through two dimerized sets of receptors, type I (blue) and type II (red). TGF-β and BMP act to phosphorylate different classes of Smad proteins, which act in the nucleus to activate transcription. *JAK-STAT pathway*: This pathway is activated by many cytokines and growth factors (yellow) that dimerize their receptors (green), leading to conformational change of the JAK proteins (purple) that convert the receptor into a tyrosine kinase. The activated receptors now phosphorylate the STAT proteins (trapezoid and rectangle) to form homodimers and heterodimers that move to the nucleus and form transcriptional complexes in association with other proteins. *Notch pathway*: Notch (green) is a transmembrane receptor activated by the membrane-bound ligands Delta, Jagged and Serrate (red) on neighboring cells, triggering the enzymatic removal of the Notch intracellular domain (NICD, star). The NICD translocates to the nucleus where it interacts with several other proteins, including HAT and RBP-Jk, to activate genes whose products repress cell differentiation. Notch is thus an important pathway for the maintenance of stem cells.

JAK-STAT, and TGF-β pathways, most of which are paracrine, but can sometimes be autocrine.

In general, these pathways use variations on a common theme to signal to the interior of the cell. Ligand binding initiates a chain of phosphorylation reactions by protein kinases that ends with the activation or suppression of transcription factors. Importantly, many small molecules have been identified that inhibit the function of ligands or receptors (called antagonists), or enhance or substitute for their function (called agonists). Antagonists and agonists are important tools for determining whether a signal/receptor pair is essential to a cellular activity.

The apoptosis and autophagy signaling pathways are also important for regeneration and fibrosis. Apoptosis is programmed cell death in which executioner caspases 3 and 7 are activated to digest cell components without evoking inflammation (Shi 2004). Apoptosis is different from necrosis, which is defined as the disintegration of a cell due to trauma with accompanying inflammation. Activation of caspases 3 and 7 results in the breakup of cells into small membrane-bound fragments that first display 'find me' signals and then an 'eat me' phosphatidylserine signal that binds to macrophage and dendritic cell surfaces (Zitvogel *et al* 2010, Nagata *et al* 2010) (figure 1.3). These are phagocytic cells that engulf and digest the fragments, while simultaneously producing anti-inflammatory cytokines

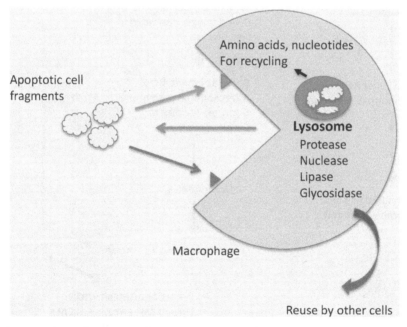

Figure 1.3. Macrophages eliminate apoptotic cell fragments. These fragments give off a 'find me' signal (green arrow) that chemoattracts macrophages (red arrow), and an 'eat me' signal (blue arrow) that triggers engulfment of the fragments by the macrophage (red arrow). Green triangle and blue triangle represent receptors for the find me and eat me signal molecules, respectively. Once engulfed, the fragments are endocytosed to the interior of the macrophage to become part of lysosomes. The lysosomes contain enzymes that digest the fragments into amino acids, nucleotides, carbohydrates and lipids for reuse by other cells.

(Ravichandran 2003, Green 2005). Adult tissues undergoing regeneration or fibrosis rely on apoptosis to regulate cell number and turnover and to sculpt the shape of regenerating organs, as well as to eliminate cells that can harm the organism such as virally infected cells and cancer cells (Nicholson and Thornberry 2003, Gilbert and Barresi 2016). Genes involved in apoptosis are remarkably conserved across vast phylogenetic distances and in fact were first discovered in the nematode worm *Caenorhabditis elegans* (Ellis and Horvitz 1986).

There are two sets of apoptotic pathways (Zitvogel *et al* 2010, Nagata *et al* 2010), one triggered by internal cell signals (intrinsic pathway), and one by external signals (extrinsic pathway) (figure 1.4). The intrinsic pathway is triggered under several circumstances: the absence of signals that inhibit apoptosis (growth factors, adhesive ECM molecules), internal damage to the cell's DNA, or the accumulation of misfolded proteins coming off the ribosomes and entering the endoplasmic reticulum. These problems lead to cell death by compromising mitochondrial function. Healthy mitochondria display the anti-apoptotic protein Bcl-2 on their surface, but in the apoptotic pathway two other proteins, Bad and Bax, now bind to Bcl-2 to block its action and allow pores to form in the mitochondrial membrane, resulting in

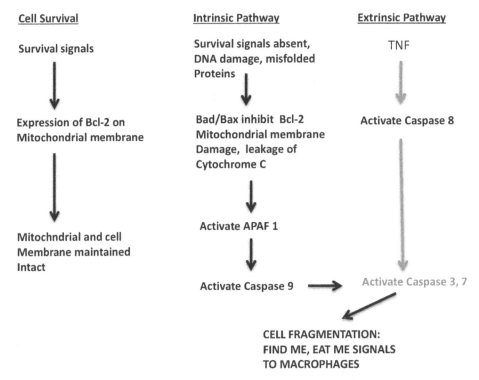

Figure 1.4. Apoptotic pathways. Left: Cell survival signals, such as growth factors, cytokines and other niche factors maintain intact cell structure. Middle: Intrinsic apoptotic pathway. Right: Extrinsic pathway, with tumor necrosis factor (TNF) acting to directly trigger cell death. Both intrinsic and extrinsic pathways ultimately activate caspases 3 and 7. The end result is cell fragmentation and production of the find me, eat me signals.

the leakage of cytochrome c. Cytochrome c binds to apoptotic protease activating factor-1 (Apaf-1) to form an 'apoptosome' complex. Apoptosomes bind and activate caspase 9, which activates caspase 3 and 7. A good example of a cell type programmed to die by this pathway is the red blood cell, which requires erythropoietin (EPO) for survival. EPO made by the liver and kidney regulates the rate of turnover of red blood cells (RBCs); it is up regulated when too many RBCs are lost and has been used illegally for maintaining high levels of RBCs ('doping') in sports requiring exceptionally high O_2 levels such as cycling.

The external pathway operates via signals that induce apoptosis directly. Two major signals that directly induce cell death are FasL and TNF. FasL is produced by cytotoxic T-cells and induces apoptosis of target cells by binding to its receptor Fas. TNF is a growth factor that triggers apoptosis by binding to the TNF receptor on target cells. In both cases, binding activates intracellular signaling pathways that activate caspase 8, which like caspase 9 leads to the activation of caspases 3 and 7 and cell digestion.

Autophagy is a different kind of cellular self-digestion that does not result in cell death, but in the re-cycling of degraded cell components into new molecules (Klionsky 2007, Mizushima et al 2008, Mizushima and Komatsu 2011). Digestion products to be recycled are packaged directly into lysosomes or first into autophagosomes that then fuse with lysosomes. In yeast, the autophagy pathway consists of ~20 genes. All cells exhibit a low basal level of autophagy that is up regulated under high-energy demand due to starvation, growth factor withdrawal, or the need to eliminate damaging cellular components that accumulate during oxidative stress or infection (Levine and Kroemer 2008). In the presence of growth factors such as insulin-like growth factor (IGF) and abundant nutrients, the autophagy pathway is inhibited by TOR (target of rapamycin) kinase (Lum et al 2005).

1.5 Approaches to the study of regeneration

1.5.1 Model organisms

Many animal (and plant) models are used to study regeneration. Regeneration by adult stem cells has been investigated largely in mammalian tissues. Invertebrates such as planaria and hydra have long been favorite research models for whole-body regeneration (Birnbaum and Sanchez-Alvarado 2008, Galliot 2012, for reviews). Among vertebrates, various species of fish, larval and adult urodele salamanders, and anuran tadpoles and adults are long-standing favorites for study of appendage and central nervous system regeneration. To this menagerie has now been added the mouse digit tip (Simkin et al 2015, Seifert and Muneoka 2017, for reviews), the MRL mouse ear (Heber-Katz et al 2013), and the skin and ear tissue of the African spiny mouse *Acomys* (Seifert et al 2012, Brant et al 2016).

Comparative analyses are useful for understanding what genetic circuits are common to regeneration-competent tissues and appendages, and for understanding the molecular differences between regeneration competence and deficiency/incompetence. For example, a direct strategy to identify the molecular differences between regenerative competence and failure is to compare and contrast the transcriptomes and proteomes of regeneration-competent versus deficient tissues. Four types of experimental models can be used for these comparisons.

- Compare wild-type tissues to the same tissues within the same species that have a genetic variation conferring a gain or loss of regenerative capacity (e.g. the MRL/lpj mouse ear tissue to wild-type *Mus* ear tissue).
- Compare the same tissues at developmental stages when they are capable of regeneration versus stages when they are not (e.g. fetal versus adult skin).
- Compare the same tissue between two species, one of which regenerates the tissue and the other does not (e.g. salamander limb versus frog limb; *Acomys* skin versus *Mus* skin).
- Compare tissue at regeneration-competent versus deficient positions (e.g. the mouse terminal phalange versus the second phalange).

Such comparisons should go a long way to reveal the mechanisms that promote or inhibit regeneration, and suggest interventions that have the potential to confer regenerative capacity on regeneration-deficient tissues and organs. Examples of each of these models will be encountered in subsequent chapters.

1.5.2 Tracing the cellular origins of regenerated tissues

The most fundamental question of regeneration concerns the origin of the cells that carry it out. Determining the origins of cells involved in maintenance or injury-induced regeneration requires marking them in some way and showing that the marked cells give rise to the regenerated tissue. Natural markers such as pigment or nuclear ploidy, or artificial markers such as DNA labeling of proliferating cells with [^{3}H]-thymidine or BrdU, or labeling groups of cells with cell membrane or cytoplasmic dyes such as DiI have been used to follow cells after grafting the marked tissue to an unmarked host and tracking the donor cells in a regenerative situation. Each of these markers, however, has a different set of limitations.

More sophisticated genetic marking of cells has refined our ability to look at the origins of regenerated tissues. Two kinds of transgenic marking schemes have been developed (figure 1.5). In the first, zygotes are injected with a construct consisting of a reporter gene driven by the promoter of a gene that is constitutive (on in every cell), or a construct driven by the promoter of a gene that is normally activated only in one specific cell type during development. In the first case, the reporter will be active in all cells. Specific cells can be tracked after injury by grafting them to an unmarked host. In the second case the reporter will be active only in the cells that express the phenotype-specific gene activity, allowing the marked cells to be followed during maintenance or injury-induced regeneration of the tissue containing them. These techniques are known as unconditional genetic marking.

In the second method, called induced or conditional genetic marking, transgenic animals are made in which the expression of a reporter gene can be activated by a stimulus at any time during development or regeneration, using a Cre/lox DNA recombination system Cre (causes recombination) is a recombinase expressed in

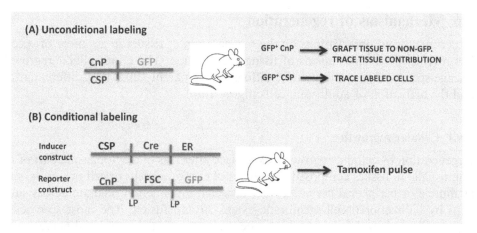

Figure 1.5. Cell labeling by transgenesis. (A) Unconditional labeling. Fertilized eggs are labeled with a construct consisting of a GFP reporter gene and either (1) a constitutive promoter (CnP) or (2) a cell-specific promoter (CSP). In the case of (1) all tissues will express GFP and specificity is achieved by grafting GFP tissues/cells to a non-GFP host. In the case of (2) GFP will be expressed only in the cells that normally activate the transcription factors that bind to the promoter. (B) Conditional labeling. Mice engineered to carry an inducer construct are mated to mice engineered to carry a reporter construct to produce mice carrying both constructs. The inducer consists of a cell-specific promoter (e.g. the insulin promoter), the gene for Cre recombinase, and an estrogen receptor gene (ER). The reporter construct consists of a constitutive promoter, a floxed stop cassette (FSC) that prevents the reporter gene from being transcribed, and the GFP reporter gene. The FSC is flanked by two loxP sites (LP), which can be cut by CRE. Labeling can be achieved at any desired stage of development by administering a tamoxifen pulse to the mouse. Tamoxifen allows the Cre/ER fusion protein made in the cytoplasm to translocate to the nucleus. Cre excises the FSC from the reporter construct, allowing the CnP to drive GFP expression. Only cells expressing transcription factors for the cell-specific promoter will express GFP.

bacteriophages that cuts DNA at loxP sequences. Two sets of transgenic animals are made. The first set carries a Cre/estrogen receptor fusion gene driven by the promoter of a gene whose activity defines a particular cell phenotype (such as the insulin gene in β-cells of the pancreas). This construct makes a Cre/estrogen receptor fusion protein only in that cell type. The second set carries a construct consisting of a constitutive ubiquitous promoter and a reporter gene separated by a sequence flanked by loxP sites (a 'floxed stop cassette', FSC). Crossing these mice yields progeny carrying both constructs. The Cre portion of the fusion protein excises the FSC of the second construct at the loxP sites to recombine the constitutive promoter with the reporter gene, activating it. However, to translocate to the nucleus and carry out this recombination, the Cre/estrogen fusion protein must first be activated, which is accomplished by administration of a pulse of tamoxifen to the animal. The Cre/ER fusion protein now enters the nucleus, where the Cre recombinase excises the FSC, recombining the constitutive promoter and reporter gene, thus marking only those cells expressing the fusion protein. Conditional genetic marking is now in wide use to differentiate the origin of regenerated tissue from stem cells or differentiated cells at different stages of development.

1.6 Mechanisms of regeneration

Experiments and observations on the regeneration of tissues in a variety of species has revealed four mechanisms of tissue regeneration. These are cellular regrowth, lineage-specific regeneration from differentiated parent cells, transdifferentiation, and the activation of adult stem cells (figure 1.6).

1.6.1 Cellular regrowth

Regeneration by cellular regrowth takes place after loss of a cytoplasmic part of the cell, usually in tissues composed of bundles of elongated cells or cell processes, such as muscle or peripheral nerves. Thus transected myofibers regenerate across small gaps by bidirectional cell membrane extension and fusion. The most spectacular single cell regeneration in vertebrates occurs in transected peripheral nerve axons, where the part of the axon still connected to the nerve cell body grows back to its target tissue (Yannas 2001, for a review).

1.6.2 Lineage-specific regeneration from differentiated cells

There are two types of lineage-specific regeneration from pre-existing cells: (1) division of cells while maintaining their differentiated state (compensatory hyperplasia) and (2) cellular dedifferentiation, proliferation and redifferentiation into their parent phenotype (dedifferentiation/redifferentiation).

Mechanism	Example
Regrowth of cell part	Transected axon
Lineage-specific regeneration from pre-existing differentiated cells	
1. Compensatory hyperplasia	Hepatocyte regeneration
2. Dedifferentiation --> redifferentiation	Cardiomyocyte regeneration
3. Epithelial → ← mesenchymal transformation	Migrating keratinocytes
Transdifferentiation	
1. Direct—genetic program changed w/o dedifferentiation or proliferation	Acinar cells of pancreas to beta cells
2. Indirect—genetic program changed by dedifferentiation and proliferation	Dorsal iris cells to lens cells during newt lens regeneration
Adult stem cell proliferation/differentiation	Epithelia, muscle, bone, blood

Figure 1.6. Summary of mechanisms of regeneration.

1.6.2.1 Compensatory hyperplasia

Compensatory hyperplasia is the proliferation of cells while they maintain their differentiated structure and function. The classic example is liver regeneration, as studied in the rat (Michalopoulos and DeFrances 1997). After partial hepatectomy the hepatocytes of the liver, as well as its non-parenchymal cell types (Kupffer, Ito, bile duct epithelial and fenestrated epithelial cells), divide while performing their functions of glucose regulation, synthesis of plasma proteins, secretion of bile, and drug metabolism, until the original mass of the liver is restored. The adult mouse pancreas, which develops from the same embryonic region as the liver, also regenerates by compensatory hyperplasia. In humans, the liver regenerates well but the pancreas does not.

1.6.2.2 Dedifferentiation/redifferentiation

Dedifferentiation is a cellular reprogramming mechanism in which chromatin remodeling leads to loss of phenotypic specialization, allowing cells to enter the cell cycle and proliferate before redifferentiating back into their parental cell type (Grafi 2004). This is a relatively common mechanism of regeneration in lower vertebrates. Teleost fish regenerate fins and barbels, and certain species of lizards regenerate tails by this mechanism. The divas of dedifferentiation are the urodele amphibians, which use this mechanism to provide progenitor cells for the regeneration of tails, jaws and limbs (Brockes and Kumar 2008). Other tissues where at least partial dedifferentiation is involved are the regeneration of cardiomyocytes of urodele amphibians, teleost fish and mice.

An interesting recent finding is that in some tissues such as stomach and airway epithelium, specific differentiated cell types can function as stem cells to regenerate all the cell types of the epithelium, including stem cells, if the normal stem cell population is depleted or ablated (Stange *et al* 2013, Tata *et al* 2013). In the stomach epithelium, these are the pepsin and chymotrypsin-secreting chief cells; in the airways, they are the secretory Clara cells.

1.6.2.3 Epithelial/mesenchymal transformation

A phenomenon that occurs during regeneration from pre-existing epithelial cell sheets is epithelial to mesenchymal transformation (EMT) and its reverse, mesenchymal to epithelial transformation (MET) (Thiery *et al* 2009). EMT endows epithelial cells with mesenchymal cell properties, allowing them to lose their adhesive contacts and migrate, whereas MET restricts migration by re-establishing an epithelium (Mani *et al* 2008). These transformations are prominent mechanisms of embryonic development and of pathologies such as fibrosis and metastasis. EMT/MET is especially prominent in wounded epithelia, such as skin epidermis, kidney tubule epithelium, and respiratory epithelium.

1.7 Transdifferentiation

Transdifferentiation is the conversion of one cell type to another. There is usually a dedifferentiation step that confers plasticity on the cell, followed by proliferation and differentiation into a new cell type (indirect transdifferentiation), but trans-differentiation can also be direct, in which the activity of the suite of genes that specifies the original cell phenotype is suppressed and immediately succeeded by activation of the suite of genes characteristic of the new cell type. Another word for transdifferentiation is metaplasia, a medical term meaning a change in cell type as a protective response to a changing environment, such as the direct transdifferentia-tion of mucus-secreting ciliated pseudostratified columnar epithelial cells of the airways to stratified squamous epithelium in response to chronic exposure to cigarette smoke. These changes are abnormal and can proceed to further patho-logical changes. Some transdifferentiation phenomena are normal (physiologic transdifferentiation), such as metaplasia of endocervical columnar cells to squamous cells as part of normal growth and development, and aging. The most spectacular transdifferentiation seen in nature is regeneration of the lens following lentectomy of the newt eye (Reyer 1977).

1.8 Regeneration by adult stem cells

Most vertebrate tissues regenerate via adult stem cells sequestered in tissue niches that maintain their stem character and regulate their proliferation for either maintenance or injury-induced regeneration. When activated, adult stem cells can divide symmetrically to produce either two lineage-committed cells or two stem cells, or they can divide asymmetrically to produce one lineage-committed cell and one stem cell (Morrison and Spradling 2008). Depending on their tissue of residence, they may be unipotential, such as epidermal stem cells, or multipotential, such as hematopoietic stem cells. Adult stem cells have signature cell surface markers and transcription factors. Their anatomical niches can be more or less extensive, depending on the tissue. For example, stem cells that regenerate hairs and epidermis reside in a restricted region of the hair follicle called the bulge, and corneal stem cells reside both in the epithelium of the cornea and in the limbus, a ring of tissue surrounding the cornea. Each niche is regionally differentiated in its molecular composition as well, and stem cells move from one region to another to accomplish different proliferative and differentiative steps (Scadden 2006).

There are two major classes of adult stem cells, epithelial and mesenchymal. Epithelial stem cells maintain and regenerate the epidermis and hairs of the skin, surface of the cornea, and the linings of airways, intestine, urogenital system, and central nervous system (Barker *et al* 2010). They are adherent to one another by desmosomes and other specialized junctions at their basolateral surfaces and to a basement membrane by hemidesmosomes. Mesenchymal stem cells (MSCs) are more individualistic cells capable of migration that reside in bone marrow, muscle, fat, and connective tissues. MSCs are ubiquitous throughout the body. This ubiquity suggests that their origin may be the perivascular cells called pericytes that adhere to, and stabilize the surface of capillaries, which form dense networks in every tissue (Da Silva Meirelles *et al* 2006, Crisan *et al* 2008). When MSCs are stimulated by specific signals, they can be released from their niche and recruited to sites of injury (Liu *et al* 2009). Bone and muscle are well-studied model systems that activate MSCs for injury repair.

ASCs have varying degrees of developmental potential, depending on the tissue they serve. For example, hematopoietic stem cells (HSCs), which reside in the bone marrow, are multipotential and give rise to erythrocytes, myeloid cells and the several cell types of the immune system, whereas epidermal stem cells appear to be unipotent and give rise only to keratinocytes. Activation of ASCs may take place differently in maintenance versus injury-induced regeneration. In maintenance regeneration, stem cells of tissues with high cellular turnover (for example, the epidermis and digestive tract) undergo continual but slow division in response to environmental signals to produce more rapidly proliferating transient amplifying cells (TACs), feeding a constant stream of progeny into a region that induces their differentiation. In tissues that normally undergo slow cellular turnover, fewer stem cells are active until injury-induced signals mobilize them.

1.9 Some tissues use multiple mechanisms of regeneration

Most tissues use one of the four mechanisms of regeneration as their primary means of regeneration, but a substantial number have redundant mechanisms. Red blood cells and immune cells of the blood and intestinal epithelium regenerate solely by adult stem cells. The liver uses compensatory hyperplasia to regenerate after surgical injury, but relies on stem cells to regenerate after chronic chemical injury. All four mechanisms of regeneration are involved in the reconstitution of an amputated urodele appendage.

References

Allan G 1977 *Life Science in the 20th Century* (Cambridge: Cambridge University Press)

Avery O T, McCleod C M and McCarty M 1944 Studies on the chemical nature of the substance inducing transformation of pneumococcal types *J. Exp. Med.* **79** 139–58

Barker N, Bartfeld S and Clevers H 2010 Tissue-resident adult stem cell populations of rapidly self-renewing organs *Cell Stem Cell* **7** 656–70

Barrientos S, Stojadinovic O, Golinko M S and Brem H 2008 Growth factors and cytokines in wound healing *Wound Rep. Reg.* **16** 585–601

Birnbaum K D and Sanchez-Alvarado A 2008 Slicing across kingdoms: regeneration in plants and animals *Cell* **132** 697–710

Boulais P E and Frenette P S 2015 Making sense of hematopoietic stem cell niches *Blood* **125** 2621–9

Brant J O *et al* 2016 Cellular events during scar-free skin regeneration in the spiny mouse *Acomys. Wound Rep. Reg.* **24** 75–88

Brockes J P and Kumar A 2008 Comparative aspects of animal regeneration *Annu. Rev. Cell Dev. Biol.* **24** 525–49

Carlson B M 2007 *Principles of Regenerative Biology* (San Diego, CA: Academic)

Coleman W 1977 *Biology in the Nineteenth Century: Problems of Form, Function and Transformation* (Cambridge: Cambridge University Press)

Crisan M, Yap S and Castiella L *et al* 2008 A peroivascular origin for mesenchymal stem cells in multiple human organs *Cell Stem Cell* **3** 301–15

Da Silva Meirelles L, Chagastelles P C and Nardi N B 2006 Mesenchymal stem cells reside in virtually all post-natal organs and tissues *J. Cell Sci.* **119** 2204–13

Deuster G 2008 Retinoic acid synthesis and signaling during early organogenesis *Cell* **134** 921–31

Dinsmore C E 1991 Lazzaro Spallanzani: concepts of generation and regeneration *A History of Regeneration Research* ed C E Dinsmore (Cambridge: Cambridge Universty Press)

Dinsmore C E 1996 Urodele limb and tail regeneration in early biological thought: an essay on scientific controversy and social change *Int. J. Dev. Biol.* **40** 621–7

Dinsmore C E 1998 Conceptual foundations of metamorphosis and regeneration: from historical links to common mechanisms *Wound Rep. Reg.* **6** 291–301

Elliot S A and Sanchez-Alvarado A 2013 The enduring contributions of planarians to the study of animal regeneration *Wiley Interdiscip. Rev. Dev. Biol.* **2** 301–26

Ellis H M and Horvitz H R 1986 Genetic control of programmed cell death in the nematode *C. elegans Cell* **44** 817–29

Engler A J, Sen S, Sweeney H L and Discher D E 2006 Matrix elasticity directs stem cell lineage specification *Cell* **126** 677–89

Galliot B 2012 *Hydra*, a fruitful model for 270 years *Int. J. Dev. Biol.* **56** 411–23

Gevers W 1984 Protein metabolism in the heart *J. Mol. Cell Cardiol.* **16** 3–32

Gilbert S F and Barresi M J F 2016 *Developmental Biology* 10th edn (Sunderland: Sinauer)

Goss R J 1969 *Principles of Regeneration* (New York: Academic)

Goss R J 1991 The natural history (and mystery) of regeneration *A History of Regeneration Research* ed C E Dinsmore (Cambridge: Cambridge University Press), pp 7–23

Grafi G 2004 How cells dedifferentiate: a lesson from plants *Dev. Biol.* **268** 1–6

Green D R 2005 Apoptotic pathways: ten minutes to dead *Cell* **121** 671–4

Guilak F *et al* 2009 Control of cell fate by physical interactions with the extracellular matrix *Cell Stem Cell* **5** 17–26

Heber-Katz E *et al* 2013 Cell cycle regulation and regeneration. New perspectives in regeneration *Curr. Top. Microbiol. Immunol.* **367** 253–77

Herrera-Rincon C *et al* 2017 The brain is required for normal muscle and nerve patterning during early *Xenopus* development *Nat. Commun.* **8** 587

Hynes R O 2002 Integrins: bidirectional, allosteric signaling machines *Cell* **110** 673–87

Ikeuchi M, Ogawa Y, Iwase A and Sugimoto K 2016 Plant regeneration: cellular origins and molecular mechanisms *Development* **143** 1442–51

Ingber D E 2006 Cellular mechanotransduction: putting all the pieces together again *FASEB J.* **20** 811–27

Klionsky D J 2007 Autophagy: from phenomenology to molecular understanding in less than a decade *Nat. Rev. Mol. Cell Biol.* **8** 931–7

Lenhoff H M and Lenhoff S G 1991 Abraham Trembly and the origins of research on regeneration in animals *A History of Regeneration Research* ed C E Dinsmore (Cambridge: Cambridge University Press), pp 47–66

Levine B and Kroemer G 2008 Autophagy in the pathogenesis of disease *Cell* **132** 27–42

Liu Z-J, Zhuge Y and Velazquez O C 2009 Trafficking and differentiation of mesenchymal stem cells *J. Cell Biochem.* **106** 984–91

Lum J J *et al* 2005 Growth factor regulation of autophagy and cell survival in the absence of apoptosis *Cell* **120** 237–48

Mani S A *et al* 2008 The epithelial–mesenchymal transition generates cells with properties of stem cells *Cell* **133** 704–15

Michalopoulos G K and DeFrances M C 1997 Liver regeneration *Science* **276** 60–6

Mizushima N, Levine B, Cuervo A and Klionsky D J 2008 Autophagy fights disease through cellular self-digestion *Nature* **451** 1069–74

Mizushima N and Komatsu M 2011 Autophagy: renovation of cells and tissues *Cell* **147** 728–58

Morrison S J and Spradling A C 2008 Stem cells and niches: Mechanisms that promote stem cell maintenance throughout life *Cell* **132** 598–611

Moskvitz K 2018 Brainless embryos suggest bioelectricity guides growth *Quanta magazine*, March 13. https://quantamagazine.org/brainless-embryos-suggest-bioelectricity-guides-growth-20180313/

Nagata S, Hanayama R and Kawane K 2010 Autoimmunity and the clearance of dead cells *Cell* **140** 619–30

Nicholson D W and Thornberry N A 2003 Apoptosis. Life and death decisions *Science* **299** 214–5

Pardo-Saganta A *et al* 2015 Parent stem cells can serve as niches for their daughter cells *Nature* **523** 597–601

Ravichandran K S 2003 Recruitment signals' from apoptotic cells: invitation to a quiet meal *Cell* **112** 817–20

Reyer R W 1977 The amphibian eye: development and regeneration *Handbook of Sensory Physiology, Vol VII/5 The Visual System in Vertebrates* ed F Crescitelli (Berlin: Springer), pp 309–90

Rompalas P, Mesa K R and Greco V 2013 Spatial organization within a niche as a determinant of stem-cell fate *Nature* **502** 513–8

Rorth P 2003 Communication by touch: role of cellular extensions in complex animals *Cell* **112** 595–8

Sanchez-Alvarado A 2000 Regeneration in the metazoans: why does it happen? *Bioessays* **22** 578–90

Scadden D T 2006 The stem-cell nice as an entity of action *Nature* **441** 1075–9

Seifert A W *et al* 2012 The influence of fundamental traits on mechanisms controlling appendage regeneration *Biol. Rev.* **87** 330–45

Seifert A W and Muneoka K 2017 The blastema and epimorphic regeneration in mammals *Dev. Biol.* **433** 190–99

Shi Y 2004 Caspase activation: revisiting the induced proximity model *Cell* **117** 855–8

Simkin J *et al* 2015 The mammalian blastema: regeneration at our fingertips *Regeneration* **2** 93–147

Singer M and Geraudie J 1991 The neurotrophic phenomenon: its history during limb regeneration in the newt *A History of Regeneration Research. Milestones in the Evolution of a Science* ed C E Dinsmore (Cambridge: Cambridge University Press), pp 101–12

Skinner D M and Cook J S 1991 New limbs for old: some highlights in the history of regeneration in Crustacea *A History of Regeneration Research* ed C E Dinsmore (Cambridge: Cambridge University Press), pp 25–46

Steward F C, Mapes M O, Kent A E and Holsten R D 1964 Growth and development of cultured plant cells *Science* **143** 20–7

Stange D E *et al* 2013 Differentiated Troy$^+$ chief cells act as reserve stem cells to generate all lineages of the stomach epithelium *Cell* **155** 357–68

Stocum D L 2012 *Regenerative Biology and Medicine* 2nd edn (San Diego, CA: Elsevier/Academic)

Sugimoto K, Gordon P and Meyerowitz E M 2011 Regeneration in plants and animals: dedifferentiation, transdifferentiation, or just differentiation? *Trends Cell Biol.* **21** 212–8

Tata P R *et al* 2013 Dedifferentiation of committed epithelial cells into stem cells *in vivo Nature* **503** 218–23

Thiery J P, Acloque H, Huang R Y J and Nieto M A 2009 Epithelial–mesenchymal transitions in development and disease *Cell* **139** 871–90

Van Doren C 1991 *A History of Knowledge* (New York: Ballantine)

Voog J and Jones D L 2010 Stem cells and the niche: a dynamic duo *Cell Stem Cell* **6** 103–15

Wachtershauser G 1990 Evolution of the first metabolic cycles *Proc. Natl Acad. Sci. USA* **87** 200–4

Watson J D and Crick F H C 1953 A structure for deoxyribose nucleic acid *Nature* **171** 737–8

Wilson E B 1925 *The Cell in Development and Heredity* (New York: Macmillan)

Yannas I V 2001 *Tissue and Organ Regeneration in Adults* (New York: Springer)

Zitvogel L, Kepp O and Kroemer G 2010 Decoding cell death signals in inflammation and immunity *Cell* **140** 798–804

IOP Publishing

Foundations of Regenerative Biology and Medicine

David L Stocum

Chapter 2

Wound repair by fibrosis

Summary

Fibrosis, or scarring, is the default state for the injured tissues of most mammalian species. Fibrosis has been best studied in skin models, where it takes place in three overlapping and integrated phases of hemostasis, inflammation and structural repair. These phases are tightly regulated by immune cells, epidermal cells, and fibroblasts of the dermis. Wounds are closed first by the formation of a plasma clot, and then by the migration of epidermal cells through the clot. Repair is accomplished by the migration into the wound of fibroblasts and regenerating capillaries to form granulation tissue, which is then remodeled into collagenous scar tissue. Aging reduces the repair capacity of skin wounds, and vascular disease can lead to chronic wounds that fail to heal without intervention. Comparison of injured fetal and adult skin has revealed differences in immune cells, ECM and growth factors that correlate with the ability of fetal skin to regenerate as opposed to scar formation in adult skin. The skin of some amphibians can regenerate, as can the skin of the African spiny mouse, *Acomys*. Comparative cellular and molecular analyses of wounded versus unwounded adult and fetal skin, and wounds of adult *Acomys* and *Mus* mouse species have revealed differences in immune cells, cytokines and growth factors, and collagens that are correlated with the ability to regenerate skin.

2.1 Fibrosis is studied primarily in skin

Scarring, or fibrosis, is a reaction of connective tissue to injury that elicits an inflammatory response by the immune system, leading to the formation of collagenous scar tissue. Even tissues capable of regeneration will repair by fibrosis if wounded to an extent that exceeds their regenerative capacity. Fibrosis that compromises tissue function, such as in the heart, central nervous system, kidneys and lungs, is a major medical problem, as is the failure of chronic venous and diabetic wounds to repair. Therefore, we seek to understand why and how tissues scar for the purpose of either accelerating this process, or devising interventions that lead to scar-free regeneration

Although fibrosis can occur in any tissue or organ, it has been most intensively studied in the skin, the largest organ of the body and the first line of defense against trauma and invasion by microorganisms. The skin functions as a barrier to dehydration, and has sensory (tactile, thermal, pain), excretory (sweating) and antibiotic functions (sebaceous glands), all of which can be disrupted by scarring. Facial skin has the important function of mediating social interactions. Major facial disfiguration by scarring can result in severe psychological trauma.

Our understanding of fibrosis comes primarily from studies on skin wounds in rodents (rats, mice, guinea pigs), lagomorphs (rabbits) and pigs. Most of these studies have been carried out *in vivo*, but the lack of standardized protocols has made comparing results difficult. *In vitro* models have recently been developed that can be easily standardized. For example, one human burn model involves harvesting a patch of skin and cutting it into pieces that are cultured with the dermal side down. When subjected to a burn on their epidermal surface, the burn heals much like it would *in vivo* (Coolen *et al* 2008). Research on mammalian skin represents a convergence of three complementary disciplines, cell biology, mouse genetics and dermatology, that have enabled insights into potential ways to regenerate skin and treat skin diseases (Watt 2014).

2.2 Structure and function of skin

The skin is a highly complex structure composed of two major layers, the epidermis and dermis, between which is a non-cellular basement membrane approximately 100 nm thick synthesized by the epidermis (Ham and Cormack 1979) (figure 2.1). The basement membrane has two layers, the lamina lucida next to the epidermis and the lamina densa next to the dermis. The primary component of the lamina lucida is the glycoprotein laminin (Ln) and that of the lamina densa is type IV collagen (Yannas 2001, for a review).

The epidermis is organized into several layers of keratinocytes, the major cell type of the epidermis. The lowest layer, the stratum basale, is anchored into the lamina lucida of the basement membrane by hemidesmosomes. Above the basal layer is the stratum germativum, stratum granulosum, and stratum corneum, which is the outermost layer of the epidermis. The keratinocytes of these layers are held together laterally by adhesion belts, desmosomes and tight junctions to form a water-impermeable barrier. A number of epidermal appendages project downward into the dermis: hair follicles, sweat glands and sebaceous glands. Non-epithelial cell types found among the keratinocytes are melanocytes, which give the skin its color, and Langerhans cells, antigen-presenting dendritic cells of the immune system.

The dermis is composed of three layers: the upper papillary layer, the subjacent reticular layer, and the deepest layer, the hypodermis, also called the fascia (Watt 2014). The papillary and reticular layers are composed of fibroblasts, whereas the reticular fibroblasts and hypodermis is composed of white adipocytes (fat cells) that differentiate from fibroblasts. The papillary layer is anchored into the lamina densa

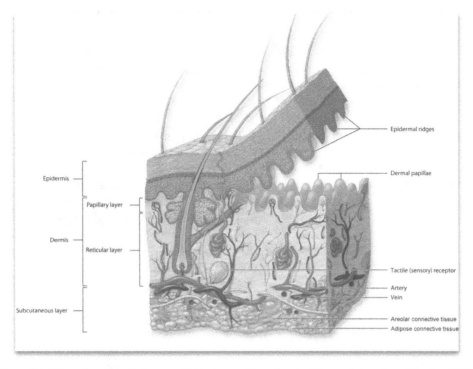

Figure 2.1. Three-dimensional diagram of skin layers. Reproduced with permission from (Mescher 2013). Copyright 2013 McGraw-Hill.

of the basement membrane by type VII collagen fibrils. A capillary network pervades the papillary layer that provides nourishment to the epidermis and acts as a heat exchanger with the external environment. Fibroblast density is greatest in the papillary layer. The reticular layer is thicker than the papillary layer and has fewer fibroblasts and capillaries. Bundles of collagen fibers anchor the reticular layer to the underlying hypodermis. Other cells present in the dermis are mast cells, which release histamine in allergic reactions, and tissue macrophages (phagocytic cells). Genetic lineage tracing has shown that the fibroblasts of the papillary layer and the fibroblasts of the reticular and hypodermal layers constitute two separate lineages (Driskell *et al* 2013). The papillary fibroblasts express the cell surface marker proteins CD26 and LRIG1, whereas the reticular and hypodermis express Dlk1 and Sca1. There is evidence for fibroblast heterogeneity that is correlated with different anatomical position, though positional markers reflecting such heterogeneity are largely undefined.

The physical properties of the dermis reflect the composition and organization of its ECM, which consists of (1) structural proteins, (2) adhesive glycoproteins, (3) proteoglycans (PGs), and (4) matricellular proteins (Schultz and Wyscki 2009).

The structural proteins are the fibrous collagens and elastins, which confer tensile strength and resiliency to the ECM. The major collagen of dermal ECM is type I (90%). Type III collagen makes up 5% and smaller amounts of other collagens the rest. The collagen and elastic fibers of the reticular layer are coarser than those of the papillary layer, but both layers are organized in a three-dimensional reticular pattern. The adhesive proteins fibronectin (Fn), vitronectin (Vn) and laminin (Ln) serve as substrates for cell migration to heal the wound (Clark 1996). The collagens and adhesive proteins have amino acid recognition sequences allowing them to bind to cell surface receptors belonging to the integrin family. Significant dermal PGs are versican, decorin and perlecan, which are linked to the large, non-sulfated GAG molecule hyaluronate, the most abundant GAG of the dermal ECM, to create hyaluronic acid:PG complexes. These complexes bind water avidly, giving the dermal ECM its property of resisting compressive force, and creating space for cell migration in injured skin (Clark 1996).

Matricellular proteins are secreted macromolecules that regulate cell function by interacting with cell surface receptors, proteases, hormones and other bioeffector molecules, as well as with structural matrix proteins such as collagens (Bornstein 2009).

The dermal ECM is a reservoir for growth factors that interact with cell receptors when released and activated (Schultz and Wyscki 2009).

The skin has several types of thermoreceptors, mechanoreceptors (tactile receptors) and nocioreceptors (receptors that detect noxious stimuli) (Zimmerman *et al* 2014). In the epidermis subsets of polymodal free nerve endings acting in all these different capacities end in the stratum granulosum, and the stratum basale contains touch receptors called Merkel cells. The dermis harbors Meissner's corpuscles that detect soft touch, Pacinian corpuscles that detect pressure, thermal receptors called bulbs of Krause, and mechanoreceptors called Ruffini endings. Each of these has a nerve ending that transmits the information to the spinal cord and/or brain.

2.3 Types of wounds

Wounds to healthy skin that heal normally on a predictable schedule are called acute wounds. Acute wounds can be due to incision, excision, and second and third-degree burns. Incisional wounds do not involve tissue loss and scarring is not extensive if the edges of the wound are tightly apposed, whereas excisional and burn wounds involve tissue loss, take longer to heal, and can result in extensive scarring. By contrast, chronic wounds develop gradually and fail to heal because of underlying circulation and metabolic problems that interfere with the repair process. Diabetic ulcers, venous ulcers and pressure ulcers are chronic wounds that are major factors in escalating health care costs, lost economic productivity, diminished quality of life and premature death.

2.4 Stages of repair in acute wounds

A wound that breaches only the epidermis heals quickly and without scar. Penetration of the dermis is required to initiate repair by fibrosis. The repair of both incisional and excisional wounds can be divided into four overlapping and tightly integrated phases, each of which initiates the next: hemostasis, inflammation, structural repair and remodeling (figure 2.2). The wound is initially hypoxic due to destruction of blood vessels, but regains blood flow as angiogenesis re-establishes

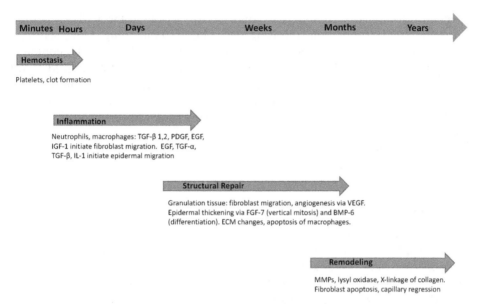

Figure 2.2. Timeline and stages of repair of a human skin wound.

normal circulation. These phases are regulated by signals among hematopoietic, immunologic and structural skin cells (Sun *et al* 2014).

Eleven cell types play major roles in the repair process. Cells of the immune system (platelets, neutrophils, macrophages, leukocytes, mast cells), and injured axons of sensory and sympathetic post-ganglionic neurons are involved in hemostasis and inflammation. Epidermal cells, dermal fibroblasts, endothelial cells and pericytes (perivascular cells that wrap around arterioles and capillaries) provide the means for structural repair. These cell types orchestrate the process of repair through their production of ECM molecules, proteases, growth factors, cytokines and chemokines that stimulate or inhibit specific cell activities (Barrientos *et al* 2008, Schultz and Wyscki 2009, Sun *et al* 2014, Bodnar *et al* 2016). Two of the most prominent growth factors regulating wound repair are platelet derived growth factor (PDGF) and transforming growth factor beta (TGFβ) (Barrientos *et al* 2008, Lichtman *et al* 2016). TGFβ has three isoforms (1, 2, 3) that have different roles in wound repair and regeneration.

2.4.1 Hemostasis

Hemostasis occurs within a matter of minutes to stop bleeding and seal off blood vessels by primary and secondary mechanisms (Clark 1996). Primary hemostasis involves the release of ADP from damaged endothelial cells, which causes platelets to clump at the site of blood vessel injury. Secondary hemostasis involves vasoconstriction and calcium-dependent clotting of the plasma by the tissue factor cascade (Lansdown 2002). The cascade is initiated by tissue factor (Factor 3) produced by the non-endothelial cells of injured blood vessels. Factor 3, in combination with Factor 7, activates prothrombinase (a complex of Factor 5 and Factor 10), which cleaves prothrombin to thrombin. Thrombin has two actions in a wound. The first is to induce platelet degranulation, releasing dense bodies containing ADP, serotonin, thromboxane A2, and Ca^{2+} that cause vasoconstriction, and α-granules containing fibrinogen, vitronectin, thrombospondin, thrombosthenin, platelet derived growth factor (PDGF) and the TGF-β1 isoform. The second action is to convert fibrinogen to fibrin, which gels into a soft clot, enmeshing platelets, red blood cells and leukocytes. The fibronectin, thrombospondin and vitronectin released by the α-granules, as well as collagens I, III and IV synthesized by blood monocytes, become part of the clot structure. Thrombosthenin contracts the clot, which then dehydrates to form a hard scab over the clot surface (Yamada and Clark 1996).

2.4.2 Inflammation

Inflammation is an adaptive response of the immune system to the cellular stress of injury (Medzhitov 2010) that is regulated by a plethora of growth factors and cytokines, including chemokines (small chemoattractant molecules), interleukins and tumor necrosis factor alpha (TNF-α) produced by immune cells, endothelial cells and fibroblasts (Eming *et al* 2007, Martins-Greene *et al* 2013).

The inflammatory response is initiated by the release of TGF-β1 and PDGF from degranulating platelets, and lasts for approximately 5–7 days (Wietecha *et al* 2014, Lichtman *et al* 2016). TGF-β1 is the only TGF-β isoform in human platelets and comprises 85% of the TGF-β in wound fluid (Roberts and Sporn 1996). These factors attract neutrophils, monocytes and T-lymphocytes into the wound. Within the clot, the monocytes differentiate into macrophages (Martin 1997). The macrophages secrete TGF-β1 and 2, PDGF and other chemoattractants that increase the number of neutrophils and macrophages entering the wound (Clark 1996). These attractions are mediated by an integrin receptor called Mac-1 (Novak and Koh 2013, Das *et al* 2015). Neutrophil influx diminishes within 3–4 days after injury. T-cell lymphocyte infiltration peaks by the end of the first week or later (Swift *et al* 2001).

Neutrophils and macrophages kill bacteria through oxygen-dependent mechanisms that generate hydrogen peroxide and hypochlorous acid, but neutrophils also kill by the production of bactericidal peptides and proteins (Clark 1996, for a review). These cells also secrete the collagenases MMP-1 and MMP-8 to degrade collagen types I, III within the wound (Nwomeh *et al* 1998). The neutrophils and macrophages phagocytize dead bacteria and cellular and molecular debris. Neutrophils and macrophages normally have a limited lifetime in the wound space and undergo apoptosis within a few hours after entering the wound. The neutrophils are ingested intact by macrophages, thus preventing their contents from being released into the wound (Haslett and Henson 1996). Macrophages also must be cleared from the wound after they have performed their functions. Apoptosis of macrophages is induced by vascular endothelial growth factor (VEGF) (Petreaca *et al* 2008), perhaps secreted by fibroblasts during structural repair, but the mechanism of their disposal is uncertain.

The macrophages entering a wound are initially pro-inflammatory (designated M1) in line with their bactericidal and phagocytic activities and their secretion of growth factors and cytokines that regulate inflammation. The macrophages then switch to an anti-inflammatory phenotype (designated M2), secreting factors that resolve inflammation and initiate structural repair by fibroblasts when activated by IL4/13 (Mescher 2017, for a review). Making the M1 to M2 switch requires that the macrophages first sense apoptotic neutrophils (Bosurgi *et al* 2017).

A number of molecules have been identified that terminate inflammation (Wietecha *et al* 2014, for a review). Lipoxins, resolvins and annexin A1 restrict further entry of neutrophils into the repairing wound from the bloodstream, hydrogen sulfide from endothelial cells stimulates neutrophils to undergo apoptosis, protectins restrict the release of pro-inflammatory molecules, and maresins stimulate tissue repair and act on nerves to relieve pain.

2.4.3 Structural repair

Structural repair begins during the inflammatory phase with the initiation of epidermal migration and continues as the fibrin clot is gradually invaded and degraded by fibroblasts. Structural repair is initiated by a variety of growth factors secreted by neutrophils and macrophages (Greenalgh 1996, Barrientos *et al* 2008, Wietecha *et al* 2014, for reviews).

2.4.3.1 Regeneration of epidermis

The epidermis closes the wound via the centripetal migration of both basal and suprabasal keratinocyes. Basal cells express keratin 14 and suprabasal cells express keratin 10 prior to wounding. After wounding, however, migrating cells express both keratin-16 and MMP-2 and -9 (Usui *et al* 2005). Epidermal cells at the edge of the wound migrate into the fibrin clot within a day or two after injury. The initiation of migration may involve a 'free edge' effect, in which a lack of neighbors on one side stimulates the cells to undergo EMT, permitting movement (Woodley 1996, for a review). The migrating cells lose their apical–basal polarity and dissolve the desmosomes holding them together laterally, as well as (in the case of basal cells) the hemidesmosomes that anchor them to the basement membrane. Simultaneously, the epidermal cells form the peripheral actin locomotory apparatus that gives them motility by the active protrusion of lamellipodia and filopodia (Fenteany *et al* 2000). The migrating cells do not themselves divide, but cells just behind the wound edge divide to produce a continuous feed of migrating cells.

Epidermal cell migration is regulated by several growth factors produced by neutrophils and macrophages, including EGF, TGF-α, TGF-β1 and TGF-β3 (Barrientos *et al* 2008, Wietecha *et al* 2014, Lichtman *et al* 2016, for reviews). The epidermal cells themselves secrete EGF, TGF-α and TGF-β3 to sustain their migration through autocrine signaling. The migrating cells secrete MMP-2 and -9 to cut a path through the fibrin clot and use fibronectin, vitronectin and collagen in the fibrin clot as adhesive substrates. Once the epidermal cells have covered the wound surface myofibroblasts of the granulation tissue induce them to divide vertically and synthesize a new basement membrane by the paracrine action of FGF-7 (keratinocyte growth factor, KGF), EGF, and granulocyte/monocyte-colony stimulating factor

(GM-CSF) produced. Migration is halted, and the cells undergo MET to re-establish normal epidermal thickness and attachment to the dermal scar tissue (Miller and Gay 1992, for a review; Ghalbzouri and Ponec 2004).

2.4.3.2 Formation of granulation tissue and wound contraction

Even before the inflammatory phase is over, fibroblasts begin migrating into the fibrin clot where they proliferate to form a tissue that under low magnification has a granular appearance, first described by the physician John Hunter in 1786. Genetic lineage tracing of wound fibroblasts indicates that they are restricted primarily to the dermis and hair follicle sheaths and contribute to dermal repair in two waves. The first wave is from the reticular and hypodermal layers, followed by a later wave from the papillary dermis (Driskell *et al* 2013). The fibroblasts proliferate and replace the clot with an ECM consisting of Fn, hyaluronate-PG complexes and a high ratio of type III to type I collagens (Weitzhandler and Bernfield 1992, for a review; Chen and Abatangelo 1999). High molecular weight hyaluronan is prominent in the early stages of repair, but as granulation tissue develops, progressively more fragmented forms of the molecule occur (Aya and Stern 2014). The hyaluronic acid–PG complexes bind water, expanding extracellular space and facilitating fibroblast migration.

The fibroblasts of the granulation tissue are not uniform in their fibrotic potential. Rinkevich *et al* (2015) identified a subpopulation of dermal fibroblasts derived from progenitors that transiently express *Engrailed1* (*En1*) in the central dermamyotome of the embryo. These fibroblasts are scattered throughout the entire dermis and are primarily responsible for the collagen deposition of structural repair. Ablating them results in a significant decrease in ECM deposition, suggesting that they are the cells involved in scar formation. Studies in which subsets of human macrophages were co-cultured with human dermal fibroblasts indicated that it is the M2 macrophages that promote the fibrogenic activities of the fibroblasts (Zhu *et al* 2017).

The size of the wound that has to be re-covered by epidermis and filled by granulation tissue is decreased by contraction of the wound edges. Wound contraction plays an important role in excisional wounds of mammals with loose skin such as rodents, but much less in other mammals (Yannas 2001, for a review). In humans, less than 50% of excisional wound closure is due to contraction; the majority is due to scar tissue formation. Wound contraction occurs by both the constriction of the skin by shortening and thickening of collagen fibrils in dermal fibroblasts around the edge of the wound and by smooth muscle actin contraction of myofibroblasts of the granulation tissue (Yannas 2001, Erlich and Hunt 2012, Darby *et al* 2014, Lichtman *et al* 2016). Myofibroblasts are cells with features of fibroblasts and smooth muscle cells that are able to contract (Darby *et al* 2014). They contract the granulation tissue by assembling fibronection into fibrils at their surfaces to form fibronexi, transmembrane connections linking their smooth muscle actin myofilaments with extracellular Fn fibrils (Newman and Tomasek 1996, Tomasek *et al* 1999).

Fibroblast proliferation and ECM synthesis are orchestrated by PDGF, TGF-β1, EGF and IGF-1 produced first by macrophages during the inflammatory phase, then by the fibroblasts themselves (Pierce 1991, Grotendorst 1992, Lawrence and Diegelmann 1994, Gottwald *et al* 1998). Animals in which wound healing is impaired by destroying macrophages have greatly reduced levels of PDGF, TGF-β1 and EGF activity (Leibovich and Ross 1975). PDGF renders fibroblasts competent to leave G_0 and enter the G_1 phase of the cell cycle (Morgan and Pledger 1992) and promotes the early synthesis of HA and the later synthesis of sulfated GAGs (Pierce 1991). TGF-β1 stimulates the early synthesis of Fn and, along with EGF, the later synthesis of type I collagen, elastin and sulfated PGs (Pierce 1991, Roberts and Sporn 1996). TGF-β also reduces collagen degradation by two complementary mechanisms, the reduction of collagenase gene transcription, and an increase in the synthesis of tissue inhibitors of metalloproteinases (TIMPs) (Jeffrey 1992, for a review).

Capillary regeneration (angiogenesis) occurs simultaneously with fibroblast migration into the wound space to provide the oxygen and nutrients required to form granulation tissue (Carmeliet and Jain 2011). Initially, there is capillary hyper-regeneration, giving the granulation tissue its characteristic red color. Later, most of the capillaries regress as the granulation tissue is remodeled into scar. Angiogenesis occurs primarily by the sprouting of new capillaries from injured venules. Endothelial cells of the venule wall lose their specialized junctions and undergo mitosis to form a cord of proliferating cells. As the cord grows, its cells tubulate to form a new capillary extending from the venule (figure 2.3). Tubulation is accomplished by the endocytotic formation of vacuoles that fuse with the endothelial cell membrane where several cells meet to form larger vessels or the fusion of vacuoles in tandem to form capillaries (Kamei *et al* 2006).

Capillary regeneration is regulated by a variety of growth factors and ECM molecules that act at different phases of the process. These factors are released from platelets and ECM and synthesized by macrophages, fibroblasts, endothelial cells and pericytes. A major function of pericytes is to stabilize endothelial cell walls of capillaries and arterioles (Bodnar *et al* 2016). FGF-2, TGF-β1, IL-8 and TNF-α activate endothelial cells to release them from the stabilizing pericytes during angiogenesis. In the presence of VEGF, which is synthesized in large quantities by

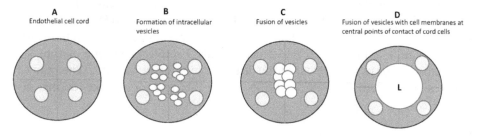

Figure 2.3. Diagrams showing the tubulation of a cord of regenerating endothelial cells. Yellow circles = nuclei. (A) cord of endothelial cells. (B) Vesicles form in the cells. (C) Vesicles fuse together and wit the plasma membrane of each cell near the central junction of the cells. (D) Vesicles incorporated into cell membranes to form the lumen (L) of the new vessel. Reprinted by permission from Macmillan Publishers Ltd: *Nature*, Mostov and Martin-Belmonte (2006). Copyright 2006.

the regenerating epidermis (Clark 1996), this leads to endothelial cell motility and proliferation (Yancopoulos *et al* 2000). The endothelial cells secrete collagenase and tissue plasminogen activator (tPA), which converts plasminogen to plasmin, helping to dissolve the fibrin clot and facilitate their migration (Madri *et al* 1996, for a review).

Migration of endothelial cells is promoted by PD-ECGF, TNF-α, FGF-1 and -2, and matrix molecules Fn, HA, Ln and Tn (Tomanek and Schatterman 2000). Migrating endothelial cells express PECAM-1 in a diffuse pattern on their surfaces, suggesting that this adhesion molecule plays a role in their movement through the fibrin matrix (Madri *et al* 1996). Proliferation is promoted by TGF-α, VEGF, PDGF and FGF-2 (Kinnaird *et al* 2004, Rehman *et al* 2004). Migration and proliferation of endothelial cells is also strongly stimulated by HGF produced by fibroblasts (Conway *et al* 2006).

2.4.3.3 Remodeling of granulation tissue

In the final phase of structural repair, the granulation tissue is remodeled into a relatively acellular fibrous scar tissue. At first, the collagens of the granulation tissue are organized in the normal reticular network. Dermal repair then begins to show visible signs of following a fibrotic rather than a regenerative pathway. Hyaluronate synthesis is decreased and synthesis of chondroitin sulfate and dermatan sulfate-PGs is increased. Much more type I collagen is synthesized than type III, and the absolute amount of collagen synthesized is greater than in uninjured skin. The scar differs from normal dermis in several ways (Linares 1996, for a review). Fibronectin and HA levels return to normal, but the level of chondroitin-4-sulfate PG is much higher, decorin PG is lower than normal, and the number of elastin fibers is reduced. Type I collagen fibers are broken down by MMPs produced by the epidermis and fibroblasts, and cross-linked by the enzyme lysyl oxidase into thick bundles oriented parallel to the surface of the wound, instead of the normal reticular organization.

As the scar matures, the fibroblasts and capillaries are eliminated by apoptosis (Lorena *et al* 2002). Apoptosis of fibroblasts is due to the down regulation of EGF, which acts as a survival signal produced by macrophages during the inflammatory and early granulation tissue formation phases (Shao *et al* 2008). Macrophages and pericytes are thought to induce the apoptosis of endothelial cells and thus regression of the capillary network (Diez-Roux *et al* 1999, Bodnar *et al* 2016). Failure to prune the capillary network and fibroblasts from the scar leads to hypertrophic scarring, which is manifested in thick scars termed keloids. Establishment of a mature, stable scar takes ~80 days in rodents, but at least six months in humans. The scar never achieves more than 70%–80% of the tensile strength of normal dermal tissue (Mast 1992).

2.4.3.4 Role of reinnervation

Peripheral sensory and sympathetic post-ganglionic nerves regenerate in healing wounds (Gottwald *et al* 1998, Kim *et al* 1998). After wounding, the portion of the nerves distal to the injury degenerates within one to two days, then regenerates over the next two weeks, resulting in hyperinnervation of the granulation tissue.

Subsequently, many of the nerve fibers regress. Thus, reinnervation follows the same pattern of advance and recession as angiogenesis, suggesting that angiogenesis and nerve regeneration are mechanistically coupled. Denervation of wounded rat skin significantly delays wound contraction and re-epithelialization (Fukai 2005), suggesting that regenerating axons provide factors essential for these processes.

2.4.4 Hair follicles and skin repair

Wounded skin undergoing fibrotic repair does not normally restore appendages such as hair follicles and sebaceous glands. Fibroblasts exposed to FGF-2 and transfected with *Bmp-2* and *Wnt-3*, however induced hair follicles from the regenerating epidermis when transplanted to back wounds of rats (Ono *et al* 2009). Sox2+ stem cells called skin precursor (SKP) cells have been isolated from the dermis and the dermal papillae of hair follicles that can differentiate into adipocytes, chondrocytes, osteoblasts and Schwann cells, and may contribute cells for dermal maintenance and repair (McKenzie *et al* 2006, Biernaskie *et al* 2009). Plikus *et al* (2017) reported that myofibroblasts are reprogrammed to adipocytes by BMP signaling from hair follicles during wound repair, and that myofibroblasts from human keloids formed adipocytes when treated *in vitro* with BMP or cultured with human hair follicles. This reprogramming may be useful in reducing keloid scarring by turning myofibroblasts into adipocytes.

2.5 Aging reduces the repair capacity of acute skin wounds

The capacity for wound healing declines with age. Aging skin shows changes in proteoglycans that decrease thickness, elasticity and hydration of the dermis, and increases in lipofuchsin deposits and benign lesions of various kinds, all of which are intensified by over-exposure to solar UV radiation (Kim *et al* 2015, for a review). In addition, there is decreased wound contraction and increased collagen remodeling by MMPs with age (Yao *et al* 2001, Ballas and Davidson 2001, Swift *et al* 2001).

Studies of the inflammatory response in skin wounds of mice indicate that during the early inflammatory phase, neutrophil content is the same in young versus old mice. Monocyte chemoattractant protein (MCP-1) is elevated in old mice leading to a 56% higher content of macrophages that, however, have a 37%–43% reduction in their phagocytic capacity. Most chemokines are reduced in old mice, and infiltration of the wound by lymphocytes during the later inflammatory phase is delayed (Swift *et al* 2001). Skin wounds in old rats exhibit a significant delay in wound closure during the early phase of repair, but old and young rats had similar rates of closure thereafter. The delay was not caused by differences in abundance of fibroblasts or myofibroblasts in granulation tissue, but skin fibroblasts at the edge of the wound expressed more MMP-2 in old rats, suggesting that proteolysis could play a role in the delay (Ballas and Davidson 2001). A comparison of rate of wound closure and growth factor profiles (VEGF, PDGF-BB) in wounds of pigs of various ages revealed statistically significant reductions in younger versus older animals, whereas there was no difference in TGF-β1 levels (Yao *et al* 2001).

2.6 Chronic wounds fail to heal

Chronic (non-healing) wounds are a major health problem the world over (Sen *et al* 2009, Frykberg and Banks 2015, Gould *et al* 2015). Treatment of chronic wounds is reported to require over $25 billion US dollars per year and the biomedical and socioeconomic burdens of these wounds are growing with the aging population in the US and Europe. The wounds are primarily ulcers of the lower extremities resulting from compromised circulation due to pressure, diabetes and venous incompetence (figure 2.4). Chronic wounds are characterized by a failure to re-epithelialize, a persistent inflammatory phase that prevents transition to the formation of granulation tissue, and by defective production and remodeling of ECM (Mast and Schultz 1996, Mustoe 2005). Hypoxia due to poor circulation is thought to be a major factor in the failure of chronic wounds to heal where there is a persistent inflammatory phase. Poor circulation in diabetic patients is correlated with impaired proliferation, adhesion and incorporation of circulating endothelial precursor cells into vascular structures (Tepper *et al* 2002) and diabetic ulcers are reported to have significantly decreased numbers of these cells (Keswani *et al* 2004).

The profiles of growth factors and inflammatory cytokines are different in acute and chronic wounds of human patients (Moor *et al* 2009). Proteolytic activity, particularly MMP-9, is much higher in chronic wounds, and TIMP levels much lower. The activity of pro-inflammatory cytokines appears to be much higher in chronic wounds, consistent with a persistent inflammatory phase. The wound fluid of human leg ulcers was reported to have significantly higher concentrations of the

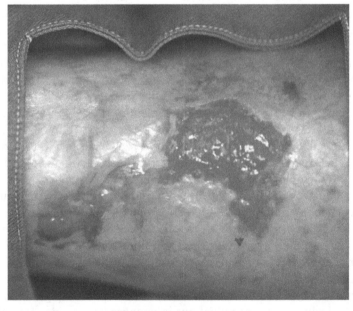

Figure 2.4. Chronic lower leg wound. Reproduced with permission from (Liu *et al* 2006). John Wiley & Sons. Copyright 2006.

pro-inflammatory cytokines IL-1, IL-6 and TNF-α, but no significant differences between healing and chronic wounds in the levels of PDGF, EGF, FGF-2 or TGF-β (Trengrove *et al* 2000).

2.7 Non-scarring skin repair models

There are several scar-free wound repair models that have given us a deeper understanding of the cellular and molecular conditions underlying a regenerative versus fibrotic response to wounding. While some data are contradictory, and there are many unanswered questions, these models strongly suggest that a regenerative response of skin to wounding requires rapid re-epithelialization, a higher rate of fibroblast migration, slower deposition of ECM components, an increased ratio of collagen III/collagen I, and a dampened inflammatory response.

2.7.1 Fetal skin

2.7.1.1 Cellular and ECM differences in fetal and adult wound repair
Fetal mammalian skin regenerates without scarring after wounding (Olutoye and Cohen 1996). Late in gestation, fetal skin changes its regenerative response to injury to the adult fibrotic response. In the rat and mouse, this transition takes place at 16–18 days of gestation, 3–5 days prior to birth (Ihara *et al* 1990). Comparison of fetal wound healing with adult wound healing is a valuable approach to elucidating the differences between regeneration and fibrosis and how we might intervene to prevent scarring in adults.

These studies have revealed a number of cellular differences in the repair of fetal and adult wounds in mice and rats that reflect differences in the injury niche composition (Lichtman *et al* 2016). The area of the wound to be re-covered by epidermis is rapidly reduced in fetal wounds by a purse-string contraction of a filamentous actin cable assembled in the epidermis at the edges of the wound, making re-epithelialization faster (McCluskey and Martin 1995). Observations on DiI-labeled fetal fibroblasts indicated that fibroblast migration is more rapid in fetal wounds (Mast *et al* 1997, Ihara and Motobayashi 1992). Re-epithelialization and fibroblast migration were also much faster in fetal wounds in an *in vitro* burn model of fetal versus adult human skin (Coolen *et al* 2010).

The ratio of collagen/total protein in the ECM of wounded adult skin is higher than in unwounded adult skin, whereas there is no significant difference in this ratio between wounded and unwounded fetal skin (Houghton *et al* 1995). Fetal skin wounds exhibit a higher ratio of type III to type I collagen (Merkel *et al* 1988) than adult wounds and sulfated PG synthesis does not accompany collagen synthesis in fetal wounds as it does in adult wounds (Whitby and Ferguson 1992). The fibroblasts of fetal wounds synthesize the same collagens as those of adult wounds but there is no excessive deposition of type I collagen fibrils in fetal wounds and the fibrils are organized in the basket-weave pattern of normal dermis (Frantz *et al* 1992, Mast *et al* 1997). Hyaluronic acid plays an important role in facilitating fetal dermal regeneration. Fetal wound fibroblasts synthesize higher levels of HA and HA receptor (Alaish *et al* 1993), facilitating cell migration while simultaneously inhibiting fetal fibroblast proliferation (Mast *et al* 1993). Treatment of wounded adult tympanic membranes with hyaluronate decreases scar formation (Hellstrom and Laurent 1987). Conversely, treatment of fetal rabbit skin wounds with hyaluronidase or HA degradation products alters the regenerative response toward fibrosis (Mast *et al* 1995).

2.7.1.2 Regeneration of fetal skin is correlated with an immature immune system

The ability of fetal wounds to regenerate is associated with an under-developed immune system (Mescher *et al* 2017, for a review). Fetal wounds of rodents exhibit a minimal inflammatory response associated with smaller numbers of platelets, neutrophils and macrophages compared to adult skin wounds (McCallion and Ferguson 1995, Yang *et al* 2003, Ferguson and O'Kane 2004). In adult wounds, PDGF induces the persistent expression of IL-6 by fibroblasts, which helps maintain an environment that promotes production and deposition of fibrotic matrix. Fetal wounds have much lower levels of PDGF and although they initially express IL-6, this expression rapidly disappears (Liechty *et al* 2000).

TGF-β1, 2 and their receptors play a prominent role in adult scarring (Walraven *et al* 2014). Lower expression of TGF-β1 and 2 and higher levels of TGF-β3 protein have been reported for both unwounded and wounded fetal skin (Ferguson and O'Kane 2004, Chen *et al* 2005). Gosiewska *et al* (2001) found that TGF-β1 transcript expression was 1.5 times higher in cultured fetal fibroblasts compared to neonatal fibroblasts. However, the level of active TGF-β1 protein was virtually undetectable in fetal fibroblasts due to low levels of the LTBP-1 protein, which is required to activate latent TGF-β1, and is readily detectable in neonatal fibroblasts, suggesting that low levels of active TGF-β1 in fetal mouse skin are the result of low levels of LTBP-1 (Walraven *et al* 2014). Consistent with these results, deep dermal burn wounds of fetal sheep exhibit only a slight increase in active TGF-β1 and heal without scarring (Fraser *et al* 2005).

2.7.1.3 Intrinsic changes in fetal fibroblasts might play a role in loss of regenerative capacity

The phenotype of fetal fibroblasts differs from that of the adult fibroblast, including production of and response to cytokines and growth factors, synthesis of matrix molecules, pericellular HA coats and antigen determinants (Moriarty *et al* 1996, Ellis *et al* 1997, Gosiewska *et al* 2001). Fetal fibroblasts appear refractory to prostaglandin E2, an important mediator of fibrosis in adult skin (Lupulescu 1975, Sandulache *et al* 2006). An interesting *in vitro* experiment suggested that the transition from fetal to adult fibroblast, and from a regenerative to scarring response is independent of immune cells and systemic factors (Chopra *et al* 1997). Mouse limb buds were isolated in culture at 12 days of development, prior to development of the immune system. After wounding, the skin of the limb buds regenerated. However, if the 12 day cultured limb buds were allowed to develop to the equivalent of an 18 day limb bud and then wounded, the response to wounding was scarring.

2.7.2 Immune-deficient mice

Further evidence that the maturity of the immune system is a determining factor in the fetal versus adult response to wounding comes from studies on immune-deficient mice. PU.1-null mice lack a hematopoietic lineage transcription factor that results in the absence of macrophages and neutrophils and die of bacterial infection within 24 h after birth unless they are given a wild-type bone marrow transplant (Dovi *et al* 2004). Antibiotics can be used to prolong the lives of neonatal PU.1 mice. Excisional wounds made in the skin of antibiotic-maintained PU.1-null mice were repaired by regeneration, not fibrosis (Martin *et al* 2003). Regeneration was associated with greatly reduced expression of IL-6 and TGF-β1, 2 mRNA (Redd *et al* 2004). Similarly, athymic nude-*nu* adult mice, which lack T-cells, repair skin wounds without scarring (Barbul and Regan 1990, 1995). As in fetal wounds, the lack of scar is associated with low levels of collagen, PDGF-B and TGF-β1, and high levels of HA (Gawronska-Kozak *et al* 2006). These mice also exhibit an enhanced ability to regenerate ear tissue lost to punch holes (Gawronska-Kozak 2004, Gawronska-Kozak *et al* 2006). Selective depletion of T-suppressor and cytotoxic lymphocytes or T-helper and T-suppressor lymphocytes shifted the wound response toward regeneration in control wild-type mice (Barbul *et al* 1989a), while reconstitution of T-cells in thymic nude-*nu* mice decreased the breaking strength of the healed wound, consistent with the presence of scar tissue (Barbul *et al* 1989b).

2.7.3 Adult oral mucosa

The structure of the oral mucosa is largely similar to that of skin, but the process of repair in this tissue is more rapid than in skin dermal wounds, and occurs without scarring. Glim *et al* (2013) have reviewed what is known about dermal versus oral wound healing (see also Boink *et al* 2016). The lack of scarring is associated with more rapid re-epithelialization and fewer blood vessels, but unlike fetal wounds there is an increased rate of fibroblast proliferation. Mast cells, neutrophils, macrophages and T-cells are reduced. Rapid re-epithelialization is associated with the down regulation of the gap junction connexins 26, 30 and 43 (Davis *et al* 2013). Collagen I and FNED-A (fibronectin splice variant extra domain A, which is thought to play a role in fibrosis) are up regulated for a shorter period of time and tenascin-C persists longer. There is a reduced inflammatory response characterized by brief expression of IL-6 and IL-8, the absence of other cytokines normally present in dermal wounds, reduced TGF-β1 and increased expression of TGF-β3.

2.7.4 Amphibian skin

Yannas (1996) reported that the skin of the anuran tadpole *Rana catesbeiana* regenerates without scarring, but that adult frog skin repairs by scar. By contrast, the skin of another adult anuran, *Xenopus laevis,* was reported to regenerate perfectly (Yokoyama *et al* 2011). Regeneration of *Xenopus* skin is associated with mononuclear cells that collect under a wound epidermis and express the paired-type homeobox-containing transcription factor Prx1 and its enhancer. The Prx1 enhancer is not activated in the excisional wounds of mice, suggesting that it is required for scar-free wound repair in skin.

The axolotl, a neotenous urodele (newts and salamanders) amphibian, also exhibits scar-free healing of skin wounds. Seifert *et al* (2012a) studied excisional flank wounds in adult axolotls and found that the lack of scarring was correlated with high expression of MMPs by migrating keratinocytes, reduced neutrophil infiltration, and a relatively long delay in the accumulation of new dermal ECM with low levels of fibronectin and high levels of tenascin-C.

2.7.5 African spiny mouse skin

The African spiny mouse, *Acomys*, is the first mammalian model of scar-free adult skin repair. The genus derives its name from the presence of spine-like hairs on its back skin. Predators of this mouse grab its neck skin, but the skin is weak and tears away easily to facilitate escape, followed by rapid and scar-free regeneration of the skin. Seifert *et al* (2012b) found that *Acomys* can restore up to 60% of its total dorsal skin area. The weakness of the skin appears due to a lack of elasticity compared to *Mus*, because of the large hair follicles that reduce the space available for dermal connective tissue. Skin regeneration in *Acomys* is characterized by faster re-epithelialization than in *Mus*, extensive contraction of the wound edges by 64%, and a more slowly deposited dermal ECM with a high collagen III/collagen I ratio. The original reticular organization of the fibrous ECM components is restored. The regeneration of new hair follicles appears to involve the same Wnt and BMP signaling pathways as in embryonic hair follicle development.

2.8 Comparative molecular analysis of wound repair and regeneration

To obtain a more comprehensive picture of the molecular differences between fibrosis and regeneration in wound repair, comparative analyses have been conducted in mice and rabbits for unwounded versus wounded adult and fetal skin, fetal wounded versus adult wounded skin, and *Mus* versus *Acomys* adult skin wounds.

2.8.1 Unwounded versus wounded adult and fetal skin

Microarray analysis of unwounded versus wounded adult mouse skin revealed that at 30 min post-injury, 3% of the genes were up regulated by two-fold or more, with most of these involved in signaling and signal transduction (Cole *et al* 2001). By one hour, the percentage of up regulated genes had declined to 1.15%, and 6.6% were down regulated two-fold or more. Analysis of a subtractive cDNA library made between unwounded and wounded mouse back skin identified several wound-regulated genes (Kaesler *et al* 2004). One of these encodes the chemokine receptor CCR1, which was barely detectable in unwounded skin, but was strongly expressed in macrophages and neutrophils after injury. However, CCR1 knockout mice healed wounds normally, indicating that there is redundancy in chemokine/receptor signaling in skin wounds. A transcriptional signature for the various stages of wound repair in mouse skin has been identified by Peake *et al* (2014).

The epidermal transcriptome has been compared in human adult skin before and at various times after wounding (Nuutila *et al* 2012). Many genes are up regulated and down regulated after wounding, particularly keratins, matrix metalloproteinases, which are involved in epithelial migration through the fibrin matrix, and serine protease inhibitors, which regulate the functions of proteases such as MMPs. An interesting finding was that transcription of growth factor genes did not change appreciably, indicating that the growth factor proteins are not regulated at the transcriptional level during epidermal regeneration. This finding is consistent with the fact that growth factors in the skin are produced as inactive precursors that are activated by up regulated MMPs.

PCR suppression subtraction of wounded versus unwounded rabbit fetal skin revealed 15 genes that were up regulated and 20 that were down regulated in the injured skin (Kathju *et al* 2006). Several of the down regulated genes were associated with cytoskeletal modulation, suggesting that scarless healing might involve alteration of fibroblast/myofibroblast action. Comparison *in vitro* of mid-gestation and late-gestation mouse skin fibroblasts revealed the differential expression of 62 genes (Wulff *et al* 2013). The most differentially expressed genes belonged to the cell proliferation category and were more highly up regulated in mid-gestational fibroblasts, consonant with their higher proliferation rates shown in other studies. The anti-inflammatory genes SOCS2 and CD109 were also up regulated in mid-gestational fibroblasts, while the pro-inflammatory gene for CO-1 was up regulated in late-gestational fibroblasts. These results suggest that the transition from scar-free to fibrotic repair of fetal skin may reflect developmental changes in dermal fibroblasts, as suggested above.

2.8.2 Fetal wounded versus adult wounded skin

Microarray studies of fetal wounded versus adult wounded rat skin (Chen *et al* 2007) found that expression of most growth factor genes did not differ between the two, except for FGF-2 and -8. TGF-β1 expression was similar in fetal and adult skin, but follistatin, an inhibitor of TGF-β1, was higher in fetal skin. The expression of the type 1α collagen chain was higher in adult skin, but that of collagen III was the same. Differential expression was noted between fetal and adult skin of genes involved in transcription, cell cycle regulation, protein homeostasis and intracellular signaling (Colwell *et al* 2008). Microarray analysis comparing cultured, serum-stimulated fibroblasts from adult oral mucosa, normal skin and chronic wounds, which show a continuum of repair from high to low in that order, showed that genes whose expression increased in the same order were associated with dysfunctional healing, whereas genes with the opposite pattern were associated with enhanced healing.

2.8.3 Adult skin wounds of *Acomys* versus *Mus*

Acomys skin regeneration has been compared with *Mus musculus* skin repair by fibrosis (Brant *et al* 2016). This study measured immune cells, wound cytokines, cell proliferation and collagenous components of the wound by microarray, targeted proteomic and histological analyses and found that *Acomys* skin regeneration shares a number of features with the regeneration of *Mus* fetal skin. There was minimal difference in cell proliferation, but very few collagens are present or up regulated in the *Acomys* wound except collagen XII, which is up regulated 10- to 30-fold. The failure to up regulate fibrotic collagens was associated with the absence or low levels of pro-inflammatory cytokines and a lack of macrophages.

These kinds of comparative studies all tie skin regeneration to a weak or absent immune response as opposed to a vigorous immune response that leads to fibrosis (Mescher *et al* 2017). Coupled with bioinformatic analysis and systems biology approaches (Rao *et al* 2014, Sood *et al* 2015), such studies have the potential to lead to a much deeper molecular understanding of the differences between scarring and regeneration of wounded skin, and thus how to shift the response of adult skin to wounding from fibrosis to regeneration.

References

Alaish S M *et al* 1993 Hyaluronate receptor expression in fetal fibroblasts *Surg. Forum* **44** 733–5

Aya K L and Stern R 2014 Hyaluronan in wound healing: rediscovering a major player *Wound Rep. Reg.* **22** 579–93

Barbul A, Breslin J R and Woodyard J P 1989a The effect of *in vivo* T helper and T suppressor lymphocyte depletion on wound healing *Ann. Surg.* **209** 479–83

Barbul A *et al* 1989b Wound healing in nude mice: a study on the regulatory role of lymphocytes in fibroplasia *Surgery* **105** 764–9

Barbul A and Regan M C 1990 The regulatory role of T lymphocytes in wound healing *J. Trauma* **30** S97–100

Barbul A and Regan M C 1995 Immune involvement in wound healing *Otolaryngol. Clin. North Am.* **28** 955–68

Ballas C B and Davidson J M 2001 Delayed wound healing in aged rats is associated with increased collagen gel remodeling and contraction by skin fibroblasts, not with differences in apoptotic of myofibroblast cell populations *Wound Rep. Reg.* **9** 223–37

Barrientos S *et al* 2008 Growth factors and cytokines in wound healing *Wound Rep. Reg.* **16** 585–601

Biernaskie J *et al* 2009 SKPs derive from hair follicle precirsors and exhibit properties of adult dermal stem cells *Cell Stem Cell* **5** 610–23

Bodnar R J, Satish L, Yates C C and Wells A 2016 Pericytes: a newly recognized player in wound healing *Wound Rep. Reg.* **24** 204–14

Boink M A *et al* 2016 Different wound healing properties of dermis, adipose, and gingiva mesenchymal stromal cells *Wound Rep. Reg.* **24** 100–9

Bornstein P 2009 Matricellular proteins: an overview *J. Cell Commun. Signal.* **3** 163–5

Bosurgi L *et al* 2017 Macrophage function in tissue repair and remodeling requires IL-4 or IL-13 with apoptotic cells *Science* **356** 1072–6

Brant J O *et al* 2016 Cellular events during scar-free skin regeneration in the spiny mouse, *Acomys Wound Rep. Reg.* **24** 75–88

Carmeliet P and Jain R K 2011 Molecular mechanisms and clinical applications of angiogenesis *Nature* **473** 298–307

Chen W Y and Abatangelo G 1999 Functions of hyaluronan in wound repair *Wound Rep. Reg.* **7** 79–89

Chen W *et al* 2005 Ontogeny of expression of transforming growth factor-β and its receptors and their possible relationship with scarless healing in human fetal skin *Wound Rep. Reg.* **13** 68–75

Chen W *et al* 2007 Profiling of genes differentially expressed in a rat of early and later gestational ages with high-density oligonucleotide DNA array *Wound Rep. Reg.* **15** 147–56

Chopra V, Blewett C J, Ehrlich H P and Krummel T M 1997 Transition from fetal to adult repair occurring in mouse forelimbs maintained in organ culture *Wound Rep. Reg.* **5** 47–51

Clark R A F 1996 Wound repair: overview and general considerations *Molecular and Cellular Biology of Wound Repair* ed R A F Clark (New York: Plenum), pp 3–50

Cole J *et al* 2001 Early gene expression profile of human skin to injury using high-density cDNA microarrays *Wound Rep. Reg.* **9** 360–70

Colwell A S, Longaker M T and Lorenz H P 2008 Identificaton of differentially regulated genes in fetal wounds during regenerative repair *Wound Rep. Reg.* **16** 450–9

Conway K, Price P, Harding K G and Jiang W G 2006 The molecular and clinical impact of hepatocyte growth factor, its receptor, activators, and inhibitors in wound healing *Wound Rep. Reg.* **14** 2–10

Coolen N A *et al* 2008 Development of an *in vitro* burn wound model *Wound Rep. Reg.* **16** 559–67

Coolen N A *et al* 2010 Wound healing in a fetal, adult, and scar tissue model: a comparative study *Wound Rep. Reg.* **18** 291–301

Darby I A, Laverdet B, Bonte F and Desmouliere A 2014 Fibroblasts and myofibroblasts in wound healing *Clin. Cosmet. Investig. Dermatol.* **7** 301–11

Das A *et al* 2015 Monocyte and macrophage plasticity in tissue repair and regeneration *Am. Pathol.* **185** 2596–606

Davis K, Phillips A and Becker D L 2013 Connexin dynamics in the privileged wound healing of the buccal mucosa *Wound Rep. Reg.* **21** 571–8

Diez-Roux G *et al* 1999 Macrophages kill capillary cells in G_1 phase of the cell cycle during programmed vascular regression *Development* **126** 2141–7

Dovi J V, Szpaderska A M and Di Pietro L A 2004 Neutrophil function in the healing wound: adding insult to injury *Thromb. Haemost.* **92** 275–80

Driskell R R *et al* 2013 Distinct fibroblast lineages determine dermal architecture in skin development and repair *Nature* **504** 277–81

Ellis I, Banyard J and Schor S 1997 Differential response of fetal and adult fibroblasts to cytokines: cell migration and hyaluronan synthesis *Development* **124** 1593–600

Eming S A, Krieg T and Davidson J M 2007 Inflammation in wound repair: molecular and cellular mechanisms *J. Investig. Dermatol.* **127** 514–25

Fenteany G, Janmey P A and Stossel T P 2000 Signaling pathways and cell mechanics involved in wound closure by epithelial cell sheets *Curr. Biol.* **10** 831–8

Ferguson M W J and O'Kane S 2004 Scar-free healing: from embryonic mechanisms to adult therapeutic intervention *Philos. Trans. R. Soc. Lond.* B **359** 839–50

Frantz F W, Diegelmann R F, Mast B A and Cohen I K 1992 Biology of fetal wound healing: collagen biosynthesis during dermal repair *J. Ped. Surg.* **27** 945–9

Fraser J F *et al* 2005 Deep dermal burn injury results in scarless wound healing in the ovine fetus *Wound Rep. Reg.* **13** 189–97

Frykberg R G and Banks J 2015 Challenges in the treatment of chronic wounds *Adv. Wound Care* **4** 560–82

Fukai T, Takeda A and Uchinuma E 2005 Wound healing in denervated rat skin *Wound Rep. Reg.* **13** 175–80

Gawronska-Kozak B 2004 Regeneration in the ears of immunodeficient mice: identification and lineage analysis of mesenchymal stem cells *Tiss. Eng.* **10** 1251–61

Gawronska-Kozak B, Bogacki M, Rim J S, Monroe W T and Manuel J A 2006 Scarless skin repair in immunodeficient mice *Wound Rep. Reg.* **14** 265–76

Ghalbzouri A and Ponec M 2004 Diffusable factors released by fibroblasts support epidermal morphogenesis and deposition of basement membrane components *Wound Rep. Reg.* **12** 359–67

Glim J E *et al* 2013 Detrimental dermal wound healing: what can we learn from the oral mucosa? *Wound Rep. Reg.* **21** 648–60

Gosiewska A *et al* 2001 Differential expression and regulation of extracellular matrix-associated genes in fetal and neonatal fibroblasts *Wound Rep. Reg.* **9** 213–22

Gottwald T, Coerper S, Schaffer M, Koveker G and Stead R H 1998 The mast cell-nerve axis in wound healing: a hypothesis *Wound Rep. Reg* **6** 8–20

Gould L, Abadir P and Brem H *et al* 2015 Chronic wound repair and healing in older adults: current status and future research *Wound Rep. Reg.* **23** 1–13

Greenalgh D G 1996 The role of growth factors in wound healing *J Trauma Injury Infect. Crit. Care* **41** 159–67

Grotendorst G R 1992 Chemoattractants and growth factors *Wound Healing: Biochemical and Clinical Aspects* ed I K Cohen, R F Diegelmann and W J Lindblad (Philadelphia, PA: WB Saunders), pp 237–47

Ham A W and Cormack D H 1979 *Histology* 8th edn (Philadelphia: JB Lippincott), pp 614–44

Haslett C and Henson P 1996 Resolution of inflammation *The Molecular and Cellular Biology of Wound Repair* ed R A F Clark (New York: Plenum), pp 143–70

Hellstrom S and Laurent C 1987 Hyaluronan and healing of typanic membrane perforations. An experimental study *Acta Otolaryngol.* **42** 54–61

Houghton P E, Keefer K A and Krummel T M 1995 The role of transforming growth factor-beta in the conversion from 'scarless' healing to healing with scar formation *Wound Rep. Reg.* **3** 229–36

Ihara S, Motobuyashi Y, Nagao E and Kistler A 1990 Ontogenetic transition of wound healing pattern in rat skin occurring at the fetal stage *Development* **110** 671–80

Ihara S and Motobayashi Y 1992 Wound closure in foetal rat skin *Development* **114** 573–82

Jeffrey J J 1992 Collagen degradation *Wound Healing: Biochemical and Clinical Aspects* ed I K Cohen, R F Diegelmann and W J Lindblad (Philadelphia, PA: WB Saunders), pp 177–94

Kaesler S *et al* 2004 The chemokine receptor CCR1 is strongly up-regulated after skin injury but dispensable for wound-healing *Wound Rep. Reg.* **12** 193–204

Kamei M, Saunders W B, Bayless K J, Dye L, Davis G E and Weinstein B M 2006 Endothelial tubes assemble from intracellular vacuoles *in vivo Nature* **442** 453–6

Kathju S *et al* 2006 Identification of differentially expressed genes in scarless wound healing utilizing polymerase chain reaction-suppression subtractive hybridization *Wound Rep. Reg.* **14** 413–20

Keswani S G *et al* 2004 Adenoviral mediated gene transfer of PDF-B enhances wound healing in type I and type II diabetic ulcers *Wound Rep. Reg.* **12** 497–504

Kim L R, Whelpdale K, Zurowski M and Pomerantz B 1998 Sympathetic denervation impairs epidermal healing in cutaneous wounds *Wound Rep. Reg.* **6** 194–201

Kim D J *et al* 2015 Cutaneous wound healing in aging small mammals: a systematic review *Wound Rep. Reg.* **23** 318–39 .

Kinnaird T *et al* 2004 Local delivery of marrow-derived stromal cell cells augments collateral perfusion through paractine mechanisms *Circulation* **109** 1543–9

Lansdown A B G 2002 Calcium: a potential central regulator in wound healing in the skin *Wound Rep. Reg.* **10** 271–85

Lawrence W T and Diegelman R 1994 Growth factors in wound healing *Clin. Dermatol.* **12** 157–69

Leibovich S B and Ross R 1975 The role of the macrophage in wound repair: a study with hydrocortisone and anti-macrophage serum *Am. J. Pathol.* **1978** 71–91

Liechty K W, Adzick N S and Cromblehome T M 2000 Diminished interleukin 6 (IL-6) production during scarless human fetal wound repair *Cytokine* **12** 671–6

Lichtman M K *et al* 2016 Transforming growth factor beta (TGF-β) isoforms in wound healing and fibrosis *Wound Rep. Reg.* **24** 215–22

Linares H A 1996 From wound to scar *Burns* **22** 339–52

Liu J Y *et al* 2006 Autologous keratinocytes on porcine gelatin microbeads effectively heal chronic venous leg ulcers Wound Rep. Reg. **12** 148–56

Lorena D, Uchio K, Monte Alto Costa A and Desmouliere A 2002 Normal scarring: importance of myofibroblasts *Wound Rep. Reg.* **10** 86–92

Lupulescu A 1975 Effect of prostaglandins on protein, RNA and DNA and collagen synthesis in experimental wounds *Prostaglandins* **10** 573–9

Madri J A, Asankar S and Romanic A M 1996 Angiogenesis *The Molecular and Cellular Biology of Wound Repair* 2nd edn ed R A F Clark (New York: Plenum), pp 355–72

Martin P 1997 Wound healing—aiming for perfect skin regeneration *Science* **276** 75–81

Martin P *et al* 2003 Wound healing in the PU.1 mouse—tissue repair is not dependent on inflammatory cells *Curr. Biol.* **13** 1122–8

Martins-Greene M, Petreaca M and Wang L 2013 Chemokines and their receptors are key players in the orchestra that regulates wound healing *Adv. Wound Care* **2** 327–47

Mast B A 1992 The skin *Wound Healing: Biochemical and Clinical Aspects* ed I K Cohen, R F Diegelmann and W J Lindblad (Philadelphia, PA: WB Saunders), pp 344–55

Mast B A, Diegelmann R F, Krummel T M and Cohen I K 1993 Hyaluronic acid modulates proliferation, collagen and protein synthesis of cultured fetal fibroblasts *Matrix* **13** 441–6

Mast B A *et al* 1995 Hyaluronic acid degradation products induce neovascularization and fibroplasias in fetal rabbit wounds *Wound Rep. Reg.* **3** 66–72

Mast B A and Schultz G S 1996 Interactions of cytokines, growth factors, and proteases in acute and chronic wounds *Wound Rep. Reg.* **4** 411–20

Mast B A *et al* 1997 Ultrastructural analysis of fetal rabbit wounds *Wound Rep. Reg.* **6** 243–8

McCallion R L and Ferguson M W J 1995 Fetal wound healing and the development of anti-scarring therapies for adult wound healing *The Molecular and Cellular Biology of Wound Repair* 2nd edn ed R A F Clark (New York: Plenum), pp 561–600

McCluskey J and Martin P 1995 Analysis of the tissue movements of embryonic wound healing—DiI studies in the limb bud stage mouse embryo *Dev. Biol.* **170** 102–14

McKenzie L A *et al* 2006 Skin-derived precursors generate myelinating Schwann cells for the injured and demyelinated nervous system *J. Neurosci.* **26** 6651–60

Medzhitov R 2010 Inflammation 2010 New adventures of an old flame *Cell* **140** 771–6

Merkel J R, DiPaolo B R, Hallok G G and Rice D C 1988 Type I and type III collagen content of healing wounds in fetal and adult rats *Proc. Soc. Exp. Biol. Med.* **187** 493–7

Mescher A L 2013 Junqueira's Basic Histology 13th edn (New York: McGraw-Hill)

Mescher A L 2017 Macrophages and fibroblasts during inflammation and tissue repair in models of organ regeneration *Regeneration* **4** 39–53

Mescher A L, Neff A W and King M W 2017 Inflammation and immunity in organ regeneration *Dev. Comp. Immunol.* **66** 98–110

Miller E J and Gay S 1992 Cllagen structure and function *Wound Healing: Biochemical and Clinical Aspects* ed I K Cohen, R F Diegelmann and W J Lindblad (Philadelphia, PA: WB Saunders), pp 130–51

Moor A N, Vachon D J and Gould L J 2009 Proteolytic activity in wound fluids and tissues derived from chronic venous ulcers *Wound Rep. Reg.* **17** 832–9

Morgan C J and Pledger W J 1992 Fibroblast proliferation *Wound Healing: Biochemical and Clinical Aspects* ed I K Cohen, R F Diegelmann and W J Lindblad (Philadelphia, PA: WB Saunders), pp 63–76

Moriarty K P, Cromblehome T M, Gallivan E K and O'Donnell C 1996 Hyaluronic acid-dependent pericellular matrices in fetal fibroblasts: implication for scar-free wound repair *Wound Rep. Reg.* **4** 346–52

Mostov K and Martin-Belmonte F 2006 Developmental biology: the hole picture *Nature* **442** 363–4

Mustoe T 2005 Understanding chronic wounds: a unifying hypothesis on their pathogenesis and implications for therapy *Am. J. Surg.* **187S** 65S–70S

Newman S A and Tomasek J J 1996 Morphogenesis of connective tissues *Extracellular Matrix, volume 2. Molecular Components and Interactions* (Amsterdam: Harwood Acadacemic), pp 335–69

Novak M L and Koh T J 2013 Macrophage phenotypes during tissue repair *J. Leukoc. Biol.* **93** 875–81

Nuutila K *et al* 2012 Human skin transcriptome during superficial cutaneous wound healing *Wound Rep. Reg.* **20** 830–9

Nwomeh C *et al* 1998 Dynamics of the matrix metalloproteinases MMP-1 and MMP-8 in acute open human dermal wounds *Wound Rep. Reg.* **6** 127–34

Olutoye O and Cohen I K 1996 Fetal wound healing: an overview *Wound Rep. Reg.* **4** 66–74

Ono I *et al* 2009 *De novo* follicular regeneration of the skin by wingless int 3 and bone morphogenetic protein 2 genes introduced into dermal fibroblasts and fibroblast growth factor-2 protein *Wound Rep. Reg.* **17** 436–48

Peake M A *et al* 2014 Identification of a transcriptional signature for the wound healing continuum *Wound Rep. Reg.* **22** 399–406

Petreaca M L, Yao M, Ware C and Martins-Green M M 2008 Vascular endothelial growth factor promotes macrophage apoptosis through stimulation of tumor necrosis factor superfamily member 14 (TNFSF14/LIGHT) *Wound Rep. Reg.* **16** 602–14

Pierce G F 1991 Tissue repair and growth factors *Encyclopedia of Human Biology* ed R Dulbecco (New York: Academic), pp 499–509

Plikus M V *et al* 2017 Regeneration of fat cells from myofibroblasts during wound healing *Science* **355** 748–52

Rao N *et al* 2014 Proteomic analysis of fibroblastema formation in regenerating hind limbs of *Xenopus laevis* froglets and comparison to axolotl *BMC Dev. Biol.* **14** 32

Redd M J, Cooper L, Wood W, Stramer B and Martin P 2004 Wound healing and inflammation: embryos reveal the way to perfect repair *Philos. Trans. R. Soc. Lond.* B **359** 777–84

Rehman J *et al* 2004 Secretion of angioogenic and antiapoptotic factors by human adipose stromal cells *Circulation* **109** 1292–8

Rinkevich Y *et al* 2015 Identification and isolation of a dermal lineage with intrinsic fibrogenic potential *Science* **348** 302

Roberts A B and Sporn M B 1996 Transforming growth factor-beta *The Molecular and Cellular Biology of Wound Repair* 2nd edn ed R A F Clark (New York: Plenum Press), pp 275–310

Sandulache V C *et al* 2006 Prostaglandin E2 differentially modulates human fetal and adult dermal fibroblast migration and contraction: implication for wound healing *Wound Rep. Reg.* **14** 633–43

Schultz G S and Wysocki A 2009 Interaction between extracellular matrix and factors in wound healing *Wound Rep. Reg.* **17** 153–62

Seifert A W *et al* 2012b Skin shedding and tissue regeneration in African spiny mice (*Acomys*) *Nature* **489** 561–5

Seifert A W, Monaghan J R, Voss S R and Maden M 2012a Skin regeneration in adult axolotls: a blueprint for scar-free healing in vertebrates *PLoS One* **7** e32875

Sen C K *et al* 2009 Human skin wounds a major and snowballing threat to public health and the economy *Wound Rep. Reg.* **17** 763–71

Sood A *et al* 2015 Targeted metabolic profiling of wounds in diabetic and non-diabetic mice *Wound Rep. Reg.* **23** 423–34

Shao H, Yi X-M and Wells A 2008 Epidermal growth factor protects fibroblasts from apoptosis via PI3 kinase and Rac signaling pathways *Wound Rep. Reg.* **16** 551–8

Swift M E, Burns A, Gray K L and Di Pietro L A 2001 Age-related alterations in the inflammatory response to dermal injury *J. Invest. Dermatol.* **117** 1027–35

Sun B K, Siprashvili Z and Khaveri P A 2014 Advances in skin grafting and treatment of cutaneous wounds *Science* **346** 941–5

Tepper O M, Galian R D and Caplan J M 2002 Human endothelial progenitor cells from type II diabetics exhibit impaired proliferation, adhesion, and incorporation into vascular structures *Circulation* **106** 2781–6

Tomanek R J and Schatterman G C 2000 Angiogenesis: new insights and therapeutic potential *Anat. Rec.* **261** 126–35

Tomasek J J, Vaughn B and Haaksma C J 1999 Cellular structure and biology of Dupuytren's disease *Hand Clin.* **15** 1–15

Trengrove N J, Bielfeldt-Ohmann H and Stacey M C 2000 Mitogenic activity and cytokine levels in non-healing and healing chronic leg ulcers *Wound Rep. Reg.* **8** 13–25

Usui M L *et al* 2005 Morphological evidence for the role of suprabasal keratinocytes in wound reepitheliazation *Wound Rep. Reg.* **13** 468–79

Verstappen J, Katsaros C, Kuijpers-Jagtman A, Torensma R and Von den Hoff J W 2011 The recruitment of bone marrow-derived dells to skin wounds is independent of wound size *Wound Rep. Reg.* **19** 260–7

Walraven M *et al* 2014 Altered TGF-beta signaling in fetal fibroblasts: what is known about the underlying mechanisms? *Wound Rep. Reg.* **22** 3–13

Watt F M 2014 Mammalian skin biology: at the interface between laboratory and clinic *Science* **346** 937–40

Weitzhandler M and Bernfield M R 1992 Protoglycan glyconjugates *Wound Healing: Biochemical and Clinical Aspects* ed I K Cohen, R F Diegelmann and W J Lindblad (Philadelphia: WB Saunders), pp 195–208

Whitby D J and Ferguson M W J 1992 Immunohistochemical studies of the extracellular matrix and soluble growth factors in fetal and adult wound healing *Fetal Wound Healing* ed N S Adzick and Longaker (New York: Elsevier), pp 161–79

Wietecha M S, Cerny W L and Di Pietro L A 2014 Mechanisms of vessel regression: toward an understanding of the resolution of angiogenesis *New Perspectives in Regeneration Current Topics in Microbiology and Immunology* vol 367 ed E Heber-Katz and D L Stocum (Berlin: Springer), pp 3–32

Woodley D T 1996 Reepithelialization *The Molecular and Cellular Biology of Wound Repair* ed R A F Clark (New York: Plenum), pp 339–54

Wulff B C, Parnt A E and Wilgus T A 2013 Novel differences in the expression of inflammation-associated genes between mid-and late-gestational dermal fibroblasts *Wound Rep. Reg.* **21** 103–12

Yamada K M and Clark R A F 1996 Provisional matrix *The Molecular and Cellular Biology of Wound Repair* 2nd edn ed R A F Clark (New York: Plenum Press), pp 51–94

Yancopoulos G D *et al* 2000 Vascular-specific growth factors and blood vessel formation *Nature* **407** 242–8

Yang G P *et al* 2003 From scarless feta lwounds to keloids: molecular studies in wound healing *Wound Rep. Reg.* **11** 411–8

Yannas I V, Colt J and Wai Y C 1996 Wound contraction and scar synthesis during development of the amphibian *Rana catesbeiana Wound Rep. Reg.* **4** 29–39

Yannas I V 2001 *Tissue and Organ Regeneration in Adults* (New York: Springer)

Yao F *et al* 2001 Age and growth factors in porcine full-thickness wound healing *Wound Rep. Reg.* **9** 371–7

Yokoyama H *et al* 2011 *Prx-1* expression in *Xenopus laevis* scarless skin wound healing and its resemblance to epimorphic regeneration *J. Invest. Dermatol.* **131** 2477–85

Zhu Z *et al* 2017 Alternatively activated macrophages derived from THP-1 cells promote the fibrogenic activities of human dermal fibroblasts *Wound Rep. Reg.* **25** 377–88

Zimmerman A, Bai L and Ginty D D 2014 The gentle touch receptors of mammalian skin *Science* **346** 950–54

IOP Publishing

Foundations of Regenerative Biology and Medicine

David L Stocum

Chapter 3

Regeneration by adult stem cells

Summary

The most common mechanism of regeneration is the activation of tissue-resident adult stem cells. There are two broad classes of adult stem cells, epithelial and mesenchymal. Epithelial stem cells renew tissues such as epidermis and hair, intestinal epithelium, and injured spinal cord in fish and amphibians. Mesenchymal stem cells renew muscle, bone, blood and adipose tissue. The stem cells of each of these tissues reside in specific anatomical niches, and differ in their morphology, molecular signatures and gene expression patterns, and respond to different suites of systemic and niche signals that regulate their proliferation and differentiation. Stem cell populations also reside in adult tissues where they back up non-stem cell mechanisms of regeneration (see chapter 4), and in mammalian tissues of limited regenerative capacity such as the retina, spinal cord and hippocampus, where their stem cell capability is normally niche-suppressed, even after injury. In this chapter, we will consider a number of mammalian tissues known to regenerate by epithelial and mesenchymal stem cells.

3.1 Epithelial stem cells

Epithelial stem cells renew the epidermis, hair follicles and cornea of the skin and eye, as well as the lining of virtually every tubular structure in the body, including the digestive system, respiratory system, central nervous system and urogenital system. Although there are many types of epithelial adult stem cells, they have been found to share the expression of the transcription factor Sox2, which is part of the embryonic core pluripotency network (Arnold *et al* 2011).

3.1.1 Maintenance regeneration of mammalian interfollicular epidermis

The mammalian epidermis includes hair follicles and associated sebaceous and sweat glands interspersed throughout fields of interfollicular epidermis (IFE). Stratum corneum cells are continually shed from mammalian skin. Maintenance regeneration of the IFE takes place by the upward migration and differentiation of epidermal stem cell (EpSC) progeny in the stratum basale to replace keratinocytes shed from the stratum corneum (Verstappen *et al* 2009), a process somewhat different from the injury-induced lateral epidermal regeneration in skin wounds described in chapter 2.

A signature feature of quiescent EpSCs of the murine IFE is their high levels of expression of the notch ligand Delta 1 (Lowell *et al* 2000) and the $\alpha_5\beta_1$, $\alpha_2\beta_1$ and $\alpha_3\beta_1$ integrin receptors for fibronectin, laminin, and collagen of the basement membrane ('Delta 1/integrin bright' cells, Jones *et al* 1995). Mouse EpSCs are activated by Wnt signaling (Blanpain *et al* 2007). The progeny of these cells down regulate Notch, Delta 1 and β_1 integrin to become 'Delta 1/integrin-dull' cells as they commit to becoming keratinocytes, and detach from the basement membrane (Blanpain and Fuchs 2006, Ezhkova *et al* 2009, Verstappen *et al* 2009). Committed cells switch from the expression of keratins 5/14 to keratins 1/10, and migrate upward to become stratum spinosum and stratum granulosum cells (Barker *et al* 2010, Wen *et al* 2010). Expression of β_1 integrin is extinguished as the cells terminally differentiate into the heavily keratinized stratum corneum (Hotchin *et al* 1995). In human epidermis, Lrig-1 (leucine-rich repeats and immunoglobulin-like domain protein-1) is a biomarker of EpSCs. Lrig-1 decreases signaling through ErbB receptors by growth factors such as EGF, TGFα, epiregulin, neuregulin and others, to help maintain their quiescence (Jensen *et al* 2009). Mouse IFE cells do not express Lrig-1, but their hair follicles do. This difference between human and mouse EpSCs may be related to the differential amount and importance of hair in these species. Analysis of EpSC proliferation and differentiation kinetics in the mouse suggests the operation of a 'three-choice' model to insure the maintenance of this population. In this model, symmetric and asymmetric divisions of EPSCs combine to produce equal numbers of stem cells and cells committed to differentiation as keratinocytes (Clayton *et al* 2007, Jones *et al* 2007). Tracing the fates of dividing EpSCs revealed that 8% of them divide symmetrically to give rise to two EpSCs, 8% divide symmetrically to give rise to two lineage-committed transit amplifying cells, and 84% divide asymmetrically to produce one EpSC and one lineage-committed transit amplifying cell. Thus, the number of stem cells and lineage-committed cells remains equal.

3.1.2 Hair follicle regeneration

The mammalian mouse hair follicle has been a favorite subject to understand how stem cells self-renew structures. Hair is cyclically regenerated by stem cells located in the hair follicles, which are epidermal invaginations that project into the dermis (figure 3.1). The tip of the projection is indented to form a cap of progenitor cells called the matrix over a condensation of dermal fibroblasts termed the dermal papilla (Hardy 1992). Together, the dermal papilla and the matrix constitute the bulb of the hair follicle. The hair shaft is formed by the proliferation and differentiation of the matrix cells into keratinocytes that form a medullary core surrounded by a cortex. Nails are variants of hair follicles in mammals, and the scales of fish and reptiles are modifications of hair follicles.

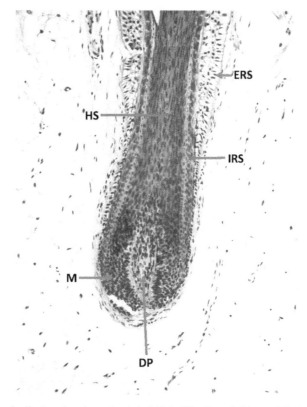

Figure 3.1. (A) Longitudinal section through a hair follicle. The distended base of the hair follicle consists of the matrix (M) of hair progenitor cells surrounding the dermal papilla (DP). ERS = external root sheath (outer root sheath); IRS = internal root sheath; HS =hair shaft differentiating from matrix progenitors. Courtesy of Dr Matt Allen, Indiana University School of Medicine.

The walls of the mouse hair follicle above the matrix form the inner and outer root sheaths. The upper third of the outer root sheath is differentiated into several regions. The lower part forms a thickening called the bulge, where the arrector pili muscles of the hair attach. Just above the bulge, a region of the outer root sheath called the isthmus evaginates to form the sebaceous gland. Above the isthmus, the outer root sheath curves to join the IFE. This curved region is called the infundibulum. A sheath of dermal cells surrounds each hair follicle.

3.1.2.1 The hair follicle contains several populations of stem cells

As shown in figure 3.2, heterogeneous populations of adult stem cells are located in distinct anatomical niches of the bulge and the isthmus of the follicle (Fuchs 2009a, 2009b, Barker *et al* 2010). Stem cells in the lower and middle bulge regenerate the hair, while stem cells in the upper bulge and isthmus renew the non-hair parts of the follicle (Brownell *et al* 2011). Bulge and isthmus stem cells together contribute to the injury-induced regeneration of the interfollicular epidermis (Taylor *et al* 2000, Snippert *et al* 2010).

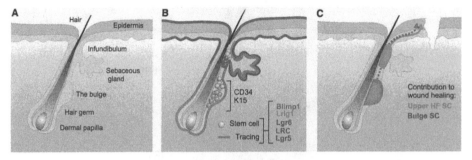

Figure 3.2. (A) Anatomical components of the hair follicle at telogen. (B) Stem cell contributions of the follicle to hair and interfollicular epidermis. Stem cells, pink; transit amplifying (TA) cells, blue. Stem and TA cells are found in both the bulge and the isthmus and can be traced by their expression of signature genes (blue and red outlines). The hair follicle is renewed by bulge TA cells (blue outline). Stem and TA cells in the isthmus renew the isthmus, sebaceous gland, and infundibulum (red outline). Both bulge and isthmus cells contribute to the renewal of the interfollicular epidermis (green outline). (C) Bulge and isthmus stem and TA cells contribute to the repair of wounded epidermis. Reproduced with permission from (Barker *et al* 2010). Copyright 2010 Elsevier.

The genes for keratins 14 and 15, and *Lgr5*, a Wnt target gene encoding an orphan G protein coupled receptor of unknown function, together constitute the molecular signature of hair germ and bulge stem cells (Morris *et al* 2004, Tumbar *et al* 2004, Jaks *et al* 2008). In ROSA 26 mice transgenic for *GFP* or *LacZ* reporter transgenes driven by bulge-specific *keratin 14* or *15* or *Lgr5* promoters, only bulge and hair germ stem cells expressed the reporters during telogen, allowing the cells to be tracked during hair regeneration. During anagen, labeled stem cell progeny moved out of the bulge and hair germ as transit amplifying progenitor cells and migrated downward in the lengthening follicle to form the new matrix. The regenerated hair was composed entirely of labeled cells, confirming that stem cells of the hair germ and bulge give rise to the new hair.

More refined analysis showed that Gli-1$^+$/K-15$^+$ cells located in the lower bulge regenerate the mouse hair follicle, while Gli-1$^+$/K-15$^-$ cells located in the upper bulge contribute to regeneration of the interfollicular epidermis. The latter cells are dependent on Shh signaling from sensory neurons for their ability to form epidermis (Brownell *et al* 2011). Bulge cells also deposit the ECM molecule nephronectin into the underlying basement membrane. Nephronectin induces the adhesion of $\alpha 8\beta 1$ integrin$^+$ mesenchymal cells to the basement membrane and their differentiation into the smooth muscle that anchors the arrector pili muscle to the hair follicle (Fujiwara *et al* 2011). Melanocyte stem cells originating in the neural crest also reside in the bulge and sub-bulge area (Nishimura *et al* 2002).

The isthmus of the mouse hair follicle harbors stem cells characterized by their specific expression of the *Lgr6* gene (Snippert *et al* 2010). Some of these *Lgr6*$^+$ cells also express the thymus epithelial progenitor marker MTS24, Lrig-1 protein and B lymphocyte-induced maturation protein-1 (Blimp-1) (Nijhof *et al* 2006, Horsley *et al* 2006). *Lrg6*$^+$ cells can differentiate into all epidermal cell types after transplantation (Jensen *et al* 2009). During embryonic development, *Lgr6*$^+$ cells establish the IFE, hair follicle and sebaceous gland, but after birth their contribution to hair lineages gradually diminishes. In adult mice, $Lgr6^+$ cells renew the isthmus, the sebaceous gland and the infundibulum (Snippert *et al* 2010).

3.1.2.2 The hair regeneration cycle

Hairs are continually cycling through three phases (Hardy 1992, Messenger 1993): (1) catagen, the cessation of hair growth and regression of the hair follicle; (2) telogen, or follicular rest; and (3) anagen, where regeneration of the follicle and formation of a new hair take place (figure 3.3). This cycle is unique to the hair follicle and is not observed in any other mammalian structure. Several functions may be served by cyclic hair loss and regeneration, including seasonal thermal adaptation, cleansing the body surface, elimination of defective hair follicles and protection from malignancy (Cotsarelis 1997).

During catagen, cells in the follicle wall undergo apoptosis and the hair follicle shortens to push the hair upward and out until the dermal papilla contacts the bulge. Genetic lineage studies have shown that at the end of catagen, a small cluster of cells derived from the bulge, the hair germ, forms between it and the dermal papilla (Greco *et al* 2009). During telogen the follicle is in a resting state with the hair still in the follicle. At anagen the hair germ cells are stimulated to proliferate, followed several days later by the activation and proliferation of bulge cells, elongating the follicle downward to reconstitute the hair bulb and matrix. A new hair shaft and inner root sheath grows and differentiates from the matrix transit amplifying cells. As it grows, the new hair pushes the old one out of the hair follicle. When a hair is

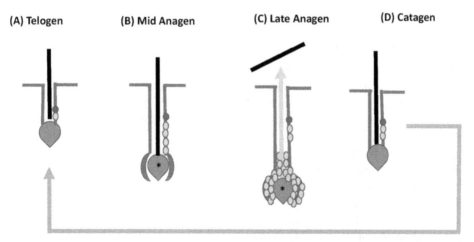

(A) Telogen **(B) Mid Anagen** **(C) Late Anagen** **(D) Catagen**

Figure 3.3. Stages of hair regeneration. (A) Telogen, with the dermal papilla (red) just below the hair germ. (B) At early to mid-anagen, signals (asterisk) from dermal papilla cells activate stem cells in the hair germ to proliferate and form the transit amplifying cell population (yellow) that migrates downward to form the matrix, increasing the length of the follicle. Bulge stem cells (purple) are then activated to self-renew and replenish the hair germ cells. (C) By late anagen, the follicle has extended to its maximum length and matrix cell progeny are proliferating and differentiating into the cells of the new hair (yellow arrow). (D) When the follicle enters catagen it shortens by apoptosis to retract the dermal papilla toward the hair germ, before again entering telogen. The black line indicates original hair in (A) and (B), being pushed out of the follicle in (C).

forcibly plucked, or the follicle damaged by chemotherapy, catagen occurs prematurely and telogen is drastically shortened to quickly bring the follicle to anagen.

Up to 90% of the hairs on the human scalp are in anagen at any one time, and this phase can occupy anywhere from 2–7 years for a given hair, enabling the growth of very long hair (a la Rapunzel!). Only 1%–2% of hairs are in catagen at a given time, and individual follicles spend 2–3 weeks in this phase. Fourteen percent of hair follicles are in telogen at any one time and this phase lasts for about three months for individual hair follicles. The lengths of these phases are different for hairs on different parts of the body. For example, anagen lasts for only 4–7 months in eyebrow hair follicles, so these hairs are much shorter than scalp hairs (Paus 1998).

During telogen, BMP (*Bmp 2, 4, 6*) and FGF (*Fgf18*) signaling maintain hair germ and bulge stem cells in a quiescent state (Blanpain *et al* 2004, Nguyen *et al* 2006, Horsley *et al* 2008). Inhibition of BMP signaling drives resting hair follicles into anagen prematurely (Botchkarev and Sharov 2004, Kobeliak *et al* 2007). Telogen is divided into two functional phases. One is refractory for hair regeneration and characterized by high BMP signaling, while the other is competent for hair regeneration and characterized by low BMP signaling (Plikus *et al* 2008). At anagen, dermal papilla cells stimulate hair germ cells and then bulge stem cells to proliferate via FGF-7 signaling. FGF-7 in turn elevates Wnt signaling in the same order to further increase proliferation (Lowry *et al* 2005, Greco *et al* 2009). Figure 3.4 illustrates the relationships of Fgf, BMP and Wnt growth factors to quiescence and activation of hair follicle cells. HGF and VEGF secreted by dermal papilla cells may also be involved in initiating anagen, since HGF stimulates the growth of mouse vibrissal follicles *in vitro* (Jindo *et al* 1994, Lachgar *et al* 1996).

3.1.3 Intestinal epithelium

The digestive system is a tubular, multilayered structure with an internal epithelial layer sitting on a basement membrane and connective tissue layers on the outside. Smooth muscle layers are interposed between the basement membrane and connective tissue layers. The epithelial cells of each part of the digestive system turn over continually and are regenerated by epithelial stem cells. The intestine has long served as a model to investigate cell turnover in the digestive system.

The intestinal epithelium is thrown into numerous absorptive villi between which are deep crypts containing a stem cell population residing at the crypt bottom that are the source of regenerated goblet, enterocyte and enteroendocrine cells (Barker *et al* 2010, Carulli *et al* 2014, for a review) (figure 3.5). Pulse labeling studies with ^{3}H-thymidine demonstrated that stem cells in the bottom of the crypt label first, with labeled differentiated cells found progressively further up the wall of the crypt at later times (Potten 1998). The intestinal stem cells are called crypt base columnar (CBC) cells. They are intercalated between supporting Paneth cells and are in fact the source of Paneth cells (Sato *et al* 2011). The molecular marker for CBC cells is the Lgr4/5 protein, which with Frz and LRP is part of a receptor complex for Wnt. Maintenance and proliferation of CBC cells is critically dependent on Notch, Wnt and EGF signals supplied by direct contact with Paneth cells, and also Il-22 from

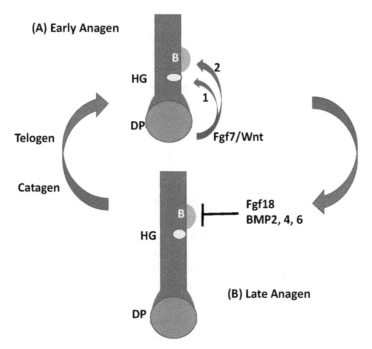

Figure 3.4. Growth factor regulation of hair stem and progenitor cells. (A) At early anagen, FGF-7 from the dermal papilla first activates the proliferation of hair germ cells (HG, yellow), and then the proliferation of bulge stem cells (B). FGF-7 elevates Wnt expression in the same order to further stimulate proliferation. (B) At late anagen, with the hair fully grown, stem cells of the hair germ and bulge are maintained in a quiescent state (bar) by Fgf18 and BMPs 2, 4, and 6. Quiescence is maintained as the hair follicle shortens during catagen and enters telogen to await the next round of stem cell activation.

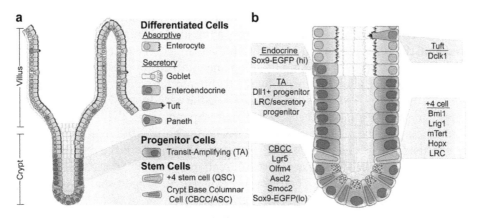

Figure 3.5. Renewal of intestinal epithelium. (a) Villous crypt showing the position of intestinal stem cells, progenitor (transit amplifying) cells and differentiated cell types. (b) Closer view of crypt bottom showing molecular signatures of TA cells, +4 cells (light green) and crypt stem cells (green), interspersed with Paneth cells (brown). Reproduced from (Carulli *et al* 2014). With permission of The Royal Society of Chemistry (Open Access).

innate lymphoid cells (van Es *et al* 2010, Sato *et al* 2011, Lindemans *et al* 2015). Paneth cells also support CBC function by providing them with lactate to sustain enhanced mitochondrial oxidative phosphorylation (Rodriguez-Colman *et al* 2017). CBC cells divide to form transient amplifying progenitor cells, which differentiate into the functional enterocyte, goblet and enteroendocrine cells of the villi.

3.1.4 Neural epithelium

3.1.4.1 Fish and amphibian spinal cord

The ependymal epithelium of the vertebrate spinal cord harbors cells capable of forming new neurons. Teleost fish and urodele amphibians regenerate spinal cord axons and neurons along with other structures after amputation of the tail (figure 3.6). In the knifefish *Apteronotus leptorhynchus*, a transient wave of apoptosis of spinal cord neurons is followed by intense proliferation of cells in the tissues surrounding the spinal cord, leading to the formation of a blastema (Sirbulescu and Zupanc 2009, Zupanc and Sirbulescu 2011). The caudal tip of the ependyma seals off to form a terminal vesicle that grows into the blastema by proliferation of ependymal cells. The blastema regenerates the musculoskeletal and connective tissue components of the caudal fin, while ependymal cells regenerate the neurons of the cord.

Pre-metamorohic *Xenopus* tadpoles regenerate spinal cord neurons after trans-ection of the trunk spinal cord, but lose regenerative capacity during metamorphosis. Regeneration is accomplished by the proliferation of Sox2+ stem cells in the subventricular zone that populate the transection gap and differentiate into new neurons. These cells decrease in metamorphosing *Xenopus*, and those Sox2+ cells that remain show only a weak response to injury (Munoz *et al* 2015). Adult urodele amphibians, which readily regenerate spinal cord axons after transection of trunk cord (see chapter 4), regenerate few new neurons (Butler and Ward 1965, Davis *et al* 1989, Tanaka and Ferretti 2009, Diaz-Quiroz and Echeverri 2013, Lee-Liu *et al* 2013).

Adult newts regenerate axons after tail amputation, but not new neurons (Duffy *et al* 1992, Benraiss *et al* 1996). By contrast, adult axolotls regenerate neurons from ependymal cells after tail amputation by a process much like that in fish (Clarke and Ferretti 1998). Neurons at the amputation plane undergo apoptosis and are removed by macrophages. A blastema forms by the dedifferentiation of cartilage, connective tissue and muscle at the amputation plane while the ependyma of the cord seals over to form a regenerating epithelial tube that expands into the blastema by cell proliferation. Differentiated neurons anterior to the amputation plane can incorpo-rate into the tube and become part of the neuronal circuitry of the regenerated cord (Zhang *et al* 2003). The blastema redifferentiates into the musculoskeletal and connective tissues of the tail. There is some evidence to suggest that FGF and Wnt signaling regulates ependymal cell proliferation.

3.1.4.2 Mammalian CNS

Neurogenesis is known to take place in two regions of the mammalian CNS, the ependymal walls of the lateral ventricles, from which NSCs migrate to the olfactory bulb at the forefront of the telencephalon, and the ventricular subgranular zone of

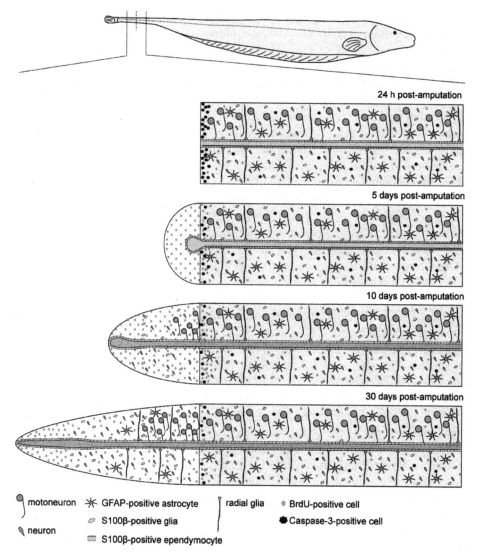

24 h post-amputation

5 days post-amputation

10 days post-amputation

30 days post-amputation

motoneuron   ✳ GFAP-positive astrocyte   radial glia   • BrdU-positive cell

         S100β-positive glia   ● Caspase-3-positive cell

neuron   S100β-positive ependymocyte

Figure 3.6. Spinal cord regeneration in the knifefish, *A. leptorhynchus* after amputation of the caudal tail at the level indicated by the dashed line. A transient wave of apoptosis (black dots, caspase-3-positive cells) is followed by proliferation of cells in the tissues surrounding the cord, which accumulate to form a blastema (red dots). The ependyma of the cord seals off and its cells proliferate to extend the cord into the blastema to its caudal tip. As the blastema and cord grow in concert, new neurons and glia differentiate from the ependymal cells and the blastemal cells differentiate into the caudal fin. Reproduced with permission from (Sirbulescu and Zupanc 2011). Copyright 2011 Elsevier.

the hippocampus, where NSCs give rise to granule cell neurons and glial cells. We will focus here on the hippocampus, an area of the brain crucial for cognition, learning and memory, particularly in humans. Hippocampal NSCs, like most NSCs, express Sox2 and are maintained in a quiescent state by BMP signaling.

Maintenance regeneration by hippocampal NSCs is crucial for the formation and retention of memory and for learning new information and tasks, and is influenced by both physical and cognitive activity. The number of hippocampal neurons declines with age, and their rate of replacement is low due to declining rate of NSC proliferation and because the majority of newborn neurons undergo apoptosis and are rapidly cleared by CNS macrophages, the microglia (Sierra *et al* 2010). However, if the mice are placed in enriched environments, NSC proliferation and differentiation into new neurons is greatly increased and cell death decreased. Surviving new neurons are functionally integrated into hippocampal neural circuitry, as shown by ultrastructural analysis and immunostaining with antibodies to synaptic proteins of labeled cells. Electrophysiological measurements indicate that the neurons function normally and receive normal excitatory input. Other studies have shown a small increase in the production of granule cell neurons in rats and gerbils after neuronal degeneration induced by focal or global ischemic injury, but not enough for functional recovery (Gould and Tanapat 1997, Jin *et al* 2001, Liu *et al* 1998).

The atomic bomb tests conducted in the 1950s produced high levels of atmospheric ^{14}C that found its way through the food chain into the DNA of stem cells throughout the human body. After cessation of the tests, ^{14}C levels in stem cell nuclei would decline due to dilution by division. This allowed neurobiologists to settle a controversy over whether the human cerebral cortex exhibits maintenance neurogenesis. Measurements of ^{14}C in the DNA of deceased persons born during the era of the tests (when the brain was forming) showed that their cerebral cortex neurons maintained their original level of ^{14}C until they died, indicating the lack of a cortical stem cell population, whereas cells known to be replaced by maintenance regeneration had undetectable levels of ^{14}C at death, indicating that they had proliferated, turned over and were replaced (Spalding *et al* 2005).

Neurogenesis does not take place from ependymal cells in the adult mammalian spinal cord. Nevertheless, the ependyma contains cells capable of forming neurons *in vitro* if they are removed from the ependyma, indicating that the cellular environment of the mammalian cord is inimical to regeneration. These facts have significance for treating spinal cord injury by NSC transplant (see chapter 8).

3.2 Mesenchymal stem cells

Mesenchymal stem cells are found throughout the body and regenerate muscle, bone, blood and fat. The archetypal MSC resides in the bone marrow stroma. Different types of tissue MSCs are identifiable by particular sets of molecular markers that overlap a great deal. They are of interest, not only for tissue regeneration, but also because they have immunomodulatory and paracrine functions. There is evidence that all of them may be derivatives of perivascular cells, the pericytes, that invest capillaries (Crisan *et al* 2008).

3.2.1 Skeletal muscle regeneration

The basic unit of skeletal muscle is the multinucleated myofiber. Myofibers originate during embryogenesis from somite-derived myoblasts that proliferate, fuse end-to-end, and form multinucleate cells that differentiate the actin–myosin contractile apparatus. Skeletal muscles are organized in much the same way as nerves (figure 3.7). Individual myofibers are surrounded by a connective tissue endomysium, which elaborates a basement membrane between it and the plasma membrane (sarcolemma) of the myofiber. The myofiber, basement membrane and endomysium together constitute the endomysial unit. Endomysial units are organized into fascicles, each surrounded by a perimysium and the fascicles are bound together by an epimysium that surrounds the muscle. Skeletal muscles are highly vascularized within and between the mysial sheaths and are heavily innervated at specialized synaptic contacts with the myofibers called neuromuscular junctions. The mysial sheaths are collectively termed muscle interstitial tissue. The ECM of the interstitial tissue consists of collagen I, and sulfated proteoglycans (PGs), including a muscle-specific sulfated PG.

The actin cytoskeleton of the myofibers is linked to the basement membrane by the dystrophin–glycoprotein complex (DGC), which prevents myofiber damage during contraction (figure 3.8). The DGC is a multisubunit complex comprised of intracellular dystrophin and syntrophins and three types of sarcolemmal proteins, the dystroglycans, sarcoglycans and sarcosan. Disruption of the DGC by mutations in dystrophin or the sarcoglycans is a feature of Duchenne muscular dystrophy that

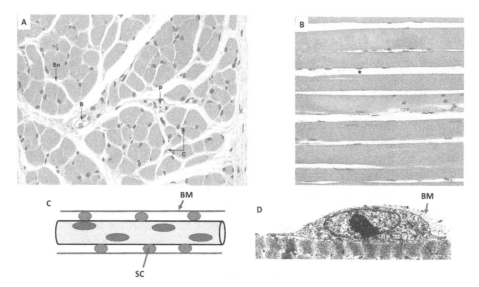

Figure 3.7. Muscle structure. (A) Cross-section showing fascicles of individual myofibers surrounded by perimysial connective tissue (P). En = delicate endmysium; B = small blood vesse; C = capillary. The blue dots under the sarcolemma on the periphery of the myofibers are myonuclei. Some nuclei appear to lie outside the myofiber; these are the nuclei of satellite cells. (B) Longitudinal section, showing individual myofibers. Blue = nuclei. Some nuclei clearly lie on the surface of the myofibers; these are satellite cell nuclei. (C) Diagram indicating the relationship of satellite cells (SC, red) to the basement membrane (BM) and sarcolemma of the myofiber. (D) Electron microscopic image of a cultured chick myofiber showing sarcomeres of myofilaments interior to the sarcolemma and a satellite cell between the sarcolemma and the basement membrane (BM). (A) and (B) Reproduced with permission from (Young *et al* 2006). Copyright 2006 Elsevier. (D) Reproduced with permission from (Konigsberg *et al* 1975). Copyright 1975 Elsevier.

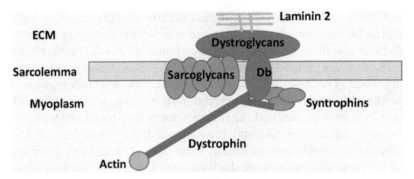

Figure 3.8. Simplified diagram of the dystrophin complex. Dystrophin is attached on one end to the actin filaments of the myofiber and on the other end is complexed to syntrophins and dystobrevin (Db). Dystobrevin and the dystroglycans form a link between dystrophin and laminin of the basement membrane through the cell membrane (sarcolemma). In the myoplasm, the dystrophin complex mechanically connects actin filaments to the sarcolemma. The complex prevents damage to the myofiber during contraction. Mutations in the structure of the components of the dystroglycan complex, particularly in dystrophin, leave the myofiber vulnerable to injury upon contraction, causing calcium influx and degeneration (dystrophy).

renders the myofibers susceptible to sarcolemmal damage and necrosis (Cohn *et al* 2002). At their ends, muscles grade into fascia or tendons, which attach them to bones.

3.2.1.1 Injured muscle is regenerated by satellite cells

Weight bearing, exercise and injury induce skeletal muscle regeneration (Grounds 1991). Astronauts lose muscle mass during long periods of weightlessness. Muscle is regenerated on return to 1g conditions (Pastoret and Partridge 1998). The effects of long-term (months, years) weightlessness on muscle structure, function and regeneration (and other structures) based on stays aboard the International Space Station are underway. This information is essential if humans are to undertake long space voyages such as missions to Mars.

Muscle does not regenerate across gaps that exceed a critical size. Small lesions in muscle evoke cytoplasmic budding of myofibers that rejoins their cut ends. More extensive lesions result in regeneration by muscle stem cells called satellite cells (SCs) (Palacios and Puri 2006). These cells are MSCs that express the SC-specific transcription factor Pax-7 and occupy a niche between the myofiber sarcolemma and its overlying basement membrane (figure 3.7). Other markers expressed by SCs are: (1) myocyte nuclear factor (Garry *et al* 1997), which inhibits expression of muscle regulatory factors that promote myogenic differentiation; (2) the c-met receptor for HGF (Cornelison and Wold 1997), which plays a key role in the activation of SCs; and (3) the cell surface marker CD34 (Zammit and Beauchamp 2001). Satellite cells form new myofibers by the same tandem fusion and differentiation process observed in embryonic development. Pax-7 expression is lost in the differentiated myofibers, but maintained in daughter stem cells.

Genetic marking experiments have revealed two subpopulations of cells in the SC niche (Kuang *et al* 2007). The majority subpopulation (90%) consists of myogenic progenitors committed to differentiate into myofibers. The minority subpopulation (10%) is the self-renewing long-term SC population. Genetic ablation of muscle fibroblasts leads to deficiencies in regeneration, including premature SC differentiation and depletion, and smaller regenerated myofibers, demonstrating that fibroblasts of the interstitial tissue provide niche factors critical for regeneration of muscle by SCs (Murphy *et al* 2011). Three other Pax-7+ stem cell populations in addition to SCs have been isolated from mouse muscle interstitial tissue that have myogenic capability *in vitro* or when engrafted into damaged muscle (Zheng *et al* 2007, Cerletti *et al* 2008, Tanaka *et al* 2009). It is not known how these three populations are related to one another and to classical SCs, or whether they normally take part in muscle regeneration. In addition, several non-SC, Pax-7 negative cell types associated with blood vessels and interstitial cells can engraft into muscle and participate in regeneration (Tedesco *et al* 2010, for a review). However, their participation requires the presence of SCs, since regeneration does not occur when they are transplanted to muscle from which Pax-7+ SCs are genetically ablated (Lepper *et al* 2011, Sambasivan *et al* 2011, Murphy *et al* 2011, McCarthy *et al* 2011). These results further emphasize that it is Pax-7+ SCs that are the indispensable cells for muscle regeneration.

Muscles subjected to overload can also mount a robust hypertrophic response in which there is an increase in protein synthesis and volume of individual myofibers, as well as the addition of SCs to existing myofibers (Green *et al* 1999, Glass 2003). Hypertrophy can take place independently of SCs, however, as shown by the hypertrophic response to mechanical overload in SC-depleted muscle (McCarthy *et al* 2011).

3.2.1.2 Cellular events of muscle regeneration

Free grafting is a widely used model to study muscle regeneration in which a muscle (usually the gastrocnemius) is removed from and then grafted back to its bed, leading to global ischemic degeneration and regeneration (Carlson 2003). The muscle regenerates in three overlapping waves that spread centripetally from the periphery to the center of the muscle (Hansen-Smith and Carlson 1979) (figure 3.9). First, a wave of myofiber necrosis takes place within their basement membranes, accompanied by a typical inflammatory response in which neutrophils and macrophages are attracted into the degenerating muscle (Grounds and Yablonka-Reuvini 1993, Pastoret and Partridge, 1998). The SCs of the myofibers survive within their basement membranes, where they secrete MMP-2 and -9 (Kherif *et al* 1999). MMP-2 partially degrades the basement membrane, allowing the detachment of SCs. MMP-9 breaks down the ECM of the epimysia to release growth factors that promote the second wave, proliferation of the SCs (Gulati *et al* 1983). The daughter cells of proliferating satellite cells take one of two pathways: commitment to myoblast progenitors or self-renewal as satellite cells. The third wave is the tandem fusion of muscle progenitors and their differentiation into myofibers. As the new myofibers differentiate, continuity of their basement membranes is restored by fibroblasts of the endomysial sheaths, and new blood vessels regenerate into the interstitial tissues (Hansen-Smith and Carlson 1979).

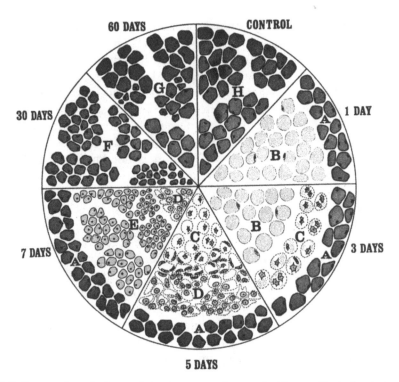

Figure 3.9. Regeneration of a free-grafted rat gastrocnemius muscle. Diagram illustrating (going clockwise) the succeeding waves of (1) myofiber degeneration (Zone B), (2) satellite cell proliferation (Zone C), (3) fusion of satellite cells within the old basement membranes to form new myofibers (Zones D, E), and maturation of muscle fibers (F, G, H). Myofibers on the periphery of the muscle (Zone A) survive. Reproduced with permission from Hansen-Smith and Carlson (1979). Copyright 1979 Elsevier.

3.2.1.3 Activation and proliferation of satellite cells

SCs are quiescent until activated by injury. How quiescence is maintained is not well understood, but a major idea is that it involves degradation of the basement membrane and interstitial ECM by MMPs to release growth factors (figure 3.10). Integrin-mediated adhesion of SCs to laminin of the basement membrane inhibits activation signals from the myofiber; detachment of SCs from the basement membrane following injury would destroy this inhibition, allowing activation signals released by injured myofibers to drive SCs into the cell cycle (Kuang *et al* 2008). Activation also involves growth factor release from ECM of the interstitial tissue. Hepatocyte growth factor (HGF) has been established as the major SC activator (Bischoff 1986, Allen *et al* 1995, Johnson and Allen 1995). As part of the activation process, the SCs themselves express HGF, which acts in an autocrine fashion. It is likely that the HGF bound to muscle ECM is activated in the same way as it is in regenerating liver, by the uPA → plasminogen activator → plasmin cascade (Miyazawa *et al* 1996). Presumably, some of the HGF produced by proliferating SCs would become bound to newly synthesized ECM of the regenerating muscle, to be available for release in a subsequent round of regeneration.

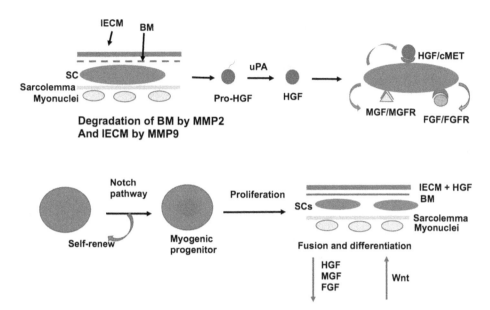

Figure 3.10. Growth factor regulation of satellite cell activation and proliferation. Upper diagrams, SC activation: IECM = interstitial ECM; BM = basement membrane; SC = satellite cell. Pro-HGF is released from the IECM as it is degraded by MMP-9 and activated by urokinase plasminogen activator (uPA). HGF is the primary activator of the SC, synergized by autocrine mechano growth factor and its receptor(MGF/ MGFR) and fibroblast growth factor and its receptor (FGF/FGFR). Satellite cells divide asymmetrically. Lower diagrams, proliferation and differentiation: Myogenic progenitors proliferate, fuse and differentiate into myofibers. Differentiation is driven by an increase in Wnt signaling, as the expression of HGF, MGF and FGF declines. Much of the HGF is re-sequestered in the IECM.

Although unable to activate SCs, PDGF, TGF-β, FGF-2 and leukemia inhibitory factor (LIF) have a synergistic effect on the proliferative activity of SCs *in vitro* (Johnson and Allen 1995, Pastoret and Partridge 1998). In uninjured muscle, they are bound to ECM components and their receptors are not expressed by quiescent SCs (DiMario *et al* 1989). They are released after injury from degrading ECM and are also secreted by macrophages to interact with their up regulated receptors on activated SCs (Pastoret and Partridge 1998, Kherif *et al* 1999). FGFs play the most important synergistic role in SC proliferation. High affinity FGF-2 receptors are not present on quiescent SCs, but are up regulated along with FGF-1, -2, -6, -7 and 13 in response to HGF *in vitro* and *in vivo* (Garrett and Anderson 1995, Zhao and Hoffman 2004). Notch signaling and heparan sulfate provided by syndecan 3 and 4 is also required for SC proliferation (Conboy and Rando 2002, Kuang *et al* 2007).

Ninety percent of the SC population has a planar (parallel) orientation of the mitotic spindle with respect to the myofiber and basement membrane and is Myf5[+] (Kuang *et al* 2007, 2008). Their daughters are thus all exposed to the same dual set of signals from myofiber and basement membrane (figure 3.11). This arrangement dictates symmetrical mitosis into two committed progenitors that express the myogenic transcription factor Myf5, or two stem cells that are negative for Myf5. The symmetric divisions of these SCs appear to be regulated through the non-

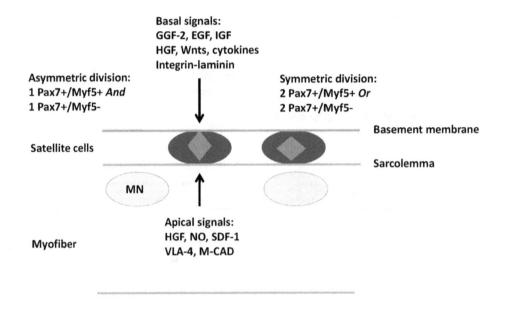

Planar vs. perpendicular stem cell division

Figure 3.11. Planar division of satellite cells is the result of exposure of the daughter cells to the same dual sets of signals, whereas perpendicular division is the result of each daughter cell being exposed to different sets of signals.

canonical Wnt planar cell polarity (PCP) pathway (LeGrand *et al* 2009). *In vitro*, Wnt7a stimulated an increase in symmetric SC divisions, whereas knockdown of its receptor Fzd7 inhibited symmetric division. The PCP protein Vangl2 was markedly up regulated and segregated into both poles of symmetrically dividing SCs. Knockdown of *Vangl2* by siRNA significantly reduced the percentage of symmetric divisions. Electroporation of *Wnt7a* into regenerating muscles *in vivo* produced a 63% increase in stem SCs, consistent with an increase in symmetric division. What determines that a symmetric division results in two committed progenitors or two stem cells is unknown.

The other 10% of SCs are Myf5$^-$ and undergo asymmetric mitosis to produce a stem cell plus a committed progenitor (figure 3.11). These cells divide perpendicularly to the basement membrane and sarcolemma so that one daughter is exposed to only the signals in the myofiber and the other daughter is exposed to only the signals in the basement membrane (Kuang *et al* 2008, for a review). The daughter in contact with the basement membrane becomes the stem cell, whereas the daughter in contact with the sarcolemma becomes the myogenic progenitor. What determines whether an SC makes a planar division versus a perpendicular one is unknown.

3.2.1.4 Differentiation of myofibers
The differentiation of myogenic progenitors is dependent on canonical Wnt signaling from myofibers (Brack *et al* 2008). Wnt signaling rises steadily during SC

proliferation and into the differentiation phase. Exposure to high concentrations of Wnt during early stages of SC proliferation *in vitro* caused premature differentiation, or even conversion to scar-forming fibroblasts. Inhibition of Wnt signaling in the differentiation phase significantly impaired myotube formation (Brack *et al* 2007). As dependency on Wnt signaling for myotube differentiation rises, dependency on Notch signaling for proliferation falls, suggesting the integration of the two signaling pathways at a convergence point, which appears to be GSK-3β. The low level of canonical Wnt signaling during the early phases of muscle regeneration allows Notch to maintain the activity of GSK-3β, resulting in the degradation of β-catenin and continued proliferation. High levels of Wnt during the later stages of regeneration inhibit GSK-3β, allowing stabilization of β-catenin and differentiation (Brack *et al* 2008). IGF-1 and -2 and TGF-β2 suppress SC proliferation and promote differentiation into myofibers (McCroskery *et al* 2003). The IGF-1Ec splice variant (mechano growth factor, MGF) is particularly important for SC differentiation (Yang and Goldspink 2002, Hill and Goldspink 2003, Musaro *et al* 2004, Philippou *et al* 2007). In the presence of growth hormone, SCs increase their production of MGF and IGF-1 receptors in response to exercise or injury (Willis *et al* 1997, Adams 1998, Owino *et al* 2001). Administration of recombinant human growth hormone in combination with strength training elevates IGF-1 transcript expression and maintains muscle mass in elderly men (Hameed *et al* 2004, Adamo and Farrar 2006).

Tension exerted by tendons is important for the normal orientation of myofibers in regenerating minced muscle. Lack of this tension results in chaotic myofiber differentiation. A normally oriented muscle can regenerate from a remaining muscle stump in the rat if a functional Achilles tendon is regenerated that connects to the muscle stump (Carlson 1970, 1974). Re-innervation of a regenerating rat gastrocnemius muscle begins during the second week post-transplantation, and the speed of spontaneous muscle contraction increases until it approaches normal at 30–40 days after transplantation (Carlson and Gutmann 1972). Muscle regeneration proceeds under conditions of denervation, but denervation prevents or retards the full structural and functional differentiation of the regenerated myofibers (Carlson and Gutmann 1975, Rogers 1982). Most likely, the nerves provide survival and proliferation factors to the regenerating muscle that restore myofiber number and size.

3.2.2 Regeneration of bone

3.2.2.1 Structure of bone
Bone is composed of osteocytes surrounded by a highly calcified organic matrix. The organic matrix is 90% collagen I, with 10% glycoproteins and proteoglycans. Hydroxyapatite crystals $[3Ca_3(PO_4)_2](OH)_2$ interspersed throughout the organic matrix give the bone its stiffness. There are two types of bones based on how they develop. Intramembranous bones, such as the flat bones of the skull, form by the direct differentiation of osteogenic progenitors to bone cells. Endochondral bones, such as the long bones of the extremities, ribs and vertebrae, first develop as a cartilage template that is then replaced by bone (Olsen 1999).

Figure 3.12 illustrates the anatomical structure of a mature endochondral long bone. Most of the length of the bone consists of a cylindrical shaft or diaphysis that grades on both ends into a short, flared metaphysis capped by a disc-shaped epiphysis surfaced with articular cartilage. The external part of the diaphysial cylinder is composed of dense cortical or compact bone, which becomes progressively thinner in the metaphysis and epiphysis. The osteocytes of cortical bone reside in small lacunae of the matrix and are organized in concentric layers around blood vessels to form Haversian systems, or osteons. The osteocytes communicate with one another and with endosteal and periosteal cells by long processes that ramify throughout a network of canaliculi in the bone matrix. A thin layer of sponge-like bone trabeculae (trabecular or spongy bone) is interior to the compact bone. Endosteal connective tissue covers the trabecular bone to line the marrow cavity, which is filled with bone marrow. The endosteum consists of fibroblasts, MSCs, pre-osteoblasts and osteoblasts. The bone marrow is a mixture of osteoblasts, fibroblasts, adipocytes, macrophages and endothelial sinusoids. These cells constitute the stroma of the bone marrow. Embedded in the stroma, and dependent upon it for survival, are hematopoietic stem cells (HSCs) and endothelial stem cells (EnSCs). The outer surface of the bone is covered with another connective tissue sheath, the periosteum. The periosteum and the linings of the Haversian canals also contain MSCs, pre-osteoblasts and osteoblasts.

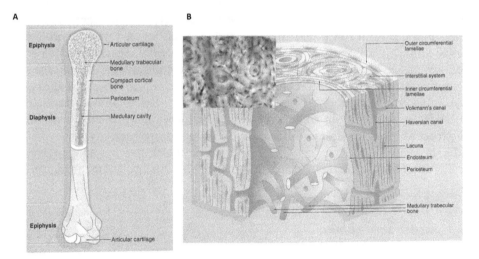

Figure 3.12. Structure of endochondral bone. (A) Bone structure as seen in a longitudinally split long bone. The bone consists of a diaphysis (shaft) and two secondary centers of ossification, epiphyses, that are covered by articular cartilage. The medullary (marrow) cavity is lined by trabecular bone. The trabecular bone is enclosed by cortical (compact) bone, which is covered by a periosteum. (B) Details of internal diaphysial structure. Compact bone is composed of inner and outer circumferential lamellae between which lie Haversian systems (osteons). Reproduced with permission from (Young *et al* 2006). Copyright 2006 Elsevier. Inset: The osteons consist of concentric rings of bone around a central Haversian canal. Blood vessels run longitudinally through the Haversian canals and transversely through Volkmann's canals. Between the osteons lie interstitial lamellae of bone matrix. Courtesy of Dr Matt Allen, Indiana University School of Medicine.

3.2.2.2 Growth of endochondral bone

Between the epiphysis and the diaphysis on both ends of a long bone is a growth plate, which allows the bone to grow proximally and distally (figure 3.13). The growth plate is a template composed of chondrocytes in several stages of differentiation leading to ossification. Chondroprogenitor cells become proliferating chondrocytes, then chondrocytes that hypertrophy and are replaced by osteoblasts invading with blood vessels from the periosteum. The activities of the growth plate are regulated by Indian hedgehog and parathyroid related protein (PTHrP) signaling (figure 3.14). Ihh and PTHrP constitute a feedback loop that controls the rate at which chondroprogenitor cells become hypertrophied (Vortkamp *et al* 1996, Vortkamp *et al* 1998). Ihh is expressed by pre-hypertrophic chondrocytes and maintains proliferation of chondroprogenitor cells. Ihh diffuses centripetally and distally to periosteal/perichondrial cells, inducing them through its downstream transcription factor Gli2 to express PTHrP, which diffuses proximally into the chondroprogenitor zone, thereby slowing the proliferation and differentiation of chondrocytes into hypertrophied chondrocytes (Kronenberg 2003). Maintaining a normal proliferation rate also requires BMP 2 in parallel with IHH signaling, and BMP 2 also modulates the expression of Ihh (Minina *et al* 2001).

3.2.2.3 Maintenance regeneration of bone

Bone is continuously degraded and regenerated throughout life. The whole skeleton is replaced once each decade by continual turnover and regeneration at an estimated two million microscopic sites throughout the skeleton (Harada and Rodan 2003).

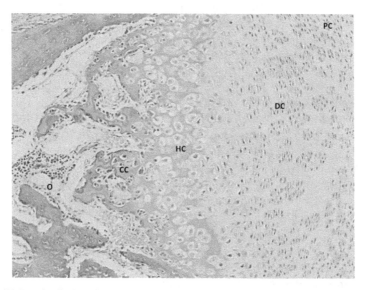

Figure 3.13. (A) Longitudinal section through the growth plate of a long bone. DC = dividing chondrocytes, arranged in columns. HC = hypertrophying chondrocytes; CC = degenerating and calcifying hypertrophied cartilage where the cartilage matrix is invaded by osteogenic cells and capillaries from the marrow cavity of the diaphysis. O = osteogenic zone where osteoblasts are secreting bone matrix. Courtesy of Dr Matt Allen, Indiana University School of Medicine.

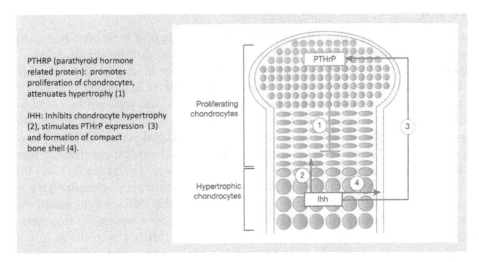

PTHRP (parathyroid hormone related protein): promotes proliferation of chondrocytes, attenuates hypertrophy (1)

IHH: Inhibits chondrocyte hypertrophy (2), stimulates PTHrP expression (3) and formation of compact bone shell (4).

Figure 3.14. Control of chondrocyte proliferation and hypertrophy in the growth plate. Ihh signaling through the periosteum induces PTHRP expression (3), inhibits chondrocyte hypertrophy (2) and induces cortical bone formation (4). PTHRP stimulates chondrocyte proliferation (1). The balance between proliferation and hypertrophy determines the rate of growth of the bone. Reproduced from (Kronenberg 2003). Copyright 2003 Nature.

At these sites, bone is removed by large multinucleated cells called osteoclasts and replaced by the bone-matrix forming cells, the osteoblasts. This maintenance regeneration is often referred to as bone remodeling. Imbalances in the rate of bone removal and replacement lead to skeletal abnormalities. When the rate of removal exceeds the rate of replacement, the result is low bone mass, or osteoporosis; the opposite results in high bone mass, or osteopetrosis. There are many genetic abnormalities associated with syndromes and disorders of the bone remodeling system (Ducy *et al* 2000, Teitlebaum 2000, Harada and Rodan 2003).

Osteoclast function and differentiation
Osteoclasts are giant cells with 4–20 nuclei derived by the fusion of monocytes and differentiated for the specialized function of bone-matrix degradation (figure 3.15). When in contact with the bone surface, the osteoclast becomes polarized, forming a ruffled membrane on one side with a peripheral 'sealing ring' that attaches the osteoclast to the bone matrix creating a microcompartment into which H^+ ions are delivered by the V-type Atp6i proton pump and Cl^- ions by chloride channels to form HCl and achieve a pH of ~4.5, which dissolves the hydroxyapatite from the matrix. The HCl also activates acid hydrolases released from lysosomes by exocytosis, such as cathepsin K, which together with MMPs degrade the organic matrix (Teitlebaum 2000). The result is a 'resorption lacuna' formed in the bone that is then filled in with new bone matrix by osteoblasts (Mundy 1999).

Osteoclast formation is regulated by factors produced by osteoblasts (Teitlebaum 2000, Martin, 2004, for reviews). Two osteoblast factors necessary and sufficient to induce osteoclast formation are macrophage colony-stimulating factor (M-CSF) and the ligand of receptor for activation of nuclear factor kappa B (RANKL).

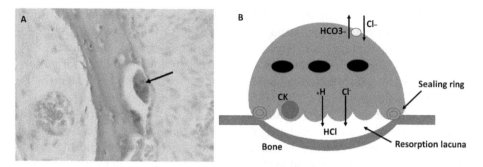

Figure 3.15. (A) Section through bone showing a resorption lacuna and an osteoclast (arrow). Courtesy of Dr Matt Allen, Indiana University School of Medicine. (B) Diagram showing osteoclast function. The osteoclast secretes lysosomal enzymes such as cathepsin K (CK) secreted via endocytosis, and hydrogen and chloride ions (to make hydrochloric acid) into the space defined by the ruffled membrane to form a resorption lacuna in the bone. The ruffled membrane at the circumference of the osteoclast forms a tight seal (sealing ring, green) on the surface of the bone to isolate the resorption lacuna.

M-CSF is soluble and binds to its receptor c-Fms on monocytes. RANKL is a juxtacrine cell surface molecule that binds to its receptor (RANK) on the monocyte cell surface. RANKL and RANK are members of the TNF and TNF receptor superfamily, respectively. Osteoblasts secrete another protein, osteoprotegerin (OPG) that competes with RANK for RANKL to negatively regulate osteoclast formation. Osteoclastogenesis is thus regulated by a balance between the concentrations of M-CSF and RANKL versus OPG (Karsenty 2003, Martin 2004). The transcription factor, c-Fos, is a key intracellular effector that drives expression of the osteoclast molecular phenotype by inducing NFATc1, the master transcription factor for osteoclastogenesis (Takayanagi *et al* 2002). Osteoclast differentiation is terminated by a negative feedback mechanism activated within the osteoclasts themselves involving the down regulation of c-fos (Boyle *et al* 2003).

Osteoblast origin and function
Osteoblasts are post-mitotic cells derived from MSCs of the endosteum, periosteum and bone marrow. Two transcription factors that play key roles in the differentiation of MSCs into osteoblasts are Runx2 (Cbfa1), which is specific for the chondro/osteo lineage, and Osterix (Osx), which is specific for pre-osteoblasts (Ducy *et al* 2000, Harada and Rodan 2003 for reviews). Together, Runx2 and Osx specify an osteoblast pattern of gene activity, with Osx acting downstream of Runx2. Mice with mutations that affect the *Osx* gene have a perfectly patterned cartilaginous skeleton, but lack osteoblasts and exhibit deficiencies in ossification.

Regulation of bone mass by systemic factors
Most of the systemic factors determining bone mass act on osteoblasts to regulate M-CSF and RANKL expression and thus the number of osteoclasts. An increase in osteoclast number results in increased bone resorption and loss of bone mass, whereas fewer osteoclasts reduces bone resorption and maintains bone mass. Non-homeostatic increases and decreases in osteoclast number result in osteoporosis and osteopetrosis,

respectively (Teitelbaum and Ross 2003). A wide variety of systemic factors determine the extent of bone mass by regulating M-CSF and RANKL expression (figure 3.16). These factors are primarily sex steroids and several non-steroidal hormones.

Sex steroids such as estrogen and testosterone inhibit osteoclast formation by reducing osteoblast expression of RANKL or increasing the expression of OPG (Ducy *et al* 2000, Boyle *et al* 2003). Studies on mice indicate that estrogen also induces osteoclast apoptosis (Nakamura *et al* 2007). The osteoporosis experienced by aging men and women is primarily due to lower production of sex steroids. The onset of osteoporosis is earlier in women because of menopause and the extent of bone loss is greater at first, but bone mass becomes more equal in men and women of the same age as time goes on (Harada and Rodan 2003, Zelzer and Olsen 2003).

Non-steroid hormones that reduce bone mass by stimulating osteoblasts to express M-CSF and/or RANKL are PTH, parathyroid hormone-related protein (PTHrP), thyroid hormone (T3) and low doses of 1, 25-dihydroxyvitamin D_3 (Teitlebaum 2000, Erben 2001). A very important non-steroidal regulator of bone mass is leptin, a hormone that functions through an inhibitory action of the

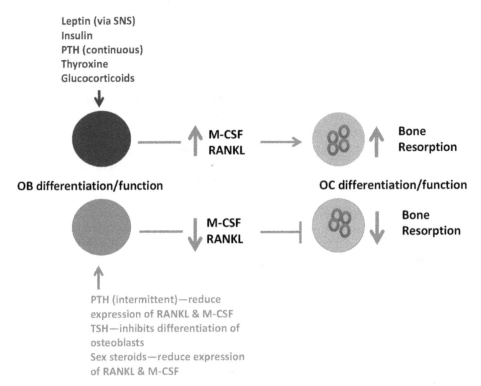

Figure 3.16. Diagram indicating the factors affecting osteoblast differentiation or function that regulate osteoclast differentiation from macrophages, and thus degree of bone resorption. The factors in blue act on the osteoblast to increase its production of M-CSF and RANKL, leading to higher numbers of osteoclasts and greater bone resorption. The factors in red act either to reduce expression of RANKL and M-CSF directly or indirectly by inhibiting the differentiation (and thus number) of osteoblasts, thus decreasing the number of osteoclasts and bone resorption.

hypothalamus on bone formation. Leptin is made by fat cells and functions to suppress appetite and inhibit bone formation by binding to receptors in the hypothalamus (figure 3.17). Mice and humans deficient in leptin or its hypothalamic receptor are obese, but have a higher than normal bone mass. Infusion of leptin into the cerebral ventricles of obese (*ob/ob*) mice rescues them from obesity and restores normal bone mass (Ducy *et al* 2000). The leptin effector pathway from hypothalamus to osteoblasts is an indirect one via sympathetic nerves that innervate osteoblasts (Elmquist and Strewler 2005, Chenu 2004). Sympathetic neurons produce noradrenaline, which binds to $\beta2$ adrenergic receptors ($\beta2$-AR) on osteoblasts (Chenu 2004), leading to the up regulation of RANKL, osteoclastogenesis and reduction of bone mass (Elefteriou *et al* 2005). Mutant mice lacking $\beta2$-AR have increased bone mass like *ob/ob* mice, but unlike *ob/ob* mice, they do not respond to leptin by reduction in bone mass (Elefteriou *et al* 2005).

Insulin is another important hormone regulating bone maintenance regeneration. A feedback linkage exists between insulin production, insulin sensitivity, and the decarboxylated form of osteocalcin, the most abundant non-collagenous protein of bone matrix (Fulzele *et al* 2010, Ferron *et al* 2010). Insulin suppresses expression of the *Twist2* gene in osteoblasts. Twist2 normally inhibits osteoblast differentiation, so its suppression allows the differentiation of higher numbers of osteoblasts, which in turn promote the production of more osteoclasts by the RANKL/RANK pathway. As a result, osteocalcin is released into the resorption lacunae and decarboxylated. The decarboxylated osteocalcin moves into the circulation and promotes insulin sensitivity in adipose tissue and muscle, as well as insulin secretion by the pancreas.

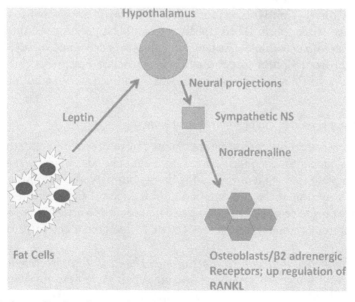

Figure 3.17. Pathway of leptin action. Leptin made by fat cells acts on the part of the hypothalamus that sends projections to neurons of the sympathetic system. Sympathetic axons innervate bone. Osteoblasts have $\beta2$ adrenergic receptors that signal to up regulate RANKL.

Regulation of bone mass by local factors

Bone mass is also regulated by local cytokines and growth factors taken up from the blood or released from the bone matrix as it is degraded by osteoclasts during bone remodeling (Harada and Rodan 2003). BMPs induce MSCs to the chondro/osteo lineage (Rosen and Thies 1992). Other growth factors stimulate proliferation and differentiation of osteoblast precursors via up regulating the collagen I gene and inhibiting transcription of the gene for MMP-13, which degrades collagen I (Wronski 2001). TNF-α and IL-1 promote the formation of osteoclasts, whereas TGF-β induces their apoptosis (Bonewald 1996).

Coupling bone removal to bone building

Osteocytes recognize when a particular region of bone is weak and needs to be replaced (Verborgt *et al* 2000). Through canaliculi, they signal the closest osteoblasts of the periosteum to retract and expose the bone surface (Hauge *et al* 2001). The retracted cells signal the nearest capillary, which sprouts into the retraction space (Parfitt 2006). The endothelial cells of this capillary, and perhaps other cell types, provide 'area code' chemoattractants that guide monocytic osteoclast precursors to bone sites in need of degradation (Parfitt 1998). Osteoblast precursors are available throughout the periosteum and marrow of bones, thus allowing resorption lacunae to become bone remodeling compartments within which osteoblasts and osteoclasts can interact. This interaction involves bidirectional ephrin signaling (Zhao *et al* 2006). When a cell expressing an ephrin receptor contacts a cell expressing an ephrin ligand there is forward signal transduction into the receptor-expressing cell, and reverse signal transduction into the ligand-expressing cell. Ephrin B2 (ligand) is expressed by osteoclasts and their precursors, whereas osteoblasts and their precursors express EphB4 (receptor). As osteoclasts resorb bone matrix, they forward signal from ephrin B2 to EphB4 in osteoblast precursors, stimulating their differentiation into osteoblasts. As new bone matrix is deposited by osteoblasts, they reverse signal from EphB4 to ephrin B2 on osteoclast precursors, inhibiting their differentiation. In this fashion, bone removal is coupled to bone building.

3.2.3 Fracture repair in endochondral long bones

Fractures of intramembranous bone are repaired by the direct differentiation of MSCs in the periosteum into osteoblasts. Fractured endochondal bones first regenerate a cartilage template that is then replaced by bone, but they also exhibit intramembranous bone formation at the proximal and distal edges of the fracture. Like skeletal muscle, bone fails to regenerate across a gap that exceeds a critical size, about 20% of the bone length in mammals (Kolar *et al* 2010, Milner and Cameron 2013).

3.2.3.1 Cellular and molecular events of endochondral fracture repair

The cellular and molecular events of fracture repair have been reviewed by Li and Stocum (2014). Figure 3.18 illustrates the repair of a fractured long bone. Regeneration is accomplished primarily by MSCs in the periosteum, with lesser contributions from the endosteum and marrow stroma. Following fracture, blood

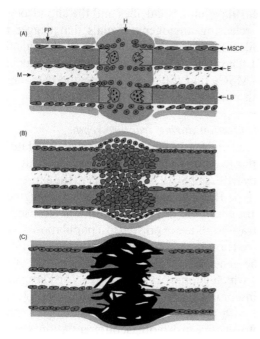

Growth factors that promote fracture repair:

BMP 2, 4, 5, 7
PDGF
FGF-2
IGF-1
TGF-β
EGF
HGF

Figure 3.18. Repair of bone fracture. (A) Shortly after fracture. A blood clot (hematoma, H) forms in the fracture space. FP = fibrous layer of the periosteum (brown); LB = living bone; E = endosteum; M= marrow; MSCP = mesenchymal stem cell layer of the periosteum. Green cells are osteoclasts that clear out dead bone fragments. Stem cells in the MSCP and endosteum (blue) migrate into the hematoma. (B) As the fibrin clot disappears, MSCs in the hematoma proliferate to form a soft callus and differentiate into hypertrophied chondrocytes. (C) Blood vessels (not shown) invade the hypertrophied cartilage template, bringing osteoblasts that replace the template with bone (black). A list of growth factors required for normal repair is given on the right. Reprinted from Stocum (2012). Copyright 2012, with permission from Elsevier.

vessels within and without the bone are torn, resulting in the formation of a fibrin clot (hematoma) in and around the break. Hypoxia results in osteocyte death for a limited distance on either side of the fracture. Platelets in the clot release PDGF and TGF-β, initiating an inflammatory phase in which the hematoma is invaded by neutrophils and macrophages (Einhorn 1998). Some of the macrophages become osteoclasts that degrade the matrix of the dead bone. Within a few days after fracture, periosteal MSCs differentiate on both edges of the fracture to osteoblasts (hard callus) in a process of direct (intramembranous) ossification. The osteoblasts secrete a bone matrix rich in type I collagen, and containing osteocalcin, the mineralization-associated glycoproteins osteonectin, osteopontin and bone sialo-protein II (BSP-II), and numerous proteoglycans (Robey 1996).

Within the fracture space itself, the phases of repair appear to recapitulate the events of embryonic endochondral bone development through a cartilage template. MSCs in the periosteum, endosteum and bone marrow proliferate to form a 'soft callus'. These precursors condense and differentiate into chondrocytes that secrete cartilage-specific matrix composed of type II and XI collagens, aggrecan, hyaluronic acid and fibronectin. The chondrocytes then up regulate collagen X and down regulate other types of collagen. The cartilage matrix calcifies and the chondrocytes hypertrophy and undergo apoptosis, releasing angiogenic signals that trigger the sprouting of periosteal capillaries, while osteoclasts excavate much of the calcified matrix (Glowacki 1998, Einhorn 1998). The periosteal capillaries invade the calcified matrix, accompanied by perivascular MSCs that differentiate into the osteoblasts that replace the cartilage template with bone.

3.2.3.2 Regulation of cartilage differentiation during fracture repair

Formation of the cartilage template and the molecular pathway that regulate its development in fracture repair are similar to embryonic skeletogenesis. Commitment to chondrogenesis requires expression by MSCs of Sox9, which induces the expression of genes for cartilage markers such as type II, X, IX and XI collagens and aggrecan (Ferguson *et al* 1999, Bi *et al* 1999). As the chondrocyte callus matures, *Ihh* transcripts are detected in chondrocytes and *Gli-1* transcripts are expressed in a population of cells on the periphery of the callus that will re-form the periosteum (Ferguson *et al* 1998). The products of these genes and of the genes for PTH and PTHrP are part of the feedback loop that controls the rate at which chondrocytes mature (Vortkamp *et al* 1996, Crowe *et al* 1999, Long and Linsenmayer 1998, Minina *et al* 2001). During replacement of the cartilage template with bone, transcripts of genes that encode signaling proteins and transcription factors active in osteoblast differentiation, such as BMPs, Runx2 and osteocalcin, are detected.

The same growth factors involved in the local control of bone remodeling are essential for fracture repair (Nikolaou and Tsiridis 2007). Members of the TGF-β family are particularly important for chondrogenesis and osteogenesis in fractured bone. BMPs, which induce MSCs to commit to the chondrogenic/osteogenic lineage, are not expressed in uninjured bone, but are released from bone matrix after fracture. BMP receptors IA and IB are expressed in the periosteal cells of uninjured bone and both they and BMPs 2, 4 and 7 are strongly expressed in periosteal mesenchymal cells in the region of hard callus formation, in the proliferating mesenchymal cells and chondroblasts of the soft callus, and in osteoblasts replacing the cartilage with bone (Bostrom 1998, Reddi 1998).

High levels of TGF-β and FGF-1 and 2 are expressed during chondrogenesis of the soft callus, but not in the region of hard callus formation (Bostrom and Asnis 1998). FGF-2 and other members of the FGF family regulate the expression of Sox9 (Murakami *et al* 2000). TGF-β is present earlier in the hematoma and periosteum, but its source appears to be platelets and release from degrading bone matrix rather than synthesis by periosteal cells. PDGF and IGF-I are expressed in the soft callus, suggesting that these growth factors are also involved in chondrogenesis (Bolander 1992, Trippel 1998).

3.2.4 Hematopoietic regeneration

3.2.4.1 Origin of HSCs

During fetal mouse development, HSCs arise from a VE-cadherin[+] CD45[+] population of cells in the ventral aortic endothelium via an endothelial to hematopoietic transition (EHT) (Boisset *et al* 2010, Bertrand *et al* 2010, Kissa and Herbome 2010). HSCs are given off into the lumen of the mouse aorta, whereas in the zebrafish they are given off adluminally. The HSCs seed the fetal liver and bone marrow at approximately similar times (Orkin and Zon 2008). In the adult, the marrow of endochondral bones becomes the principal residence of HSCs. Marrow HSCs give rise to myeloid and lymphoid cell lineages. The myeloid lineage includes erythrocytes, platelets, granulocytes (basophils, eosinophils and neutrophils), mono-cytes and myeloid dendritic cells. The lymphoid lineage includes B-cells, T-cells, natural killer cells and lymphoid dendritic cells. Mature myeloid and lymphoid cells (with the exception of memory B- and T-cells), have half-lives of only days to weeks, and their numbers are maintained by continual regeneration from HSCs.

3.2.4.2 Characteristics of adult HSCs

Long-term (LT) self-renewing HSCs residing in the bone marrow are defined by their cell surface and morphological phenotype and their ability to reconstitute the complete hematopoietic system in bone marrow transplantation assays after its elimination by lethal radiation. Serially diluted suspensions of labeled marrow cells rescued irradiated animals. Labeled myeloid and lymphoid cell colonies were found in the spleen and their derivatives were detected circulating in the blood. Serial colony-forming assays showed that a single transplanted LT-HSC is able to give rise to all the myeloid and lymphoid lineages while self-renewing (Till *et al* 1964, McCulloch 2004, Ponting *et al* 2004).

LT-HSCs are small (6 μm diameter), with dense chromatin. They represent about one in every 10^5 bone marrow cells (Berardi *et al* 1995). The LT-HSC has the cell surface phenotype CD34$^-$ c-Kit$^+$ Sca-1$^+$ VEGFR2$^+$ Thy-1^{lo} (Spangrude *et al* 1988, Ziegler *et al* 1999, Melchers and Rolink 2001). Two markers appear to be selectively expressed on the surface of LT-HSCs. The first is CD150, also called signaling lymphocyte activation molecule (SLAM) (Kiel *et al* 2005, Wagers 2005). This protein provides a simple way to isolate LT-HSCs and identify them in tissue sections. The second is CD49f (Notta *et al* 2011). Single cells expressing this marker were highly efficient in generating long-term multi-lineage engraftment in the bone marrow. LT-HSCs are cadherin low, whereas their lineage-committed progeny are cadherin-high.

3.2.4.3 Lineage maps of HSCs

Niche signals regulate specific sets of transcription factors that determine the patterns of gene activity defining each step in the differentiation of blood cell progenitors.

Hematopoietic lineage maps are based on experiments testing the developmental capacity of cell populations isolated on the basis of surface antigens (Pronk *et al* 2007) or expression levels of the transcription factors GATA-1 or PU.1 (Arinobou *et al* 2007). The details of these maps are being continually refined with the acquisition of new information, but there is general consensus that the long-term HSC gives rise to a common lymphoid progenitor, which gives rise to B and T lymphocytes, and a common myeloid progenitor, which gives rise to basophils, neutrophils, eosinophils, monocytes, macrophages and erythrocytes.

3.2.4.4 Niche regulation of LT-HSCs

The bone marrow stroma with its complex network of osteoblasts, fibroblasts, MSCs, endothelial cells of capillaries and marrow sinusoids, reticular cells associated with endothelial cells, and sympathetic noradrenergic innervation, constitutes a set of niche compartments that regulate the number of LT-HSCs and their progeny (Orkin and Zon 2008, Lin *et al* 2011). Infused LT-HSCs home to bone marrow niches where they engraft (Boulais and Frenette 2015). Homing is mediated by niche production of SDF-1, which binds to the CXCR-4 receptor on the surface of LT-HSCs and promotes migration (Spiegel *et al* 2008, Peled *et al* 1999). Antibodies to SDF-1 or CXCR-4 prevent HSC engraftment in severe combined immunodeficiency (SCID) mice.

Immunostaining for SLAM and live visualization of labeled LT-HSCs indicate that LT-HSCs engraft preferably in perivascular locations of the bone marrow (Kiel *et al* 2005, Lo Celso *et al* 2009, Itkin *et al* 2016). Different types of blood vessels maintain HSC quiescence and activation. Following destruction of marrow sinusoids by lethal irradiation, however, SDF-1 was expressed primarily in the endosteum, and LT-HSCs engrafted closer to osteoblasts (Slayton *et al* 2007). These results suggest that perivascular cells of the sinusoidal endothelium constitute the preferred niche site for quiescent LT-HSCs, but that in its absence, LT-HSCs seek refuge with endosteal osteoblasts. Only 7%–35% of cells identified immunologically as LT-HSCs are actually capable of reconstituting the hematopoietic system after irradiation (Chen *et al* 2016). These cells are positive for Hoxb5. *In situ* imaging showed that they are attached to perivascular cells that express the leptin receptor and high levels of Cxcl12. Engraftment requires not only CXCR-4, but also the guidance molecule Robo4 to specifically anchor the HSCs onto the niche cells (Acar *et al* 2015).

Maintenance of quiescence

Signals affecting osteoblast number or function are a major factor in regulating the numbers of quiescent LT-HSCs in the niche. LT-HSCs are lost after osteoblast depletion *in vivo*. The Ang-1 molecule produced by osteoblasts imposes quiescence via its binding to the Tie-2 receptor on LT-HSCs. Quiescent LT- HSCs are resistant to 5-fluorouracil (FU), a molecule that kills cycling cells. After 5-FU treatment *in vivo*, increased numbers of Tie-2R-expressing HSCs were found adhering exclusively to Ang-1-expressing osteoblasts. Bone marrow cells transfected with a retroviral Ang-1 construct dramatically increased the number of LT-HSCs in the G_0 phase of the cell cycle when injected into irradiated mice.

Other molecules such as thrombopoietin, BMPs and Wnts also help maintain LT-HSC quiescence. Thrombopoietin, a hormone made by the liver and kidney that stimulates megakaryocyte production and differentiation, is also made by osteoblasts. BMP signaling decreases osteoblast number and reduces the number of quiescent LT-HSCs. Conversely, mice with conditional inactivation of the BMP receptor type 1A (BMPR1A) exhibit an increased number of osteoblasts and an increased number of quiescent LT-HSCs. Wnt deficiency or inhibition of the Wnt pathway by DKK or axin impairs the ability of LT-HSCs to reconstitute the bone marrow (Luis *et al* 2008, Fleming *et al* 2008), while exposure to purified Wnt 3 protein or over expression of stable β-catenin expands the pool of LT-HSCs and their progeny *in vitro* (Raya *et al* 2003) and enables them to reconstitute the bone marrow in lethally irradiated mice (Willert *et al* 2003).

Mobilization of ST-HSCs

ST-HSCs must physically move from a quiescent niche to a proliferation-permissive niche. Differential arterial blood vessel permeability in the niche helps regulate the transition. Blood vessels of lower permeability and low exposure to reactive oxygen species (ROS) favor quiescence, whereas blood vessels of higher permeability that expose HSCs to higher concentrations of ROS favor HSC activation (Itkin *et al* 2016). Movement to the proliferation-permissive part of the niche 'is facilitated by stromal SDF-1-induced up regulation of MMP-9 (Levesque *et al* 2001, Heissig *et al* 2002). MMP-9 stimulates the release of Kit Ligand (KitL) from stromal cells. Kit L binds to c-Kit on the surface of the ST-HSCs, increasing their motility and allowing egress from the quiescent niche. Mutations in either c-Kit or KitL impair hematopoiesis (Huang *et al* 1992). Sympathetic innervation plays a key role in regulating the mobilization of ST-HSCs in mice via adrenergic signals. Egress of ST-HSCs from the LT-HSC niche is induced by G-CSF, which activates sympathetic noradrenergic neurons to produce norepinephrine. Pharmacologic or genetic ablation of adrenergic neurotransmission inhibits ST-HSC mobilization by G-CSF, whereas $\beta 2$ adrenergic agonists enhance mobilization.

Pericytes may play the central role in HSC niche creation

Pericytes have long been known for their ability to form osteoblasts, osteocytes and chondrocytes and are contenders for the origination of all MSCs. $CD146^+$ perivascular pericytes from human bone marrow were used to make an endochondral bone organoid that was transplanted subcutaneously into mice. Regenerated pericytes of the organoid were associated with host HSCs and sinusoidal endothelial cells, and exhibited high expression of Ang-1 (Sacchetti *et al* 2007). A CD45-negative, nestin-positive perivascular mesenchymal cell population was identified in mouse bone marrow that co-localized with LT-HSCs and could form osteoblasts, osteocytes and chondrocytes (Mendez-Ferrer *et al* 2010). These cells displayed strong expression of core LT-HSC maintenance genes, were closely associated with sympathetic nerve fibers, and expressed high levels of the β_3 adrenergic receptor and SDF-1. Consistent with these results, a population of mouse marrow cells with surface markers $CD105^+/Thy-1^-$ was identified that when implanted under the kidney capsule recruited host-derived blood vessels and produced donor-derived ectopic endochondral bone with a marrow cavity populated by host-derived LT-HSCs (Chan *et al* 2009). Because they generate features of the osteoblast, vascular and sympathetic nervous niches, perivascular cells may be a unifying feature that can reconcile divergent views about HSC niche cellular compartments.

3.3 Stem cells in non-regenerating organs

Stem cells are present in several non-regenerating mammalian organs, but their potential regenerative response to injury is suppressed due to a non-permissive niche. The spinal cord harbors NSCs that are able to form neurons and glia when isolated from the cord, and cells of the pigmented ciliary margin have proliferative potential *in vitro* and can differentiate into retinal neurons after dedifferentiation (Ahmad *et al* 2000, Tropepe *et al* 2000, Shatos *et al* 2001, Meletis *et al* 2008).

The mammalian kidney is a particularly interesting example. Although nephrons can respond to loss of kidney tissue by hypertrophy of remaining nephrons involving and increase in epithelial cell size, they are unable to regenerate. Nephrons of teleost fish and skate kidneys, which share features of the mammalian kidney in their development, anatomy and physiology, are able to regenerate nephrons from a population of stem cells resident to a special nephrogenic zone of the kidney (Drummond 2003, Diep *et al* 2011), and there are reports of epithelial stem cells that can regenerate tubular structures of the mammalian nephron (Li and Wingert 2013). Of great interest is a report that a subset of parietal epithelial cells expressing the stem cell markers $CD24^+$ $CD133^+$ and transcription factors Oct-4 and Bml-1 was isolated from the human Bowman's capsule of the kidney (Sagrinati *et al* 2006). These cells regenerated tubular structures of the nephron when injected into SCID mice with acute renal failure and significantly attenuated morphological and functional kidney damage. This report suggests that human kidneys may harbor renal stem cell populations that could be expanded in culture from a biopsy for use in intra-renal transplant or biomimetic kidney construction.

References

Adamo M L and Farrar P 2006 Resistance training and IGF involvement in the maintenance of muscle mass during the aging process *Ageing Res. Rev.* **5** 310–31

Adams G R 1998 Role of insulin-like growth factor-1 in the regulation of skeletal muscle adaptation to increased loading *Exerc. Sport Sci. Rev.* **26** 31–60

Acar M *et al* 2015 Deep imaging of bone marrow shows non-dividing stem cells are mainly perisinusoidal *Nature* **526** 126–30

Ahmad I, Tang L and Pham H 2000 Identification of neural progenitors in the adult mammalian eye *Biochem. Biophys. Res. Commun.* **270** 517–21

Allen R E, Sheehan S M and Taylor M G *et al* 1995 Hepatocyte growth factor activates quiescent skeletal muscle satellite cells *in vitro J. Cell Physiol* **165** 307–12

Arinobou Y *et al* 2007 Reciprocal activation of GATA-1 and PU.1 marks initial specification of hematopoietic stem cells into myeloerythroid and myelolymphoid lineages *Cell Stem Cell* **1** 416–27

Arnold K *et al* 2011 Sox2+ adult stem and progenitor cells are important for tissue regeneration and survival of mice *Cell Stem Cell* **8** 317–9

Barker N, Bartfield S and Clevers H 2010 Tissue-resident adult stem cell populations of rapidly self-renewing organs *Cell Stem Cell* **7** 656–70

Benraiss A *et al* 1996 Clonal cell cultures from adult spinal cord of the amphibian urodele *Pleurodeles waltl* to study the identity and potentialities of cells during tail regeneration *Dev. Dyn.* **205** 135–49

Berardi A *et al* 1995 Functional isolation and characterization of human hematopoietic stem cells *Science* **267** 104–8

Bertrand J Y *et al* 2010 Haematopoietic stem cells derive directly from aortic endothelium during development *Nature* **464** 108–11

Bi W, Deng J M, Behringer R R and de Crombrugghe B 1999 Sox 9 is required for cartilage formation *Nat. Genet.* **22** 85–9

Bischoff R 1986 A satellite cell mitogen from crushed adult muscle *Dev. Biol.* **115** 140–7

Blanpain C *et al* 2004 Self-renewal, multipotency and the existence of two cell populations within an epithelial stem cell niche *Cell* **118** 635–48

Blanpain C and Fuchs E 2006 Epidermal stem cells of the skin *Annu. Rev. Cell Dev. Biol.* **22** 339–73

Blanpain C, Horsely V and Fuchs E 2007 Epithelial stem cells: turning over new leaves *Cell* **128** 445–58

Boisset J-C *et al* 2010 *In vivo* imaging of maematopoietic cells emerging from the mouse aortic endothelium *Nature* **464** 116–20

Bolander M E 1992 Regulation of fracture repair by growth factors *Proc. Soc. Exp. Biol. Med.* **200** 165–70

Bonewald L F 1996 Transforming growth factor-β *Principles of Bone Biology* ed J P Bilezikian, L G Raisz and G A Rhodan (New York: Academic), pp 647–59

Bostrom M P G 1998 Expression of bone morphogenetic proteins in fracture healing *Clin. Orthopaed. Rel. Res.* **355S** S116–23

Bostrom M P G and Asnis P 1998 Transforming growth factor beta in fracture repair *Clin. Orthopaed. Rel. Res.* **355S** S124–31

Botchkarev V A and Sharov A A 2004 BMP signaling in the control of skin development and hair follicle growth *Differentiation* **72** 512–26

Boulais P E and Frenette P S 2015 Making sense of hematopoietic stem cell niches *Blood* **125** 26212629

Boyle W J, Simonet W S and Lacey D L 2003 Osteoclast differentiation and activation *Nature* **423** 337–42

Brack A S *et al* 2007 Increased Wnt signaling during aging alters muscle stem cell fate and increases fibrosis *Science* **317** 807–10

Brack A S *et al* 2008 A temporal switch from Notch to Wnt signaling in muscle stem cells is necessary for noemal adult myogenesis *Cell Stem Cell* **2** 50–9

Brownell I *et al* 2011 Nerve-derived sonic hedgehog defines a niche for hair follicle stem cells capable of becoming epidermal stem cells *Cell Stem Cell* **8** 552–65

Butler E G and Ward M B 1965 Reconstitution of the spinal cord following ablation in urodele larvae *J. Exp. Zool.* **160** 47–66

Carlson B M 1970 Organizational aspects of muscle regeneration *Research in Muscle Development and the Muscle Spindle Excerpta Medica International Congress Series* 240 (Amsterdam: Excerpta Medica), 3–17

Carlson B M 1974 Regeneration from short stumps of the rat gastrocnemius muscle *Experientia* **30** 275–6

Carlson B M 2003 Muscle regeneration in amphibians and mammals: passing the torch *Dev. Dyn.* **226** 167–81

Carlson B M and Gutmann E 1972 Development of contractile properties of minced muscle regenerates in the rat *Exp. Neurol.* **36** 239–49

Carlson B M and Gutmann E 1975 Regeneration in free grafts of normal and denervated muscles in the rat: morphology and histochemistry *Anat. Rec.* **183** 47–62

Carulli A J, Samuelson L C and Schnell S 2014 Unraveling intestinal stem cell behavior with models of crypt dynamics *Integr. Biol.* **6** 243–57

Cerletti M *et al* 2008 Skeletal muscle precursor grafts in dystrophic muscle *Cell* **135** 998–9

Chan C K F *et al* 2009 Endochondral ossification is required for hematopoietic stem-cell niche formation *Nature* **457** 490–4

Chen J Y *et al* 2016 Hoxb5 marks long-term hematopoietic stem cells and reveals a homogeneous perivascular nuche *Nature* **457** 887–91

Chenu C 2004 Role of innervation in the control of bone remodeling *J. Musculoskel. Neuronal. Interact.* **4** 132–4

Clarke J D W and Ferretti P 1998 CNS regeneration in lower vertebrates *Cellular and Molecular Basis of Regeneration* ed P Ferretti and J Geraudie (New York: Wiley), pp 255–72

Clayton E *et al* 2007 A single type of progenitor cell maintains normal epidermis *Nature* **446** 185–9

Cohn R D *et al* 2002 Disruption of *Dag1* in differentiated skeletal muscle reveals a role for dystroglycan in muscle regeneration *Cell* **110** 639–48

Conboy I M and Rando T A 2002 The regulation of Notch signaling controls satellite cell activation and cell fate determination in postnatal myogenesis *Dev. Cell* **3** 397–409

Cornelison D D W and Wold B J 1997 Single-cell analysis of regulatory gene expression in quiescent and activated mouse skeletal muscle satellite cells *Dev. Biol.* **191** 270–83

Cotsarelis G 1997 The hair follicle: dying for attention *Am. J. Pathol.* **151** 505–9

Crisan M *et al* 2008 A perivascular origin for mesenchymal stem cells in multiple human organs *Cell Stem Cell* **3** 301–13

Crowe R, Zikherman J and Niswander L 1999 Delta-1 negatively regulates the transition from prehypertrophic to hypertrophic chondrocytes during cartilage formation *Development* **126** 987–98

Davis B M, Duffy M T and Simpson S B Jr 1989 Bulbospinal and intraspinal connection in normal and regenerated salamander spinal cord *Exp. Neurol.* **103** 41–51

Diaz-Quiroz J F and Echeverri K 2013 Spinal cord regeneration: where fish, frogs and salamanders lead the way, can we follow? *Biochem. J.* **451** 353–64

Diep C Q *et al* 2011 Identification of adult nephron progenitors capable of kidney regeneration in zebrafish *Nature* **470** 95–100

DiMario J, Buffinger N, Yamada S and Strohman R C 1989 Fibroblast growth factor in the extra cellular matrix of dystrophic (mdx) mouse muscle *Science* **244** 688–90

Ducy P, Schinke T and Karsenty G 2000 The osteoblast: a sophisticated fibroblast under central surveillance *Science* **289** 1501–4

Duffy M T *et al* 1992 Axonal sprouting and frank regeneration in the lizard tail spinal cord:correlation between changes in synaptic circuitry and axonal growth *J. Comput. Neurol.* **316** 363–74

Drummond I 2003 The skate weighs in on kidney regeneration *J. Am. Soc. Nephrol.* **14** 1704–5

Einhorn T A 1998 The cell and molecular biology of fracture healing *Clin. Orthopaed. Related. Res.* **355S** 7–21

Elefteriou F *et al* 2005 Leptin regulation of bone resorption by the sympathetic nervous system and CART *Nature* **434** 514–20

Elmquist J K and Strewler G J 2005 Do neural signals remodel bone? *Nature* **434** 447–8

Erben R G 2001 Vitamin D analogs and bone *J. Musculoskel. Neuronal. Interact.* **2** 59–69

Ezhkova E *et al* 2009 Ezh2 orchestrates gene expression for the stepwise differentiation of tissue-specific stem cells *Cell* **136** 1122–35

Ferguson C M *et al* 1998 Common molecular pathways in skeletal morphogenesis and repair *Ann. N. Y. Acad. Sci.* **857** 33–42

Ferguson C M, Alpern E, Miclau T and Helms J A 1999 Does adult fracture repair recapitulate embryonic skeletal formation? *Mech. Dev.* **87** 57–66

Ferron M *et al* 2010 Insulin signaling in osteoblasts integrates bone remodeling and energy metabolism *Cell* **142** 296–308

Fleming H E *et al* 2008 Wnt signaling in the niche enforces hematopoietic stem cell quiescence and is necessary to preserve self-renewal *in vivo Cell Stem Cell* **2** 274–83

Fuchs E 2009a Finding one's niche in the skin *Cell Stem Cell* **4** 499–502

Fuchs E 2009b The tortoise and the hair: slow-cycling cells in the stem cell race *Cell* **137** 811–9

Fujiwara H *et al* 2011 The basement membrane of hair follicle stem cells is a muscle cell niche *Cell* **144** 577–89

Fulzele K *et al* 2010 Insulin receptor signaling in osteoblasts regulates postnatal bone acquisition and body composition *Cell* **142** 309–19

Garrett K L and Anderson J E 1995 Colocalization of bFGF and the myogenic regulatory gene myogenenin in dystrophic mdx muscle precursors and young myotubes *in vivo Dev. Biol.* **169** 596–608

Garry D J, Yang Q, Bassel-Duby R and Williams R S 1997 Persistent expression of MNF identifies myogenic stem cells in postnatal muscles *Dev. Biol.* **188** 280–94

Glass D J 2003 Molecular mechanisms modulating muscle mass *Trends Mol. Med.* **9** 345–50

Glowacki J 1998 Angiogenesis in fracture repair *Clin. Orthopaed. Rel. Res.* **355S** S82–9

Gould E and Tanapat P 1997 Lesion-induced proliferation of neuronal progenitors in the dentate gyrus of the adult rat *Neuroscience* **80** 427–36

Greco V *et al* 2009 A two-step mechanism for stem cell activation during hair regeneration *Cell Stem Cell* **4** 155–69

Green H *et al* 1999 Regulation of fiber size, oxidative potential and capillarization in human muscle by resistance exercise *Am. J. Physiol.* **276** R591–6

Grounds M D 1991 Towards understanding skeletal muscle regeneration *Pathol. Res. Pract.* **187** 1–22

Grounds M D and Yablonka-Reuvini Z 1993 Molecular and cell biology of skeletal muscle regeneration *Molecular and Cell Biology of Muscular Dystrophy* ed T Partridge (London: Chapman and Hall), pp 210–56

Gulati A K, Reddi A H and Zalewski A A 1983 Changes in the basement membrane zone components during skeletal muscle fiber degeneration and regeneration *J. Cell Biol.* **97** 957–62

Hameed M *et al* 2004 The effect of recombinant human growth hormone and resistance training on IGF-I mRNA expression in the muscles of elderly men *J. Physiol.* **555** 231–40

Hansen-Smith F M and Carlson B M 1979 Cellular responses to free grafting of the extensor digitorus longus muscle of the rat *J. Neurol. Sci.* **41** 149–73

Harada S-I and Rodan G 2003 Control of osteoblast function and regulation of bone mass *Nature* **423** 349–54

Hardy M H 1992 The secret life of the hair follicle *Trends Genet.* **8** 55–61

Hauge E M *et al* 2001 Cancellous bone remodeling occurs in specialized compartments lined by cells expressing osteoblastic markers *J. Bone Miner. Res.* **16** 1575–82

Heissig B *et al* 2002 Recruitment of stem and progenitor cells from the bone marrow niche requires MMP-9 mediated release of Kit-ligand *Cell* **109** 625–37

Hill M and Goldspink G 2003 Expression and splicing of the insulin-like growth factor-I gene in rodent muscle is associated with muscle satellite (stem) cell activation following local tissue damage *J. Physiol.* **549** 409–18

Horsley V *et al* 2008 NFATc1 balances quiescence and proliferation of skin stem cells *Cell* **132** 299–310

Hotchin N A, Gandarillas A and Watt F M 1995 Regulation of cell surface beta₁ integrin levels during keratinocyte terminal differentiation *J. Cell Biol.* **128** 1209–19

Huang E J, Nocka K H, Buck J and Besmer P 1992 Differential expression and processing of two cell associated forms of the kit-ligand: KL-1 and KL-2 *Mol. Biol. Cell* **3** 349–60

Itkin T *et al* 2016 Distinct bone marrow blood vessels differentially regulate haematopoiesis *Nature* **532** 323–8

Jaks V *et al* 2008 Lgr5 marks cycling, yet long-lived, hair follicle stem cells *Nat. Genet.* **40** 1291–9

Jensen K B *et al* 2009 Lrig1 expression defines a distinct multipotent stem cell population in mammalian epidermis *Cell Stem Cell* **4** 427–39

Jin K *et al* 2001 Neurogenesis in dentate subgranular zone and rostral subventricular zone after focal cerebral ischemia in the rat *Proc. Natl Acad. Sci. USA* **98** 4710–5

Jindo T *et al* 1994 Hepatocyte growth factor/scatter factor stimulates hair growth of mouse vibrissae in organ culture *J. Invest. Dermatol.* **103** 306–9

Johnson S E and Allen R E 1995 Activation of skeletal muscle satellite cells and the role of fibroblast growth factor receptors *Exp. Cell Res.* **219** 449–53

Jones P H, Harper S and Watt F M 1995 Stem cell patterning and fate in human epidermis *Cell* **80** 83–93

Jones P H, Simons B D and Watt F M 2007 Sic Transit Gloria: farewell to the epidermal transit amplifying cell? *Cell Stem Cell* **1** 371–81

Karsenty G 2003 The complexities of skeletal biology *Nature* **423** 316–8

Kherif S *et al* 1999 Expression of matrix metalloproteinases 2 and 9 in regenerating skeletal muscle: a study in experimentally injured and *mdx* muscles *Dev. Biol.* **205** 158–70

Kiel M J *et al* 2005 SLAM family receptors distinguish hematopoietic stem and progenitor cells and reveal endothelial niches for stem cells *Cell* **121** 1109–21

Kissa K and Herbome P 2010 Blood stem cells emerge from aortic endothelium by a novel type of cell transition *Nature* **464** 112–5

Kobeliak K *et al* 2007 Loss of a quiescent niche but not follicle stem cells in the absence of bone morphogenetic protein signaling *Proc. Natl Acad. Sci. USA* **104** 10063–8

Kolar P *et al* 2010 The early fracture hematoma and its potential role in fracture healing *Tissue Eng. Part B* **16** 427–34

Konigsberg U R *et al* 1975 The regenerative response of single mature muscle fibers isolated *in vitro* *Dev. Biol.* **45** 260–75

Kronenberg H M 2003 Developmental regulation of the growth plate *Nature* **423** 332–6

Kuang S, Kuroda K, Le Grand F and Rudnicki M A 2007 Asymmetric self-renewal and commitment of satellite stem cells in skeletal muscle *Cell* **129** 999–1010

Kuang S, Gillespie M A and Rudnicki M A 2008 Niche regulation of muscle satellite cell self-renewal and differentiation *Cell Stem Cell* **2** 22–31

Lachgar S *et al* 1996 Vascular endothelial growth factor is an autocrine growth factor for hair dermal papilla cells *J. Invest. Dermatol.* **106** 17–23

Lee-Liu D, Edwards-Faret G, Tapia V S and Larrain J 2013 Spinal cord regeneration: Lessons for mammals from non-mammalian vertebrates *Genesis* **51** 529–44

LeGrand F *et al* 2009 Wnt7a activates the planar cell polarity pathway to drive the symmetric expansion of satellite stem cells *Cell Stem Cell* **4** 535–47

Lepper C, Partridge T A and Fan C-M 2011 An absolute requirement for Pax7-poitive satellite cells in acute injury-induced skeletal muscle regeneration *Development* **138** 3639–46

Levesque J P *et al* 2001 Vascular cell adhesion molecule-1 (CD106) is cleaved by neutrophil proteases in the bone marrow following hematopoietic progenitor mobilization by granulocyte colony-stimulating factor *Blood* **98** 1289–97

Li J and Stocum D L 2014 Fracture healing *Basic and Applied Bone Biology* ed D B Burr and M R Allen (New York: Academic), pp 205–23

Li Y and Wingert R A 2013 Regenerative medicine for the kidney: stem cell prospects and challenges *Clin. Transl. Med.* **2** 11

Lin K K, Challen G A and Goodell M A 2011 Hematopoietic stem cell properties, markers, and therapeutics *Principles of Regenerative Medicine* 2nd edn ed A Atala, R Lanza, J A Thomson and R Nerem (San Diego, CA: Elsevier/Academic), pp 273–84

Lindemans C A *et al* 2015 Iterleukin-22 promotes intestinal-stem-cell-mediated epithelial regeneration *Nature* **528** 560–4

Liu J, Solway K, Messing R O and Sharp F R 1998 Increased neurogenesis in the dentate gyrus after transient global ischemia in gerbils *J. Neurosci.* **18** 7768–78

Lo Celso C *et al* 2009 Live-animal tracking of individual haematopoietic stem/progenitor cells in their niche *Nature* **457** 92–6

Long F and Linsenmayer T F 1998 Regulation of growth region cartilage proliferation and differentiation by perichondrium *Development* **125** 1067–73

Lowell S *et al* 2000 Stimulation of human epidermal differentiation by Delta-Notch signaling at the boundaries of stem-cell clusters *Curr. Biol.* **10** 491–500

Lowry W E *et al* 2005 Defining the impact of beta-catenin/Tcf transactivation on epithelial stem cells *Genes Dev.* **19** 1596–611

Luis T C *et al* 2008 Wnt3a deficiency irreversibly impairs hematopoietic stem cell self-renewal and leads to defects in progenitor cell differentiation *Blood* **113** 546–54

Martin T J 2004 Paracrine regulation of osteoclast formation and activity: Milestones in discovery *J. Musculoskel. Neuronal. Interact.* **4** 243–53

McCarthy J J *et al* 2011 Effective fiber hypertrophy in satellite cell-depleted skeletal muscle *Development* **138** 3657–66

McCroskery S *et al* 2003 Myostatin negatively regulates satellite cell activation and self-renewal *J. Cell Biol.* **162** 1135–47

McCulloch E A 2004 Normal and leukemic hematopoietic stem cells and lineages *Stem Cells Handbook* ed S Sell (Towota, NJ: Humana), pp 119–32

Melchers F and Rolink A 2001 Hematopoietic stem cells: lymphopoiesis and the problem of commitment versus plasticity *Stem Cell Biology* ed D R Marshak, R Gardner and D Gottleib (Cold Spring Harbor, NY: Cold Spring Laboratory Press), pp 307–27

Meletis K *et al* 2008 Spinal cord injury reveals multilineage differentiation of ependymal cells *PLoS Biol.* **6** e182

Mendez-Ferrer S *et al* 2010 Mesenchymal and haematopoietic stem cells form a unique bone marrow niche *Nature* **66** 829–34

Messenger A G 1993 The control of hair growth: an overview *J. Invest. Dermatol.* **101** 45–95

Milner D J and Cameron J A 2013 Muscle repair and regeneration: Stem cells, scaffolds, and the contribution of skeletal muscle to amphibian limb regeneraton *New Perspectives in Regeneration. Current Topics in Microbiology and Immunology* vol 367 ed E Heber-Katz and D L Stocum (Berlin: Springer), pp 133–62

Minina E *et al* 2001 BMP and Ihh/PTHrP signaling interact to coordinate chondrocyte proliferation and differentiation *Development* **128** 4523–34

Miyazawa K, Shimomura T and Kitamura N 1996 Activation of hepatocyte growth factor in injured tissues is mediated by hepatocyte growth factor activator *J. Biol. Chem.* **271** 3615–8

Morris R J *et al* 2004 Capturing and profiling adult hair follicle stem cells *Nat. Biotechnol.* **22** 411–7

Mundy G R 1999 Bone remodeling *Primer on the Metabolic Bone Diseases and Disorders of Mineral Metabolism* 4th edn ed M J Favus (Philadelphia: Lippincott Williams and Wilkins), pp 30–8

Munoz R *et al* 2015 Regeneration of *Xenopus laevis* spinal cord requires Sox2/3 expressing cells *Dev. Biol.* **408** 229–43

Murakami S, Kan M, McKeehan W L and de Crombrugghe B 2000 Up-regulation of the chondrogenic *Sox9* gene by fibroblast growth factors is mediated by the mitogen-activated protein kinase pathway *Proc. Natl Acad. Sci. USA* **97** 1113–8

Murphy M M *et al* 2011 Satellite cells, connective tissue fibroblasts and their interactions are crucial for muscle regeneration *Development* **138** 3625–37

Musaro A *et al* 2004 Stem cell-mediated muscle regeneration is enhanced by local isoform of insulin-like growth factor 1 *Proc. Natl Acad. Sci. USA* **101** 1206–10

Nakamura T *et al* 2007 Estrogen prevents bone loss via estrogen receptor α and induction of Fas ligand in osteoclasts *Cell* **130** 811–23

Nguyen H, Rendl M and Fuchs E 2006 Tcf3 governs stem cell features and represses cell fate determination in skin *Cell* **127** 171–83

Nikolaou V S and Tsiridis E 2007 Minisymposium: Fracture healing: pathways and signaling molecules *Curr. Orthopaed.* **21** 249–57

Nijhof J G W *et al* 2006 The cell-surface marker MTS24 identifies a novel population of follicular keratinocytes with characteristics of progenitor cells *Development* **133** 3027–37

Nishimura E K *et al* 2002 Dominant role of the niche in melanocyte stem-cell fate determination *Nature* **416** 854–60

Notta F *et al* 2011 Isolation of single human hematopoietic stem cells capable of long-term multilineage engraftment *Science* **333** 218–21

Olsen B R 1999 Bone morphogenesis and embryologic development *Primer on the Metabolic Bone Diseases and Disorders of Mineral Metabolism* 4th edn ed M J Favus (Philadelphia: Lippincott Williams and Wilkins), pp 11–4

Orkin S H and Zon L I 2008 Hematopoiesis: An evolving paradigm for stem cell biology *Cell* **132** 631–44

Owino V, Yang S Y and Goldspink G 2001 Age-related loss of skeletal muscle function and the inability to express the autocrine form of insulin-like growth factor-1 (MGF) in response to mechanical overload *FEBS Lett.* **505** 259–63

Palacios D and Puri P L 2006 The epigenetic network regulating muscle development and regeneration *J. Cell Physiol.* **207** 1–11

Parfitt A M 1998 Osteoclast precursors as leukocytes: importance of the area code *Bone* **23** 491–4

Parfitt A M 2006 Misconceptions V-activation of osteoclasts is the first step in the bone remodeling cycle *Bone* **39** 1170–2

Pastoret C and Partridge T A 1998 Muscle regeneration *Cellular and Molecular Basis of Regeneration* ed P Ferretti and J Geraudie (New York: Wiley), pp 309–34

Paus R 1998 Principles of hair cycle control *J. Dermatol.* **25** 793–802

Peled A *et al* 1999 Dependence of human stem cell engraftment and repopulation of NOD/SCID mice on CXCR4 *Science* **283** 845–8

Philippou A, Halapas A, Maridaki M and Koutsilieris M 2007 Type I insulin-like growth factor receptor signaling in skeletal muscle regeneration and hypertrophy *J. Musculoskel. Neuronal. Interact.* **7** 208–18

Plikus M V *et al* 2008 Cyclic dermal BMP signaling regulates stem cell activation during hair regeneration *Nature* **451** 340–4

Ponting I, Zhao Y and Anderson W F 2004 Hematopoietic stem cells *Stem Cells Handbook* ed S Sell (Towota NJ: Humana), pp 155–62

Potten C S 1998 Stem cells in gastrointestinal epithelium: numbers, characteristics and death *Philos. Trans. Soc. Lond.* B **353** 821–30

Pronk C J H, Rossi D J, Mansson R, Attema J L, Norddahl G L, Chan C K F, Sigvardsson M, Weissman I L and Bryder D 2007 Elucidation of the phenotypic, functional, and molecular topography of a myeloerythroid progenitor cell hierarchy *Cell Stem Cell* **1** 428–42

Raya A *et al* 2003 Activation of Notch signaling pathway precedes heart regeneration in zebrafish *Proc. Natl Acad. Sci USA* **100** 11889–95

Reddi A H 1998 Initiation of fracture repair by bone morphogenetic proteins *Clin. Orthopaed. Rel. Res.* **355S** S66–72

Robey P 1996 *Bone matrix proteoglycans and glycoproteins Principles of Bone Biology* ed J P Bilezikian, L G Raisz and G A Rhodan (New York: Academic Press), pp 155–65

Rodriguez-Colman M J *et al* 2017 Interplay between metabolic identities in the intestinal crypt supports stem cell function *Nature* **543** 424–7

Rogers S L 1982 Muscle spindle formation and differentiation in regenerating rat muscle grafts *Dev. Biol.* **94** 265–83

Rosen V and Thies R S 1992 The BMP proteins in bone formation and repair *Trends Genet.* **8** 97–102

Sacchetti B *et al* 2007 Self-renewing osteoprogenitors in bone marrow sinusoids can organize a hematipoietic microenvironment *Cell* **131** 324–36

Sagrinati C *et al* 2006 Isolation and characterization of multipotent progenitor cells from the Bowman's capsule of adult human kidneys *J. Am. Soc. Nephrol.* **17** 2443–56

Sambasivan R *et al* 2011 Pax7-expressing satellite cells are indispensable for adult skeletal muscle regeneration *Development* **138** 3647–56

Sato T *et al* 2011 Paneth cells constitute the niche for Lgr5 stem cells in intestinal crypts *Nature* **469** 415–8

Shatos M A *et al* 2001 Multipotent stem cells from the brain and retina of green mice *J. Reg. Med.* **2** 13–5

Sierra A *et al* 2010 Microglia shape adult hippocampal neurogenesis through apoptosis-coupled phagocytosis *Cell Stem Cell* **7** 483–95

Sirbulescu R F and Zupanc G K H 2009 Dynamics of caspase-3-mediated apoptosis during spinal cord regeneration in the teleost fish, *Apteronotus leptorhynchus Brain Res.* **1304** 14–25

Sirbulescu R F and Zupanc G K H 2011 Spinal cord repair in regeneration-competent vertebrates: adult teleost fish as a model system Brain Res. Rev. **67** 73–93

Slayton W B *et al* 2007 The role of the donor in the repair of the vascular niche following hematopoietic stem cell transplant *Stem Cells* **25** 2945–55

Snippert H J *et al* 2010 *Lgr6* marks stem cells in the hair follicle that generate all cell lineages of the skin *Science* **327** 1385–9

Stocum D L 2012 *Regenerative Biology and Medicine* 2nd edn (Elsevier)

Spalding K L *et al* 2005 Retrospective birth dating of cells in humans *Cell* **122** 133–43

Spangrude G J, Heimfeld S and Weissman I L 1988 Purification and characterization of mouse hematopoietic cells *Science* **241** 58–62

Spiegel A *et al* 2008 Stem cell regulation via dynamic interactions of the nervous and immune systems with the microenvironment *Cell Stem Cell* **3** 484–92

Takayanagi H *et al* 2002 RANKL maintains bone homeostasis through c-Fos-dependent induction of *interferon-β Nature* **416** 744–9

Tanaka E M and Ferretti P 2009 Considering the evolution of regeneration in the central nervous system *Nat. Rev. Neurosci.* **10** 713–23

Tanaka K K *et al* 2009 Syndecan-4-expressing muscle progenitor cells in the SP engraft as satellite cells during muscle regeneration *Cell Stem Cell* **4** 217–25

Taylor G *et al* 2000 Involvement of follicular stem cells in forming not only the follicle but also the epidermis *Cell* **102** 451–61

Tedesco F S *et al* 2010 Repairing skeletal muscle: regenerative potential of skeletal muscle stem cells *J. Clin. Invest.* **120** 11–9

Teitlebaum S L 2000 Bone resorption by osteoclasts *Science* **289** 1504–8

Teitelbaum S L and Ross F P 2003 Genetic regulation of osteoclast development and function *Nature Rev. Genet.* **4** 638–49

Till J E, McCullough E A and Siminovitch L 1964 A stochastic model of stem cell proliferation, based on the growth of spleen colony-forming cells *Proc. Natl Acad. Sci. USA* **51** 29–36

Trippel S B 1998 Potential role of insulin-like growth factors in fracture healing *Clin. Orthopaed. Rel. Res.* **355S** S301–13

Tropepe V *et al* 2000 Retinal stem cells in the adult mammalian eye *Science* **287** 2032–6

Tumbar T *et al* 2004 Defining the epithelial stem cell niche in skin *Science* **303** 359–63

Van Es J H *et al* 2010 Intestinal stem cells lacking the Math1 tumor suppressor are refractory to Notch inhbitors *Nat. Commun.* **1** 1–5

Verborgt O, Gibson G J and Schaffler M B 2000 Loss of osteocyte integrity in association with microdamage and bone remodeling after fatigue *in vivo J. Bone Miner. Res.* **15** 60–7

Verstappen J, Katsaros C, Torensma R and Von den Hoff J W 2009 A functional model for adult stem cells in epithelial tissues *Wound Rep. Reg.* **17** 296–305

Vortkamp A *et al* 1996 Regulation of rate of cartilage differentiation by Indian hedgehog and PTH-related protein *Science* **273** 613–22

Vortkamp A *et al* 1998 Recapitulation of signals regulating embryonic bone formation during postnatal growth and in fracture repair *Mech. Dev.* **71** 65–76

Wagers A J 2005 Stem cell grand SLAM *Cell* **121** 967–70

Wen T *et al* 2010 Integrin $\alpha3$ subunit regulates events linked to epithelial repair, including keratinocyte migration and protein expression *Wound Rep. Reg.* **18** 325–34

Willert K *et al* 2003 Wnt proteins are lipid-modified and can act as stem cell growth factors *Nature* **423** 448–52

Willis P E, Chadan S, Baracos V and Parkhouse W S 1997 Acute exercise attenuates age-associated resistance to insulin-like growth factor-1 *Am. J. Physiol.* **272** E397–404

Wronski T 2001 Skeletal effects of systemic treatment with basic fibroblast growth factor *J. Musculoskel. Neuronal. Interact.* **2** 9–14

Yang S Y and Goldspink G 2002 Different roles of the IGF-I Ec peptide (MGF) and mature IGF-I in myoblast proliferation and differentiation *FEBS Lett.* **522** 156–60

Young B *et al* 2006 Wheater's Functional Histology 5th edn (London: Elsevier)

Zammit P and Beauchamp J 2001 The skeletal muscle satellite cell: stem cell or son of stem cell? *Differentiation* **68** 193–204

Zelzer E and Olsen B R 2003 The genetic basis for skeletal diseases *Nature* **423** 343–8

Zhang J *et al* 2003 Identification of the haemopoietic stem cell niche and control of the niche size *Nature* **425** 836–41

Zhao P and Hoffman E P 2004 Embryonic myogenesis pathways in muscle regeneration *Dev. Dyn.* **229** 380–92

Zhao C *et al* 2006 Bidirectional ephrinB2-EphB4 signaling controls bone homeostasis *Cell Metabol.* **4** 111–21

Zheng B *et al* 2007 Prospective identification of myogenic endothelial cells in human skeletal muscle *Nat. Biotechnol.* **25** 1025–34

Ziegler B *et al* 1999 KDR receptor: a key marker defining hematopoietic stem cells *Science* **285** 1553–8

Zupanc G K and Sirbulescu R F 2011 Adult neurogenesis and neuronal regeneration in the central nervous system of teleost fish *Eur. J. Neurosci.* **34** 917–29

IOP Publishing

Foundations of Regenerative Biology and Medicine

David L Stocum

Chapter 4

Non-stem cell regenerative mechanisms

Summary

There are three primary mechanisms of regeneration that do not involve adult stem cells. These are cellular re-growth, regeneration from pre-existing differentiated cells, and transdifferentiation. Cellular re-growth is used by neurons to regenerate axons. Regeneration from pre-existing differentiated cells is accomplished either by compensatory hyperplasia, in which a proliferative program of gene transcription is superimposed on the differentiation program of the cell, as in hepatocyte or β-cell regeneration, or by dedifferentiation of the cell to a progenitor transcriptional state, proliferation, and lineage-specific redifferentiation (cardiomyocytes). Both the liver and pancreas use adult stem cells in the biliary and pancreatic ducts as a secondary mechanism of regeneration when the mechanism of compensatory hyperplasia is compromised. Epithelial to mesenchymal transformation is a form of lineage-specific dedifferentiation/redifferentiation that allows epithelial cells to take on a migratory phenotype, as in the movement of epidermal cells across a wound, and the regeneration of urodele and fish spinal cord. Transdifferentiation is a process that converts the transcriptional profile of a cell to that of a different cell type. This is done either directly, with no intermediate cell state, or most often indirectly, in which the cell dedifferentiates, proliferates and redifferentiates into a new pheno-type. The classic example of transdifferentiation is lens regeneration by pigmented epithelial cells of the iris and neural retina regeneration from retina pigmented epithelial cells in newts.

4.1 Cellular re-growth: axon regeneration

4.1.1 Peripheral nerve axons

Peripheral nerves consist of cell bodies with projecting short dendrites and an extended axon with branching terminals that synapse to the dendrites and cell bodies of other neurons and to target tissues such as skin and muscle. Axons are encased in wraps of Schwann cell membrane (myelin) separated by nodes of Ranvier (figure 4.1(A)). The fundamental structure of peripheral nerves is the endoneurial unit composed of an axon, its associated Schwann cell sheath, and a surrounding connective tissue sheath, the endoneurium (Ham and Cormack 1979) (figure 4.1(B), (C)). The endoneurium and the Schwann cells synthesize a basement membrane between them. Endoneurial units are organized into fascicles, each of which is surrounded by a connective tissue perineurium.

The fascicles are bundled into the nerve trunk, and are encased in an epineurium. The nerve trunks are richly vascularized by epineurial, intrafascicular and perineurial arteries and arterioles, and a capillary network in the endoneurium.

Mammalian spinal nerves are comprised of mixed sensory and motor axons (figure 4.1(D)). The cell bodies of sensory neurons are located in the series of spinal ganglia just outside the spinal cord and the cell bodies of motor neurons reside in the ventral horn of the cord gray matter. The spinal ganglia send long axons to peripheral sensory receptors and extend a shorter axon into the dorsal side of the white matter of the cord to synapse with neurons that form the ascending sensory

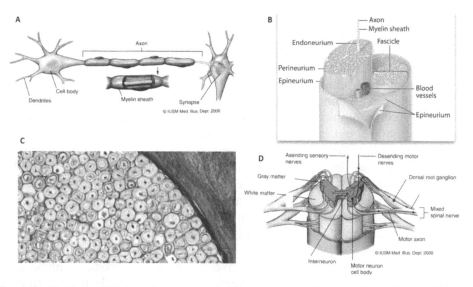

Figure 4.1. (A) Nerve cell body and axon wrapped with myelin derived from the cell membranes of Schwann cells. (B) Anatomy of a nerve. Individual axons with their Schwann sheath and basement membrane invested with a thin connective tissue endoneurium, constitute the endoneurial unit. These units are bundled into fascicles invested with a perineurium. The outside of the nerve is invested with epineurial connective tissue. (C) Cross-section of a nerve showing the axons (dark spots) surrounded by myelin sheaths and bundled into fascicles. (D) Diagram of cross-section of spinal cord. Gray matter (butterfly shape) consists of neurons organized into dorsal and ventral horns. Dorsal horn neurons relay signals to the brain (ascending nerve tracts) from sensory neurons in the spinal ganglia. These same sensory ganglia neurons project afferent axons to sensory organs such as the skin. Ventral horn neurons relay efferent motor signals from the brain to the muscles of the body. The afferent axons and efferent axons combine to form mixed sensory nerves just distal to the spinal ganglia. (A, D) Reproduced with permission of Indiana University School of Medicine Medical Illustration Department. (B) Reproduced with permission from (Mescher 2013). Copyright 2013 McGraw-Hill. (C) Courtesy of Dr Matt Allen, Indiana University School of Medicine.

tracts to the brain. The motor neurons send long axons that are bundled together with the long sensory axons just distal to the spinal ganglia to form the spinal nerve. The motor neurons of the cord synapse with motor axons from the motor centers of the brain to form the descending motor tracts. Some of these motor axons belong to the autonomic nervous system. These autonomic efferents (preganglionc fibers) synapse with a second motor neuron in a series of autonomic paravertebral ganglia outside the vertebral column. The axons of the second neuron (postganglionic fibers) innervate the viscera and glands. Interneurons within the cord synapse with both a branch of the sensory root axons and the motor cell bodies to integrate sensory and motor information and mediate reflexes. Sensory and motor axons are myelinated by Schwann cells for most of their length. Myelin consists of expanded phospho-lipid-rich Schwann cell membranes wrapped around the axons.

The axons in the part of a peripheral nerve distal to a crush injury or transection degenerates, leaving empty endoneurial tubes, whereas the part of the nerve proximal to the lesion survives. Peripheral nerves regenerate moderately well after crush injury in most vertebrates, including mammals (Yannas 2001, for a review). Regeneration after transection is more problematic because the distal endoneurial tubes are often not in register with their proximal counterparts, causing the extending axons to form painful tangles called neuromas. Surgeons attempt to minimize this problem by realigning the cut nerve ends as closely as possible. Axon regeneration is initiated by the actin-driven formation of a growth cone (McQuarrie 1983, Sincropi and McIlwain 1983). As the growth cone advances, it elongates the axon by pulling out a thin cylinder of the proximal axon stump that is stabilized by the polymerization of microtubules. Phospholipids, actin, and tubulin for building new cell membrane and cytoskeleton are synthesized by the neuron cell body and transported distally. When a growth cone contacts its target, the axon enlarges radially to its mature diameter and re-elaborates neurofilaments (McQuarrie 1983).

Schwann cells play a pivotal role in axon regeneration by recycling myelin degradation products and providing neurotrophic factors and adhesion proteins necessary for neuron survival and axon elongation (Fu and Gordon 1997). The myelin is degraded both proximal and distal to the injury in a process called Wallerian degeneration, leaving behind nucleated Schwann cell bodies that undergo dedifferentiation and proliferation. The myelin degradation products inhibit axon regeneration and are removed by macrophages, which degrade them further to cholesterol and free fatty acids. These are complexed to apolipoprotein E to form lipoprotein particles that are released from the macrophages and taken up by Schwann cells via low-density lipoprotein (LDL) receptors. The cholesterol and fatty acids are reused to synthesize new myelin (Goodrum and Bouldin 1996).

In an injured nerve, Schwann cells spread into the lesion space to form a bridge between the endoneurial tubes proximal and distal to the lesion. Regenerating axons grow across this bridge and through the distal endoneurial tubes to restore the distal part of the nerve. The Schwann cells are initially mixed with fibroblasts from the injured endoneurium and sort out from them to form cords of cells called the bands of Bungner (Parrinello *et al* 2010). There is evidence that the Schwann cells in different regions of the injury have different behaviors (Clements *et al* 2017).

The Schwann cells on the proximal side of the injury (closest to the cell body) maintain their epithelial adhesive characteristics, whereas those on the distal side of the injury, where the axon is degenerating, activate stem cell gene expression and undergo a partial EMT, becoming less adhesive and more migratory. Schwann cells in the bridge region exhibit a strong EMT in response to elevated levels of TGF-β4, and become increasingly mesenchymal and motile. These are the cells that promote extension of axons across the transection gap.

Axotomized neurons, platelets, macrophages and dedifferentiated Schwann cells provide neurotrophic, growth factor and adhesion molecules essential for neuron survival and axon extension (Gordon 2009) (figure 4.2). Important neurotrophins include nerve growth factor (NGF), brain-derived neural growth factor (BDNF), and neurotrophins 3 and 4 (NT-3, NT-4). NGF supports the survival of sensory and sympathetic axons, whereas BDNF, NT-3 and NT-4 promote survival of regenerating motor neurons. Ciliary neurotrophic factor (CNTF) and glial derived neurotrophic factor (GDNF) are also important to the survival of axotomized neurons and axon elongation. CNTF and GDNF are released from Schwann cells after injury, whereas macrophages are the primary source of IL-6. Neurotrophins exert their effects by binding to specific receptor tyrosine kinases (Trks) on the neuron surface as well as to a common low-affinity receptor p75^{NTR} that maintains a GTP signaling protein called Rho A in an inactive state. In the absence of neurotrophins, Rho A is activated, leading to the depolymerization of actin and growth cone collapse (Davies 2000). CNTF and IL-6 use JAK/STAT signaling pathways. The growth cones of the elongating axons are guided to their targets by adhering to fibronectin, laminin and tenascin C on the basement membranes of the endoneurial units, and to L1, NCAM and N-cadherin on the Schwann cell surfaces.

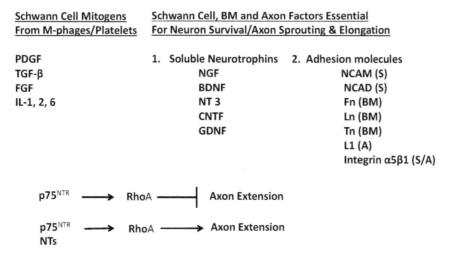

Schwann Cell Mitogens From M-phages/Platelets	Schwann Cell, BM and Axon Factors Essential For Neuron Survival/Axon Sprouting & Elongation	
PDGF	1. Soluble Neurotrophins	2. Adhesion molecules
TGF-β	NGF	NCAM (S)
FGF	BDNF	NCAD (S)
IL-1, 2, 6	NT 3	Fn (BM)
	CNTF	Ln (BM)
	GDNF	Tn (BM)
		L1 (A)
		Integrin α5β1 (S/A)

p75^{NTR} $\longrightarrow$ RhoA $\longrightarrow\!|$ Axon Extension

p75^{NTR} $\longrightarrow$ RhoA $\longrightarrow$ Axon Extension
NTs

Figure 4.2. Left: Mitogens for Schwann cells expressed by macrophages and platelets. Right: Factors provided by Schwann cells, basement membrane (BM) and axons themselves that are essential for neuron survival, and axon sprouting and elongation after peripheral nerve crush or transection. Neurotrophins activate the p75^{NTR} receptor, which promotes axon extension via Rho A.

4.1.2 CNS axons

Axons of the mammalian central nervous system (CNS) fail to regenerate after injury, for reasons that will be discussed in chapter 7. However, the spinal cord axons and optic nerve axons of urodele amphibians and teleost fish regenerate well, and can provide insights into the mechanisms of successful CNS regeneration.

4.1.2.1 Spinal cord

The transected spinal cord of the urodele amphibian regenerates by an epithelial/mesenchymal transformation (EMT/MET) of the ependyma that lines the lumen of the spinal cord (figure 4.3). The ependymal cells lose their specialized junctions and become mesenchymal cells that proliferate to bridge the transection gap and then undergo MET (Holder *et al* 1989, Ferrett *et al* 2003). The cells of the restored epithelium put out processes with expanded endfeet that constitute the glia limitans under the pia mater and form channels that promote the sprouting and elongation of axons in the spinal tracts (Holder 1990).

The axons of the transected cord in teleost fish also regenerate well. After transection of the zebrafish cord a Fgf-dependent glial scar of hypertrophied astrocytes is formed in the transection gap that promotes the growth of axons as the ends of the ependyma grow through the scar to restore the continuity of the spinal canal. The scar appears to provide niche adhesion and growth promoting factors necessary for axon elongation (Sirbulescu and Zupanc 2010, for a review). The adhesive immunoglobulins zfNLRR (Bormann *et al* 1999) and L1.1 (Becker *et al* 1998) are up regulated in neurons undergoing axon regeneration. The myelin protein zero (P0) is up regulated in glial cells and contactin1a is up regulated in neurons and oligodendrocytes around the lesion site (Schweitzer *et al* 2003, 2007).

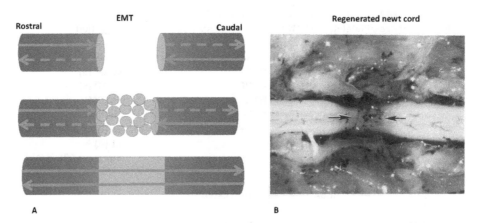

Figure 4.3. Regeneration of the urodele spinal cord. (A) Top: transection of the cord results in axon degeneration (dashed lines) caudal to the transection (descending tracts), and rostral to the transection (ascending tracts). Middle: Ependymal cells undergo EMT and proliferate to bridge the transection gap. Bottom: The mesenchymal cells undergo MET back to an ependymal epithelium that facilitates axon regeneration caudally and rostrally. (B) Image of the regenerated spinal cord of the newt, *Notophthalmus viridescens*. Reproduced with permission from (Davis *et al* 1990). Copyright 1990 Elsevier.

Inhibiting the expression of these adhesion proteins greatly diminishes axon regeneration, synapse formation and functional recovery (Becker *et al* 2004). In the knifefish, *A. leptorhynchus*, a network of radial glia cells provides a support system for axon regeneration of the transected cord, but no glial scar has been observed in this species (Vitalo *et al* 2016).

The growth-associated protein 43 (GAP-43) is up regulated in the neurons of zebrafish regenerating spinal cord axons, and sonic hedgehog is up regulated in ependymoglial cells of the central canal. Non-regenerating axons such as those of the Mauthner neuron do not up regulate GAP-43 (Becker *et al* 1998). Signaling molecules that may foster motor neuron regeneration are sonic hedgehog, retinoic acid and Fgf-3, acting through the transcription factors Nkx6.1 and Pax-6 (Reimer *et al* 2009). A genome-wide screen for secreted factors up regulated during zebrafish cord regeneration revealed induction of the connective tissue growth factor a (*ctgf a*) gene in and around glial cells participating in the initial bridging events of regeneration (Mokalled *et al* 2016). Regeneration was disrupted by mutations in the gene, whereas its overexpression, as well as delivery of human CTGF recombinant protein, accelerated regeneration.

Regeneration of the mammalian spinal cord is inhibited by myelin-associated inhibitory factors (MAFs) released by myelin (oligodendrocyte membrane) break-down (chapter 7). Three such MAFs are Nogo-A, myelin-associated glycoprotein (MAG) and oligodendrocyte-myelin glycoprotein (OMgp). Nogo and MAG act through the Nogo receptor (NgR). Nogo is present in the injured zebrafish cord but lacks an inhibitory N-terminal region (Diekmann *et al* 2005). MAG is rapidly removed in the injured adult newt spinal cord, probably by macrophages ingesting fragments of myelin (Becker *et al* 1999). Nogo-A, MAG and NgR are present at high levels in the regenerating axolotl cord, but are not inhibitory to regeneration; in fact, their expression appears to be correlated with high regenerative capacity. The reason for this may be related to the locus of expression of these molecules. They are expressed on the neuron cell bodies and axons, whereas in the injured non-regenerating mammalian cord their locus of expression shifts to oligodendrocytes (O'Neill *et al* 2004).

4.1.2.2 Optic nerve

Transected optic nerve axons of urodeles also regenerate through an astrocytic scar that bridges the transection (Turner and Singer 1974, Bohn *et al* 1982). Optic nerve regeneration in the newt is associated with removal of the inhibitory myelin proteins tenascin-R and MAG (Becker *et al* 1999). Goldfish optic nerves were reported to produce a factor that is cytotoxic to myelin-producing oligodendro-cytes (Cohen *et al* 1990). Antibodies to the cytokine IL-2 reacted with the factor, indicating that it might be a dimer of IL-2 (Eitan and Schwartz 1993). Consistent with this hypothesis, an enzyme identified as a nerve transglutaminase was purified from regenerating fish optic nerves that dimerized human IL-2. The dimerized IL-2 was cytotoxic to cultured rat brain oligodendrocytes. Dimerization of IL-2 thus might be a mechanism in the fish optic nerve that prevents oligodendrocyte inhibition of regeneration.

4.2 Regeneration from pre-existing differentiated cells

4.2.1 Compensatory hyperplasia of liver and pancreas

4.2.1.1 Liver

The liver is the largest internal organ of the body. It circulates 25% of the total cardiac output within its vessels at any given time and has secretory and metabolic functions that affect virtually every physiological process of the body. The liver converts glycogen to glucose, has a major role in carbohydrate, ammonia and triglyceride metabolism, and secretes a large number of blood serum proteins, including albumin, prothrombin, fibrinogen, the protein component of lipoproteins, and bile, which contains bile salts required for intestinal absorption. The liver also detoxifies metabolic byproducts and toxic substances.

The structural organization of the liver reflects its secretory and metabolic functions (figure 4.4). Hepatocytes, which constitute 80% of the liver, have tremendous numbers of mitochondria and a large volume of rough endoplasmic reticulum and Golgi stacks for protein synthesis and secretion. The hepatocytes are arranged as two-layered trabeculae. The two layers enclose biliary canaliculi that convey bile to the bile ductules via short canals of Hering, which are lined by a mixture of hepatocytes and bile duct cells. The trabeculae are separated by vascular sinusoids lined by fenestrated endothelial cells and macrophages called Kuppfer cells. Blood vessels and bile ducts all have basement membranes, but these are lacking in the sinusoids. Between the fenestrated endothelium of the sinusoids and the trabeculae is the space of Disse. The fenestrated endothelium and the space of Disse provide the hepatocytes with maximum exposure to hepatic blood flow. The hepatocytes have numerous microvilli projecting from their surface into the space of

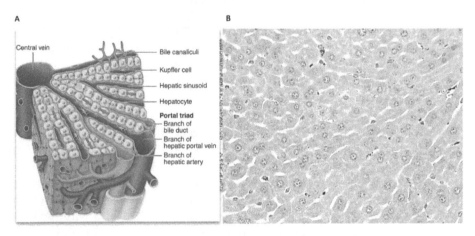

Figure 4.4. (A) Structure of the liver. Liver tissue is organized around a central vein as plates of cells, each consisting of two layers of hepatocytes facing each other with a space, the bile canaliculus, between them for access to bile ducts. Each plate is sandwiched between endothelial hepatic sinusoids for access to the blood. See the text for further details. Reproduced with permission from (Mescher 2013). Copyright 2013 McGraw-Hill. (B) Section of liver tissue showing plates of hepatocytes. Courtesy of Dr Matt Allen, Indiana University School of Medicine.

Disse, providing them with a large surface area for absorption. Interspersed among the hepatocytes and also abutting on the space of Disse are Ito cells (also called stellate cells), pericytes that store vitamin A. In mammals such as the rat and pig, but not humans, the liver is organized into roughly hexagonal lobules consisting of portal areas, each with a central vein and joined by connective tissue septae.

The best-studied model for liver regeneration is the two-thirds partial hepatectomy (PH) model developed for the rat by Higgins and Anderson (1931). The rat liver has two large and two small lobes. Excision of the two large lobes removes two-thirds of the mass. The excised lobes do not grow back, but the remaining two small lobes grow rapidly to attain the original mass of the liver within 14 days after operation. There is no cell damage when discrete lobes of the liver are removed; consequently, cell proliferation occurs in the absence of complicating factors such as cell death or inflammatory events (Fausto and Webber 1994). Hepatocytes depend on lipids stored in lipid droplets as a readily available source of energy and new membrane synthesis during proliferation (Farrell 2004). Each hepatocyte divides once or twice to restore the liver mass, but an increase in hepatocyte size (hypertrophy) also contributes to the restoration (Miyaoka and Miyajima 2013).

Damage to the liver by cutting through the lobular tissue itself as opposed to removing lobes involves an inflammatory phase. If the injury is not too severe, the liver regenerates without scarring. Major disruption of the liver stroma, however, by making a complete cut through the liver of neonatal or adult mice, results in inflammation and stellate cell activation that leads to scarring characterized by collagen deposition (Masuzaki et al 2013). The difference between regeneration and scarring is associated with differential expression of the stromal-derived factor 1 (SDF-1) receptors CXCR7 and CXCR4 by LSECs. Under conditions in which liver injury leads to scarring (cirrhosis, hepatotoxin injection, bile duct ligation) CXCR4 expression predominates over that of CXCR7, resulting in a pro-fibrotic vascular niche, whereas under regenerative conditions CXCR7 and CXCR4 collaborate to deploy pro-regenerative angiocrine factors (Ding et al 2014).

The liver regenerates by compensatory hyperplasia, in which a proliferative pattern of gene activity is superimposed on the hepatocyte-specific transcription pattern (Michalopoulos and DeFrances 1997, Trembly and Steer 1998, Fausto 2004, Otu et al 2007). This mechanism has been confirmed by genetic marking experiments showing that regenerated hepatocytes are derived from pre-existing hepatocytes (Malato et al 2011, Yanger et al 2014). Hepatocytes turn over slowly, with an average life span estimated at 200–300 days (Bucher and Malt 1971). Cell labeling studies indicate that turnover is matched by mitosis of existing hepatocytes (Grompe and Finegold 2001, Zaret and Grompe 2008). The liver has a remarkable ability to regulate its size to the metabolic needs of the body. Humans can tolerate loss of 70% of the liver. In a young person, this loss is made good in a matter of 3–4 weeks; regeneration is slower in older people. Livers transplanted from a child donor to an adult, or from an adult donor to a child adjust their size upward by mitosis or downward by apoptosis, respectively.

Recently, a restricted population of hepatocytes located in the periportal regions of the liver around the central vein was shown to proliferate in response to liver damage. The dividing hepatocytes remained in the periportal region while their descendants migrated to replenish up to 40% of the liver mass outside the periportal zone (Wang *et al* 2015). The periportal hepatocytes are the only liver cell population to express genes activated by Wnt signals generated by the endothelial cells of the adjacent central vein. This raises the question of whether the behavior of hepatocytes in different parts of the liver is differentially dependent on niche composition.

Hepatocytes have tremendous proliferative potential (Fausto 2004, Power and Rasko 2008). From serial transplants of hepatocytes to cure mice with hereditary tyrosinemia type I (a fatal recessive liver disease involving the accumulation of a hepatotoxic metabolite of tyrosine) it was calculated that a single hepatocyte is able to divide at least 70 times (Overturf *et al* 1997). All the cell types of the liver proliferate after PH, although the kinetics of DNA synthesis for each cell type is different (Michalopoulos and DeFrances 1997). Hepatocyte DNA synthesis is initiated 10–12 h after partial hepatectomy and peaks at 24 h, whereas DNA synthesis in biliary duct cells, Kupffer cells and Ito cells begins later and peaks later. The endothelial cells of the sinusoids begin to divide last, reaching a peak of DNA synthesis at four days. In a young adult rat, up to 95% of hepatocytes divide at least once in restoring the original liver size. In older animals, liver regeneration is slower, less complete, and involves the proliferation of fewer hepatocytes.

Liver cells are maintained in a mitotically quiescent state by the transcription factor CAAT/enhancer binding protein alpha (C/EBPα), which is expressed at high levels in the intact liver and acts to prevent entry into the cell cycle by inhibiting cyclin dependent kinases (cdks) (Wang *et al* 2001, 2002). Liver cells are 'primed' to respond to growth factor and cytokine signals that drive them from G_0 into G_1. This involves reduction of C/EBPα levels and the activation of transcription factors regulating a suite of over 70 'immediate-early' genes (Taub 1996), followed by induction of a set of 'delayed-early' genes encoding cell cycle proteins that allow the cells to progress through G_1 into S (Fausto and Webber 1994, Trembly and Steer 1998).

What signals regulate proliferation during liver regeneration? Mitogenic signals appear in the blood after PH, as evidenced by experiments showing that liver tissue or hepatocytes transplanted to ectopic locations replicate DNA following partial hepatectomy of the host liver, and that hepatectomy of one member of a parabiosed pair of rats induces growth of the intact liver of the other member of the pair (Michalopoulos and DeFrances 1997). Three heavily redundant signaling pathways regulate the cell cycle during liver regeneration. The mitogenic signals are produced by cells of the liver other than hepatocytes, as well as by non-liver cells elsewhere in the body (Fausto and Webber 1994, Michalopoulos and DeFrances 1997).

TNF-α/IL-6 pathway

TNF-α binds to its receptor, TNFR1 to induce IL-6 expression in Kuppfer cells and plasma concentrations of IL-6 reach high levels by 24 h after PH (Michalopoulos and DeFrances 1997). IL-6 activates the JAK-STAT and MAPK pathways, leading to transcription of immediate-early genes. Experiments deleting the TNFR1 or IL-6 genes, or inhibiting TNF-α with antibodies reduce the ability to regenerate mouse livers, suggesting a signaling pathway in which partial hepatectomy induces expression of TNF-α followed by activation of NF-κB, which induces IL-6, causing activation of STAT3.

HGF pathway

The stellate and endothelial cells of the liver produce pro-HGF, the inactive precursor of HGF. Pro-HGF is released from the hepatocyte ECM and activated by the proteases HGFA (hepatocyte growth factor activator) and urokinase plasminogen activator (uPA) (Miyazawa et al 1996, Miyazawa 2010). HGFA itself is activated by thrombin, thus HGF is functionally and structurally linked to the blood clotting system. Activated pro-HGF forms a heterodimer that binds to its receptor, c-met. Within one hour after PH the plasma concentration of active HGF rises over 20-fold (Mars et al 1995).

The hepatocyte vascular niche regulates the increase in HGF. Liver sinusoidal endothelial cells (LSECs) stimulate hepatocyte division by the release of angiocrine factors, including HGF, Wnt2 and angiopoietin 2 (Hu et al 2014). VEGF-A stimulates hepatocyte division when injected in vivo, but does so in vitro only in the presence of LSECs (LeCouter et al 2003). Injury activates two distinct pathways in liver endothelial cells that lead to elevated levels of HGF and IL-6 by up regulating hepatocyte VEGF-A production. VEGF-A binds to both the VEGFR-1 and 2 receptors on endothelial cells. Binding to VEGR-1 results in enhanced secretion of HGF and IL-6. Binding to VEGF-2 enhances endothelial cell proliferation, thus increasing the number of cells producing the growth factors. Genetic ablation of VEGFR2 in LSECs impairs the initial burst of hepatocyte proliferation by inhibiting an endothelial cell-specific transcription factor, Id1. Mice that are Id1-deficient exhibit defective liver regeneration, owing to diminished expression by LSECs of HGF (Ding et al 2010).

EGF/TGF-α pathway

EGF transcripts increase by ten-fold in the liver within 15 min after PH and EGF protein is expressed in hepatocytes, endothelial cells, Kupffer cells and Ito cells during G_1 (Michalopoulos and DeFrances 1997). The number of EGF receptors doubles in the first three hour. TGF-α also binds to the EGF receptor and is induced in hepatocytes within 2–3 h after PH, rising to a peak between 12 and 24 h. However, there is only a small increase in plasma TGF-α protein after PH, despite a large increase in TGF-α mRNA.

Importantly, disruption of any of these pathways may delay liver regeneration, but regeneration goes to completion nevertheless, indicating that the other pathways can compensate. This strong redundancy in the regulation of regeneration reflects the physiological importance of the liver. DNA synthesis in regenerating rat liver is complete by 72 h, but little is known about the mechanism by which proliferation is terminated. TGF-β1, which is normally made by Ito cells, may play a role in preventing and cessation of proliferation. In vitro, TGF-β1 inhibits the mitosis of hepatocytes. However, liver regeneration, although slowed, proceeds to completion in mice transgenic for increased TGF-β1 gene expression, suggesting that other factors act synergistically with TGF-β1 to terminate hepatocyte proliferation

(Michalopoulos and DeFrances 1997). When proliferation ceases, hepatocytes exist as clusters of 10–14 cells lacking sinusoids and ECM (Martinez-Hernandez and Amenta 1995). The normal organization of the liver is then re-established via regulation of cell number by apoptosis, cell movements and ECM synthesis.

Regeneration of the liver by stem cells

There is substantial evidence that the liver contains a population of stem cells enabling it to regenerate when the ability of hepatocytes to proliferate is compromised by compounds such as D-galactosamine (D-galN), 2-acetylaminofluorine (2-AAF) and retrosine (Dabeva and Shafritz 2003). Liver regeneration under these circumstances is achieved by the proliferation and differentiation of small cells with oval nuclei that are unaffected by the treatments. These 'oval cells' arise from epithelial stem cells (cholangiocytes) in the canals of Hering (Grompe and Finegold 2001, Bonner-Weir *et al* 2004, Raven *et al* 2017). The oval cells are thought to enzymatically digest the basement membrane of the epithelium and pass through it into the periportal tissue. Small hepatocytes with high growth potential capable of differentiating into hepatocytes or bile duct cells have been isolated from the regenerating liver by FACS (Tateno *et al* 2000). These cells may be early stages of oval cell differentiation (Gordon *et al* 2000a, 2000b). Remarkably, hepatocytes may be able to serve as their own stem cells, as it has been found that hepatocytes are able to regenerate under conditions of chronic liver damage by reverting to a stem-like state (Yimiamai *et al* 2014).

4.2.1.2 Pancreas

The pancreas consists of tubules of exocrine cells called acini, and clusters of endocrine cells called islets of Langerhans (figure 4.5). The acini secrete a large

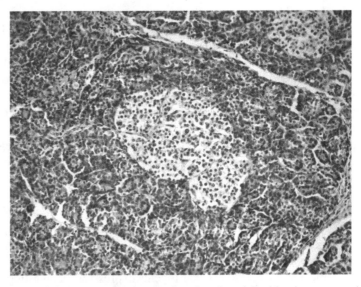

Figure 4.5. Structure of the pancreas. Section showing the pale staining islet tissue surrounded by purple-stained acinar tissue. Courtesy of Dr Matt Allen, Indiana University School of Medicine.

number of digestive enzymes into small intercalary ducts that connect to larger intralobular ducts and thence to still larger interlobular ducts. The interlobular ducts empty into a single pancreatic duct that joins with the duodenum. The islets of Langerhans constitute only 1%–2% of the pancreatic mass and are interspersed among the acini. They consist of cords and irregular clumps of cells and capillaries. The average human pancreas has approximately one million islets. The islets are supplied with extensive vasculature into which they secrete their hormones directly. Islets consist of four functional cell types, each of which secretes a different hormone. Insulin and glucagon, the main hormones involved in the regulation of blood glucose, are produced by the β-cells and α-cells, respectively. Insulin lowers blood glucose concentration whereas glucagon raises it. The δ-cells secrete somatostatin, which regulates the secretion of insulin and glucagon. The F-cells secrete pancreatic polypeptide, which affects the exocrine secretion of pancreatic enzymes, water and electrolytes, but the exact physiological role of this protein has not been defined.

Despite a few reports that the human pancreas can regenerate, most surgeons who have resected diseased pancreatic tissue say that it does not (Tsiotos *et al* 1999). Most of our knowledge of pancreatic regeneration comes from the mouse, which exhibits a low level of maintenance regeneration and is able to regenerate after injury (Slack 1995, Zaret and Grompe 2008). Partial pancreatectomy (PP) evokes the regeneration of both acinar and islet tissue, an observation that dates back to the 1920s. Different mouse models allow cell type-specific destruction and regeneration of pancreatic tissue. Temporary ligation of the pancreatic duct or ethionine treatment selectively destroys acini, which then regenerate rapidly. Streptozotocin (STZ) treatment selectively ablates the β-cell population. The β-cells do not regenerate unless STZ treatment is followed by partial pancreatectomy.

Genetic marking studies strongly suggest that β-cells are regenerated by compensatory hyperplasia after partial PP (figure 4.6). This has been shown by selectively marking fully differentiated β-cells in transgenic mice carrying a tamoxifen-inducible Cre/estrogen receptor fusion gene driven by the insulin promoter and a reporter construct consisting of the human placental alkaline phosphatase gene (HPAP) driven by a ubiquitous CMV/β-actin promoter separated by a floxed stop

1.   *IP:Cre:ER* ⟶ Cre/ER Fusion Protein expressed selectively in beta cell cytoplasm.

   CMV/ beta actin:FSC:HPAP ⊣ HPAP Expression

2.   Tamoxifen pulse → Cre:ER FusionPprotein enters nucleus

3.   *FSC* is excised to give *CMV/beta actin:HPAP.* HPAP is now expressed, selectively labeling beta cells. New beta cells that appear during growth or regeneration of the pancreas indicates derivation of these cells from the labeled pre-existing beta cells. This was the actual result.

Figure 4.6. Experiment designed to determine whether regenerated β-cells of the mouse pancreas are derived from pre-existing β-cells or stem cells.

cassette (Dor *et al* 2004). A pulse of injected tamoxifen allows the CreER protein to translocate into the nucleus and excise the *lac Z* stop cassette, initiating expression of the HPAP reporter. The marked β-cells can be tracked during growth of the pancreas or during regeneration after PP. If new β-cells arise from non-insulin producing stem cells after PP, they would be unmarked and dilute the percentage of HPAP+ cells over time, but if they arose from pre-existing β-cells, the percentage of HPAP+ cells would not change. In the intact and growing pancreas, all islets contained HPAP+ β-cells after a tamoxifen pulse. The percentage of marked cells did not change significantly over time, although their absolute number increased 6.5-fold with growth of the pancreas between 3 and 12 months of age, showing that the new β-cells were generated from pre-existing β-cells. Following a 70% PP (done 14 days after tamoxifen pulse to mark β-cells) BrdU was administered for two weeks to label dividing cells. Any new β-cells that arose from stem cells should have been BrdU+/HPAP−. Instead, the new β-cells were BrdU+/HPAP+, showing that they were derived from pre-existing β-cells. These results have been confirmed by other lineage tracing strategies (Teta *et al* 2007).

Like β-cells, acinar cells regenerate from pre-existing acinar cells after PP. Mouse acinar cells genetically marked with a tamoxifen-inducible cre:ER fusion gene driven by the acinar-specific elastase I promoter and a lacZ reporter regenerated only new acinar cells after a 70%–80% PP (Desai *et al* 2007).

The mouse pancreas, like the liver, is thought to harbor a stem cell population in the ductal epithelium (Bonner-Weir and Weir 2005). Neurogenin 3 (Ngn3) is the earliest islet cell-specific transcription factor expressed during pancreatic embryogenesis. By tracing the expression of Ngn3 in the adult mouse pancreas injured by ductal ligation, Xu *et al* (2008) identified the activation of progenitor cells that gave rise to all islet cell types, including glucose-responsive β-cells. Recently, human pancreatic duct cells expressing the cell surface purinergic receptor P2RY, which can act as a surrogate marker to identify cells expressing the pancreatic progenitor marker PDX1, were reported to be expandable by BMP-7 treatment and to differentiate into multiple pancreatic lineages after BMP withdrawal (Qadir *et al* 2018). This raises the possibility that BMP-7 could be used, in conjunction with treatments to inhibit autoimmunity, to restore β-cell mass.

4.2.2 Regeneration of myocardium by dedifferentiation

Cardiomyocytes of urodele amphibians, fish and mammals regenerate from pre-existing cardiomyocytes by a process of partial dedifferentiation to an immature state, proliferation and redifferentiation into mature cardiomyocytes (Ausoni and Sartori 2009, for a review).

4.2.2.1 Urodele myocardium

The amphibian heart is three-chambered, having two atria and one ventricle. The myocardium is thin, highly trabeculated and avascular. It is nourished by incoming blood that flows into the many sinuses that permeate the muscle trabeculae. The myocardium is covered externally by a thin epicardium that has two layers, an external mesothelial layer and an internal fibroblastic connective tissue layer. The inner surface of the myocardium is lined by an endocardium.

Both the newt and axolotl can regenerate the tip of the ventricle (figure 4.7). The epicardium regenerates first, under the fibrin clot, followed by the partial dedifferentiation and proliferation of uninjured cardiomyocytes at the cut edge of the myocardium (Oberpriller and Oberpriller 1971, 1974, 1983). The regenerated newt myocardium does not beat in synchrony with the rest of the myocardium, suggesting that electrical circuits are not regenerated. Histological, DNA labeling and ultrastructural studies of dedifferentiating cardiomyocytes showed that these cardiomyocytes resembled embryonic cardiomyocytes. Compact myofibrils were disrupted, myofibrillar patches were scattered throughout the cytoplasm, and intercalated disks were lost, resulting in the formation of mononucleate cells that re-enter the cell cycle, but maintain their contractility (Nag *et al* 1979, Rumyantsev 1992, Flink 2002). Immunostaining and RT-PCR analysis revealed strong down regulation of sarcomeric proteins and transcripts. Similar results were obtained in a system of mechanical damage to the newt heart inflicted by squeezing with forceps (Laube *et al* 2006).

Newt cardiomyocytes may not be equal in their ability to proliferate. Although the majority of them enter into S-phase *in vitro*, more than half of these stably arrest at G_2 or during cytokinesis. About one-third of the cells complete mitosis, some of which undergo successive mitoses (Bettencourt-Dias *et al* 2003). These observations

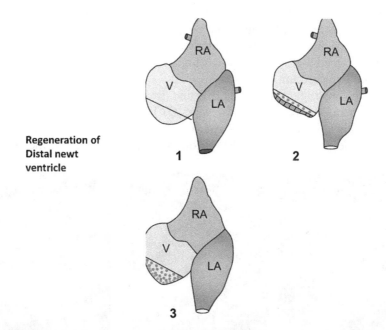

Figure 4.7. (A) Diagrams showing regeneration of the ventricular tip of the newt heart from pre-existing cardiomyocytes. 1. Heart morphology. RA = right atrium; L = left atrium; V = ventricle. Line indicates plane of amputation. 2. The epicardium (green), followed by partial dedifferentiation of underlying cardiomyocytes. 3. The dedifferentiated cells proliferate and redifferentiate into new cardiomyocytes. Reproduced with permission from (Stocum 2012a). Copyright 2012 Elsevier.

suggest that myocardial regeneration *in vivo* is carried out by a restricted subset of cardiomyocytes. What might be different about these cardiomyocytes is unknown.

The adult anuran myocardium, too, can mount a regenerative response to injury. Cardiomyocytes in the injured frog ventricular myocardium undergo DNA synthesis and proliferate to regenerate new myocardium (Borisov 1998). An interesting natural analogue of myocardial cell loss is observed in the South American toad, *Bufo arenarum*. These toads lose 30% of their cardiomyocytes during winter hibernation, but the loss is fully made up during the summer, although the mechanism by which this is achieved is unknown (Aoki *et al* 1989).

4.2.2.2 Zebrafish myocardium

The zebrafish heart is two chambered, with an atrium and a ventricle. The ventricular myocardium is organized into two layers of cardiomyocytes, a thin, vascularized external layer of compact cells, and an internal layer of cells organized into trabeculae. The myocardium is covered externally by an epicardium and lined internally by an endocardium. The zebrafish heart regenerates well, making this fish a preferred model for myocardial regeneration (Lien *et al* 2012, for a review). Removal of 20%–30% of the apex of the zebrafish ventricle is followed by the successive regeneration of the epicardium and myocardium (Poss *et al* 2002). The new myocardial tissue then expands to complete the restoration of the ventricle. Continuous BrdU labeling showed that DNA synthesis peaked at 32% of cells by 14 days after amputation. Incorporation decreased to 7% by 60 days, when regeneration was complete. BrdU pulse labeling experiments revealed showed that proliferation was diffuse during regeneration of the initial ventricular wall, but subsequently formed a gradient that was highest in the cardiomyocytes subjacent to the regenerated epicardium.

Genetic labeling (figure 4.8) for GFP and BrdU incorporation studies showed that the regenerated cardiomyocytes were GFP/BrdU-positive and thus derived from pre-existing cardiomyocytes (Raya *et al* 2003, Kikuchi *et al* 2010, Jopling *et al* 2010). The same proportion of GFP cardiomyocytes was seen in the regenerated

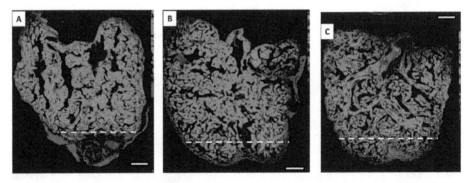

Figure 4.8. Ventricular regeneration in zebrafish with cardiomyocytes conditionally labeled with GFP. (A) As the fibrin clot (blue) is resorbed, the myocardium is replaced by cardiomyocytes expressing GFP, indicating their origin from pre-existing cardiomyocytes (B, C). White dashed line indicates level of amputation. Reproduced with permission from Macmillan Publishers Ltd (Jopling *et al* 2010). Copyright 2010.

myocardium as in the uninjured heart. These cells were fully differentiated and functionally integrated with the remaining myocardium as assessed by conduction velocity assays (Kikuchi *et al* 2010). Electron microscopy of regenerating tissue labeled for phosphorylated histone 3 (PH3), a G_2 cell cycle protein, showed that dividing cardiomyocytes disassembled their sarcomeric structure and detached from one another (Jopling *et al* 2010). Polo-like kinase (Plk1), which releases cells from G_2 arrest, was up regulated in regenerating hearts during the first week post-amputation. The frequency of regeneration was drastically reduced by treatment with cyclapolin 9, a Plk1 antagonist.

Shh, Fgf and PDGF-BB are important regulators of zebrafish heart regeneration. Epicardial regeneration requires Shh signaling associated with the bulbus arteriosus (BA, the outflow tract of the heart), as shown by the results of *ex vivo* experiments on zebrafish genetically for induced epicardial ablation (Wang *et al* 2015). The ablated epicardium regenerates robustly, but fails to do so if the BA, which expresses Shh, is removed. Replacement of the BA restores regeneration, as does transplantation of Shh-soaked microbeads at the base of the ventricle in the absence of the BA.

FGF regulates the initiation of cardiomyocyte regeneration (cf lens regeneration, below). This was shown using a transgenic fish that expressed a dominant-negative FGF receptor under the control of a heat shock gene promoter (Kikuchi *et al* 2010). Ventricles were injured and expression of the receptor was induced by heat shock for 30 days, causing inhibition of FGF signaling, regenerative arrest and scarring. Upon removal of the heat shock stimulus the resulting restoration of FGF signaling allowed myocardial regeneration to take place.

DNA synthesis in cardiomyocytes close to the wound requires PDGF-BB signaling (Lien *et al* 2006). PDGF-BB is up regulated at 7 days post-amputation, coincident with the observed increase in DNA synthesis in cardiomyocytes, and promotes DNA synthesis in partially dedifferentiated cardiomyocytes *in vitro*. Abolishing PDGF signaling by treatment of cultured cardiomyocytes with the PDGFR inhibitor AG1295 prevented DNA synthesis.

The zebrafish heart appears to have a mechanism to degrade scar tissue in the injured heart. Cryoinjury of up to ~25% of the zebrafish ventricle led to cardiomyocyte and vascular cell death within 24 h and massive scar formation by three weeks after injury. Subsequently, the scar tissue was degraded and replaced by abnormally thick myocardial tissue that exhibited asynchronous ventricular contraction (Gonzalez-Rosa *et al* 2011). The mechanism of scar degradation is not known, but is relevant to the possibility of removing scar tissue from myocardial infarcts in humans.

4.2.2.3 Mammalian myocardium

Fetal mouse myocardial tissue can regenerate and this capacity is maintained for a short time after birth. The one-day old mouse heart regenerates the ventricular apex in a manner similar to the zebrafish heart over a three-week period (Porrello *et al* 2011). The wound is first sealed by a blood clot, followed by an inflammatory response. Cardiomyocytes at the edge of the wound disassemble their sarcomeres and undergo mitosis with cytokinesis, as assayed by their expression of the mitotic markers pH3 and aurora kinase B. The regenerated ventricular apex was derived from pre-existing cardiomyocytes, as shown by labeling with BrdU and a conditional Cre/Lac Z set of constructs. The ability to regenerate is lost by seven days, coincident with the normal developmental withdrawal of cardiomyocytes from the cell cycle.

Fibroblasts are the most numerous cells of the mammalian heart. They proliferate rapidly into an infarct area to patch it with scar tissue, diminishing contractile function. Nevertheless, there is evidence that the adult myocardium initiates a regenerative response. This was first noted in histological studies on the hearts of human patients who died 4–12 days after ventricular myocardial infarction (Beltrami et al 2001, Anversa and Nadal-Ginard 2002). Four percent (2×10^6 cells) of the cardiomyocytes in the border zone surrounding the infarcts were in mitosis, 84 times the value for the similar area of the myocardium in control patients who had died of other causes, and had no major risk factors for heart disease. Maintenance regeneration of the adult mammalian myocardium appears to depend on a special population of cardiomyocytes. A rare population of hypoxic cardiomyocytes has been identified in the mouse heart that displays characteristics of proliferative neonatal cardiomyocytes, including smaller size, mononucleation and lower oxidative DNA damage, and which contributed widely to new cardiomyocyte formation during maintenance regeneration in the adult heart (Kimura et al 2015). Hypoxia (7% oxygen) for two weeks increased the proliferation of cardiomyocytes in both uninjured and injured hearts of mice and improved cardiac function in the infarcted mice. Genetic labeling and BrdU incorporation studies showed that the new cardiomyocytes were derived from pre-existing cardiomyocytes (Nakada et al 2012).

The MRL/MpJ mouse, which can regenerate ear tissue in punch wounds (Heber-Katz 2004), has also been reported to regenerate cardiac muscle after cryoinjury (Leferovich et al 2001, Leferovich and Heber-Katz 2002). Proliferation of cardiomyocytes at levels 7–10× higher than controls was indicated by the co-localization of BrdU labeling and antibody staining for α-actinin in the injured myocardium. Revascularization was also more evident at the injury site in the MRL mice. By day 60, the MRL/mpj hearts were virtually scar-free and echocardiograph measurements indicated a return of heart function to normal. However, other results have suggested that both coronary ligation (Oh et al 2003, Robey and Murry 2008, Cimini et al 2008, Grisel et al 2008) and cryoinjury (Grisel et al 2008) in the MRL/lpj mouse result in myocardial scarring, not regeneration. In still another investigation, it was reported that cardiomyocytes proliferated after cryoinjury in non-MRL mice (Van Amerongen et al 2008). These discrepancies remain to be resolved.

Two stem cell types may also regenerate mammalian myocardium (Wu et al 2008). The first type is a Lin⁻ [c-Kit Sca-1⁺] cell that in the rat is distributed throughout the uninjured myocardium at a frequency of approximately one in every 1×10^4 cardiomyocytes (Beltrami et al 2003). Similar [c-Kit Sca-1⁺] cells that could self-renew and differentiate into the same cell types were found in human hearts (Messina et al 2004, Smith et al 2007). After isolation from the myocardium by

FACS, they could differentiate *in vitro* into immature cardiomyocytes, smooth muscle cells and endothelial cells. Both rat and human cells expanded *in vitro* were reported to regenerate cardiac muscle when injected into the borders of myocardial infarcts in rats or mice. The second phenotype was Sca-1$^+$ [c-Kit$^-$ Lin$^-$] (Oh *et al* 2003). These cells express most cardiogenic transcription factors (GATA-4, MEF-2C, TEF-1) but not cardiac structural protein genes. They differentiated *in vitro* into cardiomyocytes expressing cardiac structural proteins when exposed to 5-azacytidine, and entered into infarcted hearts and differentiated into cardiac muscle when injected intravenously.

4.2.2.4 *Vascularization of regenerating myocardium*

The regenerating myocardium requires concomitant revascularization. In the zebrafish heart, this is accomplished by the rapid expansion of epicardial tissue to cover the exposed cardiomyocytes. A subpopulation of these epicardial cells undergoes EMT and invades the wound to provide new vasculature to regenerating muscle (Lepilina *et al* 2006). The EMT is induced by the up regulation of *fgf17b* in the myocardium and up regulation of *fgfr2, 4* in epicardial cells. Inhibition of Fgf signaling by expression of a dominant-negative *fgfR* results in failure of EMT and coronary revascularization, putting a halt to cardiac regeneration. In the mouse, the peptide thymosin β1 stimulates the epicardium to invade the myocardium to make coronary vessels (Smart *et al* 2007). Using a transgenic mouse genetically labeled with GFP or YFP, resident stem cells were identified that expressed the epicardial embryonic gene Wt1 (Wilm's tumor) in the epicardium of myoinfarcted hearts that responded to thymosin β4 stimulation by invading the myocardium and differentiating into cardiomyocytes (Smart *et al* 2011).

4.3 Transdifferentiation: newt lens and retina

4.3.1 Structure of the eye

Figure 4.9 illustrates the structure of the eye. Light passes through the pupillary opening and is focused by the lens onto the photoreceptors of the neural retina, which tranduces photonic energy to electrical signals. These signals go to the ganglion cells, the axons of which bundle together and pass through the retina to the brain as the optic nerve.

The lens consists of three components, anucleate transparent lens fibers characterized by their synthesis of α, β and γ crystallin proteins, an anterior lens epithelium, and a covering capsule that is a thick basement membrane laid down during eye development by lens epithelial cells (LECs). During development, the LECs of the posterior half of the epithelium are the first to differentiate into a core of lens fibers. The anterior LECs proliferate at a germinal zone in the anterior epithelium of the growing lens to add lens fibers to the original core. The embryos of several vertebrate species, including fish, amphibians, birds and rats, can regenerate the lens. Frogs can regenerate the lens as tadpoles, but newts and a cobotid fish (the Japanese weather loach) are the only vertebrates able to regenerate the lens as adults (Henry and Tsonis 2010, Henry *et al* 2013, for reviews).

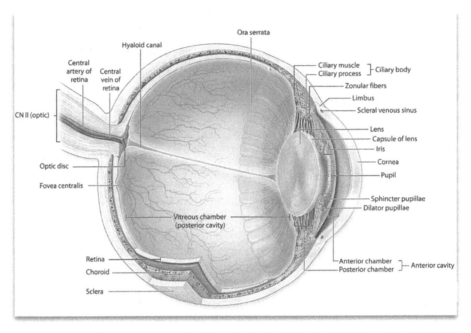

Figure 4.9. Structure of the eye. Reproduced with permission from (Mescher 2013). Copyright 2013 McGraw-Hill.

4.3.2 Lens regeneration

Newt lens regeneration was first described by Colucci in 1891 and shortly thereafter by Wolff in 1895, whose name it has borne since. Wolffian regeneration of the newt lens takes place by the dedifferentiation and proliferation of dorsal iris pigmented epithelial cells (PECs) and their transdifferentiation into lens fibers (Reyer 1977, Mitashov 1996, Del-Rio Tsonis and Eguchi 2004, Hayashi *et al* 2008, for reviews). Lens regeneration occurs in two major steps (figure 4.10). In the first step PECs undergo dedifferentiation around the whole circumference of the iris, but only dorsal dedifferentiated PECs form a lens vesicle. Step two is the differentiation of the lens vesicle posterior epithelium into transparent lens cells. The interplay between FGF, Wnt, BMP and Shh signaling has been shown to regulate lens regeneration in the newt eye (Barbosa-Sabanero *et al* 2012, Henry *et al* 2013, for reviews).

4.3.2.1 Dedifferentiation of PECs and formation of a lens vesicle
Studies to identify the epigenetic changes involved in shifting the transcriptional program during PEC dedifferentiation and proliferation have been initiated, but are still in their infancy. The genes *klf4, Sox-2* and *c-myc* are up regulated in dedifferentiating PECs (Maki *et al* 2009). Analysis of expressed sequence tags (ESTs) of iris cells during dedifferentiation has revealed the expression of a number of cancer-associated genes, apoptosis genes, and genes for DNA and chromatin-modifying enzymes (Maki *et al* 2010). Dedifferentiation of PECs appears to involve a high level of microRNA 124a (miR-124a) expression. The differential regulation

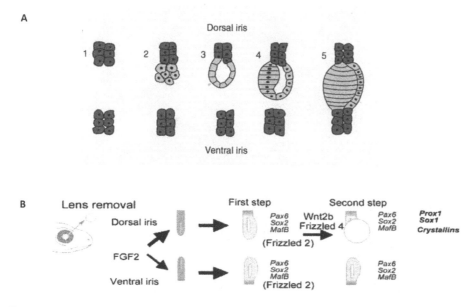

Figure 4.10. Regeneration of newt lens. (A) Cellular events of lens regeneration from the dorsal iris. 1. Immediately after lentectomy; the pigmented iris epithelium is a bilayer. 2. Depigmentation and proliferation of dorsal iris pigmented epithelial cells. 3. Formation of the lens vesicle. 4. Formation of lens fibers from progenitor cells derived from the anterior lens epithelium. 5. Maturation of lens fibers. Reproduced with permission from (Stocum 2012b). Copyright 2012 Elsevier. (B) Initiation of lens regeneration by Fgf-2 and subsequent activation of lens transcription factors in two steps. Step 1 is activated around the whole circumference of the iris. Step 2 results in the expression of Prox-1, Sox-1 and lens crystallins, and proceeds only in the dorsal iris due to stronger Wnt2b expression, reinforced by the expression of Frizzled 4 in addition to Frizzled 2. In the diagram, the dorsal iris in the first step is represented by a bilayer of cells with a space between them, the lens vesicle is represented by the circle in the second step. Reproduced with permission from (Hayashi *et al* 2008). Copyright 2008 Japanese Society of Developmental Biologists.

of two other miRNAs, miR-148 and let-7b, has been implicated in the control of proliferation genes in dedifferentiating PECs (Makarev *et al* 2006).

There is substantial evidence that Fgf-2 is necessary and sufficient for the initiation and progression of the events leading to lens vesicle formation. First, Fgf-2 is up regulated, along with early lens genes, in both the dorsal and ventral iris after lentectomy (Hayashi *et al* 2008). Second, the FGFR1 and 3 receptors are expressed at high levels in the dorsal iris (McDevitt *et al* 1997, Del-Rio Tsonis *et al* 1998). Third, injection of a soluble form of FGFR3 (isoform IIIc) to titrate Fgf-2 inhibited all molecular and morphological changes in PECs associated with lens regeneration (Hayashi *et al* 2004, 2006), and inhibition of tyrosine kinase activity by the small molecule inhibitor SU5402 inhibits FGFRs and lens regeneration (Del-Rio Tsonis *et al* 1998). Fourth, injection of Fgf-2 into the anterior chamber of the intact eye results in the same molecular changes as does lentectomy and leads to the formation of a secondary lens from dorsal iris PECs. This secondary lens replaces the primary lens, which degenerates (Hayashi *et al* 2004).

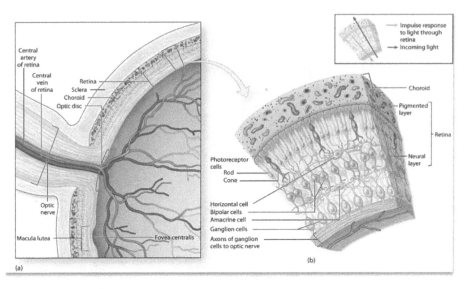

Figure 4.11. Anatomy of the retina, optic nerve and vasculature. Reproduced with permission from (Mescher 2013). Copyright 2013, McGraw-Hill.

Lens vesicle formation requires the interaction of dorsal iris cells with the neural retina (Reyer 1977, Del-Rio Tsonis and Eguchi 2004, for reviews). The presence of the retina is critical during a four-day window from day 11–15 post-lentectomy (Yamada *et al* 1973, Connelly *et al* 1986). The nature of the retinal signal is unknown, nor is it known whether this signal is part of the mechanism that evokes Fgf-2 expression, or has some other function essential to regeneration. PECs immersed in solutions of the cell cycle inhibitor SU9516 were able to dedifferentiate and form a small lens vesicle, but its cells were unable to proliferate to form a new lens (Tsonis *et al* 2004). This result suggests that dedifferentiation and cell cycle re-entry by PECs are not mechanistically linked.

4.3.2.2 Dorsal localization of lens vesicle formation
Formation of a lens vesicle occurs only in the dorsal PECs due to a preferential dorsal accumulation of FGF-2 protein and its receptor, which appears to induce stronger expression of Wnt2b and its receptor Frizzled 4 (Hayashi *et al* 2008). Addition of Wnt3A to the medium in the presence of FGF-2 induced the regeneration of lenses from cultured dorsal iris and modest lens regeneration from the ventral iris. Lens vesicle formation may thus depend on the synergistic action of FGF-2 and Wnt3A.

Macrophages invade the dorsal iris epithelium after lentectomy (Reyer 1990a, 1990b). Imokawa and Brockes (2003) have suggested that the macrophages are attracted to the dorsal iris because a clot forms there by the activation of tissue factor in the non-endothelial cells of torn dorsal ciliary vessels. The macrophages would express PDGF and TGF-β, resulting in the induction of FGF-2 protein expression by dorsal PECs (Godwin *et al* 2010). Depletion experiments have shown macrophages to be obligatory for the regeneration of amputated axolotl limbs (chapter 5). Such experiments have not been carried out for regenerating lenses, but if lens regeneration were inhibited under such circumstances, it would provide evidence for this hypothesis of dorsal localization.

BMP signaling may also have a role in restricting lens vesicle formation to the dorsal iris. Treating the ventral iris with the BMP inhibitor Chordin, or using a BMP competitor for BMPR-1A permits lens regeneration. Likewise, activating the BMP pathway by treating the dorsal iris with BMP4 or 7 reduces the capacity of the dorsal iris for lens regeneration (Grogg *et al* 2005). This suggests that BMP signaling negatively regulates the participation of ventral iris PECs in lens regeneration, thus insuring that only dedifferentiated PECs of the dorsal iris form a lens vesicle.

Another interesting asymmetry involved in localizing lens regeneration to the dorsal iris is the removal of proteoglycans from the surface of the dorsal PECs following lentectomy (Zalik and Scott 1972, 1973). A cell surface glycoprotein, 2NI-36, was identified that disappears from the surface of dorsal but not ventral, iris cells after lentectomy (Eguchi 1988). Ventral iris treated *in vitro* with antibody to this antigen and implanted into a lentectomized eye can regenerate a lens. These observations suggest that the shedding or inhibition of 2NI-36 is necessary to trigger the events of PEC dedifferentiation. 2NI-36 is also found in many other tissues of the newt and has been postulated to exert a general stabilizing effect on differentiation (Imokawa *et al* 1992, Imokawa and Eguchi 1992).

4.3.2.3 Differentiation of lens fibers

Shh signaling plays an important role in lens cell differentiation. The gene for the Shh receptor *Ptc1* is expressed in the intact lens, but not Shh itself. The suite of hedgehog/patched genes expressed during lens development (Shh, Ihh, Ptc1, Ptc 2) is also expressed during lens regeneration. Inhibition of the Shh pathway by KAAD, which prevents the action of Smoothened, or HHIP (hedgehog interacting protein), which inhibits the action of hedgehog, results in smaller regenerated lenses in nearly one-third of cases (Tsonis *et al* 2004). Hedgehog pathway inhibition reduces cell proliferation and synthesis of β-crystallin, suggesting the importance of the hedge-hog pathway for lens fiber differentiation. The hedgehog pathway does not appear to be active in the earlier stages of lens regeneration.

Transcription factors involved in lens fiber differentiation are Prox-1, Pax-6, Sox-1, Sox-2, Six-3 and MafB. All are expressed only in dorsal iris PECs. *Sox-2* and *MafB* expression begin at 8 days post-lentectomy, whereas *Prox-1* and *Sox-1* are expressed at 16 days and are thus considered late lens genes (Del-Rio Tsonis *et al* 1999, Hayashi *et al* 2004). Retinoic acid (RA) is a small molecule critical to eye development and lens cell differentiation. RA binds to and activates retinoic acid receptors (RARs), transcription factors that bind to retinoic acid response elements (RAREs) on the 5' side of target genes. An important target gene for lens differentiation is the αB crystallin gene, regulated by binding of the RAR-δ isoform to RARES in the promoter of this gene (Tsonis *et al* 2000). After lentectomy, the *RAR-δ* gene is expressed in the dedifferentiated PECs that form the new lens vesicle, and the level of expression increases to its highest level during new lens fiber

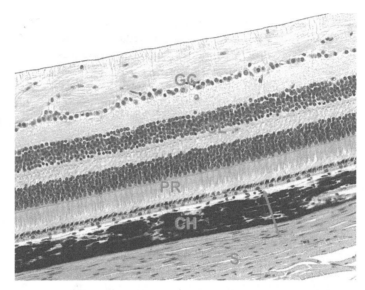

Figure 4.12. Section through the retina. GC = ganglion cell layer; ICL = intermediate cell layers; PR = photoreceptors; CH = choroid; S = sclera. The retina pigmented layer (arrow) is a thin layer of cells between the photoreceptors and the choroid. Courtesy of Dr Matt Allen, Indiana University School of Medicine.

differentiation. Inhibition of RA synthesis by disulfram, or inhibition of RAR function by the RAR antagonist 193109, results in the inhibition, retardation or abnormal morphogenesis of the regenerating lens, although in some cases ectopic lenses regenerate from the ventral iris or the cornea (Tsonis *et al* 2000).

4.3.2.4 Fidelity of newt lens regeneration
A remarkable finding is that neither repeated lentectomy nor aging reduces the capacity of newts to regenerate the lens (Eguchi *et al* 2011). Lenses of individual Japanese newts (*Cynops pyrrhogaster*) lentectomized 18 times over a period of 16 years regenerated perfectly each time! By then the newts, which had been captured in the wild, were at least 30 years old. The cellular and biochemical properties of the lenses regenerated after the 17th and 18th lentectomies had the same cellular and biochemical properties as younger newts that had never undergone lentectomy.

4.3.2.5 Mammals have some capacity for lens regeneration
Rabbits, cats, mice and rats can regenerate an imperfect lens if the lens is removed from the lens capsule, leaving the capsule behind (Gwon *et al* 1993, Call *et al* 2004, Huang and Xie 2010). A new lens is formed by the proliferation and differentiation of residual lens epithelium cells that remain adherent to the lens capsule. On a molecular level, mammalian lens regeneration appears to duplicate the gene activity of the developing lens after it has been induced to form (Huang and Xie 2010). Humans do not regenerate a lens from PECs of the dorsal iris. Nevertheless, human PECs can form lenses under the right conditions. Cells of the human H80HrPE-6

dedifferentiated PEC line, derived from an 80-year-old man, formed clear lentoid-like aggregates that expressed crystallins after four days of culture on Matrigel (Eguchi 1998, Tsonis *et al* 2001). Thus, if all the factors involved in lens regeneration from the dorsal iris in newts become known, it may be possible to intervene to regenerate a lens from the dorsal iris in humans.

4.3.3 Neural retina regeneration

The anatomical and histological structure of the retina is highly conserved across vertebrate species (figures 4.11 and 4.12). The outermost layer of the retina consists of rod and cone photoreceptor cells and the innermost layer consists of the ganglion cells, the axons of which form the optic nerve. Sandwiched between these layers are other layers that vary in number depending on species, but which always include an inner nuclear layer next to the ganglion cells and an outer nuclear layer closer to the photoreceptors. The outer nuclear layer is actually the portion of the rods and cones where their nuclei reside. The inner nuclear layer contains horizontal, bipolar and amacrine cells that relay impulses from the photoreceptors to the ganglion cells. Outer and inner plexiform layers show up as synapses between the photoreceptors of the outer nuclear layer and the bipolar cells of the inner nuclear layer, and synapses between neurons of the inner nuclear layer and the ganglion cell layer, respectively. External to the neural retina is the retinal pigmented epithelium (RPE). The RPE and neural retina are nourished by the highly vascular choroid layer that lies between the RPE and the tough connective tissue layer surrounding the eye, the sclera.

Although stem cells in the ciliary margin of the newt eye add cells throughout growth of the eye and can be mobilized for regeneration, the newt neural retina regenerates by the dedifferentiation and transdifferentiation of RPE cells into the layers of the new retina (Reyer 1977, Mitashov 1996, Del-Rio Tsonis and Tsonis 2003, Barbosa-Sabanero *et al* 2012). The dedifferentiated RPE cells facing the vitreous chamber transdifferentiate into the layers of the retina, and those next to the choroid layer redifferentiate as RPE cells. Figure 4.13 illustrates stages in the regeneration of the newt retina.

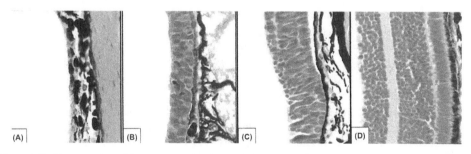

Figure 4.13. Regeneration of the newt neural retina. (A) Pigmented epithelium after removal of neural retina. (B) Neuroepithelial progenitor cells formed by dedifferentiation of pigmented epithelial cells. (C) Stratification of the differentiating progenitors into the different layers of the neural retina. (D) Fully regenerated retina. Photo courtesy of Dr Natalia Vergara and Dr Katia Del-Rio Tsonis, Miami University of Ohio. Reproduced with permission from (Stocum 2012b). Copyright 2012 Elsevier.

Loss of contact with the choroid layer while maintaining contact with the vascular membrane triggers the dedifferentiation/transdifferentiation process. The important component for initiation in the vascular membrane is contact with laminin. Proliferating newt RPE cells express the antigen RPE-1 until transdifferentiation takes place, when expression is extinguished. Re-entry into the cell cycle involves the up regulation of Musashi, a RNA binding protein characteristic of neural progenitor cells, and the Notch signaling pathway (Chiba and Mitashov 2007, Nakamura and Chiba 2007). Studies on retina regeneration in early *Xenopus* tadpoles indicated that FGF-2 can induce the whole process of retina regeneration and requires activation of the MAPK pathway (Vergara and Del-Rio Tsonis 2009). FGF-2 applied to the RPE of avian embryonic eyes after retinectomy has also been shown to induce transdifferentiation to neural retina. It would appear that in the normal urodele eye, the neural retina inhibits the RPE from transdifferentiating into neural cells. Separation of the two allows the RPE to undergo transdfferentiation to neural cells.

Species other than urodeles also have the capacity for retinal regeneration as adults. These species harbor stem cells in the ciliary marginal zone (CMZ) that contribute to regeneration. A major contribution is made, however, by Muller cells, retinal glial cells that can be reprogrammed to a progenitor state (Hamon *et al* 2016, for a review). Adult teleost fish can completely regenerate the retina after injury, whereas regeneration of adult bird and mammalian retina is very limited. However, retinal stem cells have been identified in the ciliary margin of the human eye (Haynes and Del-Rio Tsonis 2004, Coles *et al* 2004). These cells can form all the types of retinal cells *in vitro* or when transplanted into developing chick embryo eyes. Muller glia have also been identified in the avian and human retina that have been reported to have neurogenic potential (Hamon *et al* 2016, for a review).

References

Anversa P and Nadal-Ginard B 2002 Myocyte renewal and ventricular remodeling *Nature* **415** 240–3

Aoki A, Maldonado C A and Forsmann W G 1989 Seasonal changes of the endocrine heart *Functional Morphology of the Endocrine Heart* ed W-G Forsmann, D W Scheuermann and J Alt (New York: Steinkopff/Springer), pp 61–8

Ausoni S and Sartori S 2009 From fish to amphibians to mammals: in search of novel strategies to optimize cardiac regeneration *J. Cell Biol.* **184** 357

Barbosa-Sabanero K *et al* 2012 Lens and retina regeneration: new perspectives from model organisms *Biochem. J.* **447** 321–34

Becker T *et al* 1998 Readiness of zebrafish brain neurons to regenerate a spinal axon correlates with differential expression of specific cell recognition molecules *J. Neurosci.* **18** 5789–803

Becker C G, Becker T, Meyer R L and Schachner M 1999 Tenascin-R inhibits the growth of optic fibers *in vitro* but is rapidly eliminated during nerve regeneration in the salamander *Pleurodeles waltl J. Neurosci.* **19** 813–27

Becker C G *et al* 2004 L1.1 is involved in spinal cord regeneration in adult zebrafish *J. Neurosci.* **24** 7837–42

Beltrami A P *et al* 2001 Evidence that human cardiac myocytes divide after myocardial infarction *New Eng. J. Med.* **344** 1750–6

Beltrami A P *et al* 2003 Adult cardiac stem cells are multipotent and support myocardial regeneration *Cell* **114** 763–76

Bettencourt-Dias M, Mittnacht S and Brockes J P 2003 Heterogeneous proliferative potential in regenerative adult newt cardiomyocytes *J. Cell Sci.* **116** 4001–9

Bohn R, Reier P J and Sourbeer E B 1982 Axonal interactions with connective tissue and glial substrata during optic nerve regeneration in *Xenopus* larvae and adults *Am. J. Anat.* **165** 397–419

Bonner-Weir S and Weir G C 2005 New sources of pancreatic b-cells *Nat. Biotech.* **23** 857–61

Bonner-Weir S and Weir G C 2005 New sources of pancreatic β-cells *Nat. Biotechnol.* **23** 857–61

Borisov A B 1998 Cellular mechanisms of myocardial regeneration *Cellular and Molecular Basis of Regeneration* ed P Ferretti and J Geraudie (New York: Wiley), pp 335–54

Bormann P *et al* 1999 zfNLRR, a novel leucine-rich repeat protein is preferentially expressed during regeneration in zebrafish *Mol. Cell Neurosci.* **13** 167–79

Bucher N L R and Malt R A 1971 *Regeneration of the Liver and Kidney* (Boston: Little, Brown and Co)

Call M K, Grogg M W, Del-Rio Tsonis K and Tsonis P A 2004 Lens regeneration in mice; Implications in cataracts *Exp. Eye Res.* **78** 297–9

Chiba C and Mitashov V I 2007 Cellular and molecular events in the adult newt retinal regeneration *Strategies for Retinal Tissue Repair and Regeneration in Vertebrats: From Fish to Human* ed C Chba (India: Research Signpost)

Cimini M *et al* 2008 The MRL mouse heart does not recover ventricular function after a myocardial infarction *Cardiovasc. Pathol.* **17** 32–9

Clements M P *et al* 2017 The wound microenvironment reprograms Schwann cells to invasive mesenchymal-like cells to drive peripheral nerve regeneration *Neuron* **96** 98–114

Cohen A, Sivron T, Duvdedani R and Schwartz M 1990 Oligodendrocyte cytotoxic factor associated with fish optic nerve regeneration: implications for mammalian CNS nerve regeneration *Brain Res.* **537** 24–32

Coles B L K *et al* 2004 Facile isolation and the characterization of human retinal stem cells *Proc. Natl Acad. Sci. USA* **101** 15772–7

Connelly T G, Green M S, Sahijdak W M and Loyd R M 1986 Role of the neural retina in newt (*Notophthalmus viridescens*) lens regeneration in vitro *J. Exp. Zool.* **240** 343–51

Dabeva M D and Shafritz D A 2003 Hepatic stem cells and liver repopulation *Seminars in Liver Disease* ed P D Berk, S Thorgeirsson and J W Grisham (New York: Thieme Medical), pp 349–61

Davies A M 2000 Neurotrophins: neurotrophic modulation of neurite growth *Curr. Biol.* **10** R198–200

Davis B M *et al* 1990 Time course of salamander spinal cord regeneration and recovery of swimming HRP retrograde pathway tracing and kinematic analysis Exp. Neurol. **108** 198–213

Del-Rio Tsonis K, Trombley M T, McMahon G and Tsonis P A 1998 Regulation of lens regeneration by fibroblast growth factor receptor 1 *Dev Dyn.* **213** 140–6

Del-Rio Tsonis K, Tomarev S I and Tsonis P A 1999 Regulation of Prox 1 during lens regeneration *Invest. Opthalmol. Visual Sci.* **40** 2039–45

Del-Rio Tsonis K and Tsonis P A 2003 Eye regeneration in the molecular age *Dev. Dyn.* **226** 211–24

Del-Rio Tsonis K and Eguchi G 2004 Lens Regeneration *Development of the Ocular Lens* ed F J Lovicu and M L Robinson (Cambridge: Cambridge University Press), pp 290–311

Desai B M *et al* 2007 Preexisting pancreatic acinar cells contribute to acinar cell, but not islet β-cell, regeneration *J. Clin. Invest.* **117** 971–7

Diekmann H *et al* 2005 Analysis of the reticulon gene family demonstrates the absence of the neurite growth inhibitor Nogo-A in fish *Mol. Biol. Evol.* **22** 1635–48

Ding B-S *et al* 2010 Inductive angiocrine signals from sinusoidal endothelium are required for liver regeneration *Nature* **468** 310–4

Ding B-S *et al* 2014 Divergent angiocrine signals from vascular niche balance liver regeneration and fibrosis *Nature* **505** 97–102

Dor Y *et al* 2004 Adult pancreatic beta cells are formed by self-duplication rather than stem cell differentiation *Nature* **429** 41–6

Eguchi G 1988 Cellular and molecular background of Wolffian lens regeneration *Cell Differ. Dev.* Suppl. **25** 147–58

Eguchi G 1998 Transdifferentiation as the basis of eye lens regeneration *Cellular and Molecular Basis of Regeneration* ed P Ferretti and J Geraudie (New York: Wiley), pp 207–29

Eguchi G *et al* 2011 Regenerative capacity in newts is not altered by repeated regeneration and aging *Nat. Commun.* **2** 384

Eitan S and Schwartz M 1993 A transglutaminase that converts interleukin-2 into a factor cytotoxic to oligodendrocytes *Science* **261** 106–8

Farrell G C 2004 Probing Prometheus: fat fueling the fire? *Hepatology* **40** 1252–5

Fausto N 2004 Liver regeneration and repair: hepatocytes, progenitor cells and stem cells *Hepatology* **39** 1477–87

Fausto N and Webber E M 1994 Liver regeneration *The Liver: Biology and Pathobiology* 3rd edn ed I M Arias, J L Boyer, N Fausto, W B Jakoby, D A Schachter and D A Shafritz (New York: Raven), pp 1059–84

Ferretti P, Zhong F and O'Neill P O 2003 Changes in spinal cord regenerative ability through phylogenesis and development: lessons to be learnt *Dev. Dyn.* **226** 245–56

Flink I L 2002 Cell cycle reentry of ventricular and atrial cardiomyocytes and cells within the epicardium following amputation of the ventricular apex in the axolotl *Ambystoma mexicanum*: confocal microscopic immunofluorescent image analysis of bromodeoxyuridine-labeled nuclei *Acta Embryol.* **205** 235–44

Fu S Y and Gordon T 1997 The cellular and molecular basis of peripheral nerve regeneration *Mol. Neurobiol.* **14** 67–116

Godwin J W, Liem K F Jr and Brockes J P 2010 *Mech. Dev.* **127** 321–8

Goodrum J F and Bouldin T W 1996 The cell biology of myelin degeneration and regeneration in the peripheral nervous system *J. Neuropathol. Exp. Neurol.* **55** 943–53 .

Gonzalez-Rosa J M *et al* 2011 Extensive scar formation and regression during heart regeneration after cryoinjury in zebrafish *Development* **138** 1663–74

Gordon G J, Coleman W B, Hixson D C and Grisham J W 2000a Liver regeneration in rats with retrorsine-induced hepatocellular injury proceeds through a novel cellular response *Am. J. Pathol.* **156** 607–19

Gordon G J, Coleman W B and Grisham J W 2000b Temporal analysis of hepatocyte differentiation by small hepatocyte-like progenitor cells during liver regeneration in retrorsine-exposed rats *Am. J. Pathol.* **157** 771–86

Gordon T 2009 The role of neurotrophic factors in nerve regeneration *J. Neurosurg.* **26** E3

Grisel P *et al* 2008 The MRL Mouse repairs both cryogenic and ischemic myocardial infarcts with scar *Cardiovasc. Pathol.* **17** 14–22

Grogg M W *et al* 2005 BMP inhibition-driven regulation of *six-3* underlies induction of newt lens regeneration *Nature* **438** 858–62

Grompe M and Finegold M 2001 Liver stem cells ed D R Marshak, R L Gardner and D Gottlieb *Stem Cell Biology* (Cold Spring Harbor, NY: Cold Spring Harbor Laboratory Press), pp 455–97

Gwon A, Gruber L J and Mantras C J 1993 Restoring lens capsule integrity enhances lens regeneration in New Zealand albino rabbits and cats *Cataract Refract. Surg.* **19** 735–46

Ham A W and Cormack D H 1979 *Histology* 8th edn (Philadelphia: JB Lippincott), pp 614–44

Hamon A, Roger J E, Yang X-J and Perron M 2016 Muller glial cell-dependent regeneration of the neural retina: an overview across vertebrate model systems *Dev. Dyn.* **245** 727–38

Hayashi T *et al* 2004 FGF2 triggers iris-derived lens regeneration in newt eye *Mech. Dev.* **121** 519–26

Hayashi T *et al* 2006 Determinative role of Wnt signals in dorsal iris-derived lens regeneration in newt eye *Mech. Dev.* **123** 793–800

Hayashi T, Mizuno N and Kondoh H 2008 Determinative roles of FGF and Wnt signals in iris-derived lens regeneration in newt eye *Dev. Growth Differ.* **50** 279–87

Haynes T and Del-Rio Tsonis K 2004 Retina repair, stem cells and beyond *Curr. Neurovasc. Res.* **1** 1–8

Heber-Katz E 2004 Spallanzani's mouse: a model of restoration and regeneration *Regeneration: Stem Cells and Beyond (Current Topics in Microbiology and Immunology* ed E Heber-Katz vol 280 (Berlin: Springer), pp 165–89

Henry J J and Tsonis P A 2010 Molecular and cellular aspects of amphibian lens regeneration *Prog. Retin Eye Res.* **29** 543–55

Henry J J, *et al* 2013 Cell signaling pathways in vertebrate lens regeneration *New Perspectives in Regeneration* ed E Heber-Katz and D L Stocum (Berlin: Springer), pp 75–100

Higgins G M and Anderson R M 1931 Experimental pathology of the liver: I. Restoration of the liver of the white rat following partial surgical removal *Arch. Pathol.* **12** 186–202

Holder N *et al* 1989 Mechanisms controlling directed axon regeneration in the peripheral and central nervous systems of amphibians *Recent Trends in Regeneration ResearchNATO ASI Series* vol 172 ed V Kiorstsis, S Koussoulakos and H Wallace (Berlin: Springer), pp 179–90

Hu J *et al* 2014 Endothelial cell-derived angiopoietin-2 controls liver regeneration as a spatiotemporal rheostat *Science* **343** 416–9

Huang Y and Xie L 2010 Expression of transcription factors and crystalline proteins during rat lens regeneration *Mol. Vis.* **16** 341–52

Imokawa Y and Eguchi G 1992 Expression and distribution of regeneration-responsive molecule during normal development of the newt, *Cynops pyrrhogaster Int. J. Dev. Biol.* **36** 407–12

Imokawa Y, Ono S-I, Takeuchi T and Eguchi G 1992 Analysis of a unique molecule responsible for regeneration and stabilization of differentiated state of tissue cells *Int. J. Dev. Biol.* **36** 399–405

Imokawa Y and Brockes J P 2003 Selective activation of thrombin is a critical determinant for vertebrate lens regeneration *Curr. Biol.* **13** 877–81

Jopling C *et al* 2010 Zebrafish heart regeneration occurs by cardiomyocyte dedifferentiation and proliferation *Nature* **464** 606–9

Kikuchi K *et al* 2010 Primary contribution to zebrafish heart regeneration by *gata4*+ cardiomyocytes *Nature* **164** 601–9

Kimura W *et al* 2015 Hypoxia fate mapping identifies cycling cardiomyocytes in the adult heart *Nature* **523** 226–30

Laube F *et al* 2006 Re-programming of newt cardiomyocytes is induced by tissue regeneration *J. Cell Sci.* **119** 4719–29

LeCouter J *et al* 2003 Angiogenesis-independent endothelial protection of the liver: role of VEGFR-1 *Science* **299** 890–3

Leferovich J *et al* 2001 Heart regeneration in adult MRL mice *Proc. Natl Acad. Sci. USA* **98** 9830–5

Leferovich J M and Heber-Katz E 2002 The scarless heart *Cell Dev. Biol.* **13** 327–35

Lepilina A *et al* 2006 A dynamic epicardial injury response supports progenitor cell activity during zebrafish heart regeneration *Cell* **127** 607–19

Lien C-L *et al* 2006 Gene expression analysis of zebrafish heart regeneration *PLoS Biol.* **4** e260

Lien C-L *et al* 2012 Heart repair and regeneration: recent insights from zebrafish studies *Wound Rep. Reg.* **20** 638–46

Makarev E, Spence J R, Del Rio-Tsonis K and Tsonis P A 2006 Identification of microRNAs and other small RNAs from the adult newt eye *Mol Vis.* **12** 1386–91

Maki N *et al* 2009 Expression of stem cell pluripotency factors during regeneration in newts *Dev. Dyn.* **238** 1613–6

Maki N *et al* 2010 Expression profiles during dedifferentiation in newt lens regeneration revealed by expressed sequence tags *Mol Vis.* **16** 72–8

Malato Y *et al* 2011 Fate tracing of mature hepatocytes in mouse liver homeostasis and regeneration *J. Clin. Invest.* **121** 4850–60

Mars M *et al* 1995 Immediate early detection of urokinase receptor after partial hepatectomy and its implications for initiation of liver regeneration *Hepatology* **21** 1695–701

Martinez-Hernandez A and Amenta P 1995 The extracellular matrix in hepatic regeneration *FASEB J.* **9** 1401–10

Masuzaki R *et al* 2013 Scar formation and lack of regeneration in adult and neonatal liver after stromal injury *Wound Rep. Reg.* **21** 122–30

McDevitt D S, Brahma S K, Courtois Y and Jeanny J-C 1997 Fibroblast growth factor receptors and regeneration of the eye lens *Dev Dyn.* **208** 220–6

McQuarrie I G 1983 Role of the axonal cytoskeleton in the regenerating nervous system *Nerve, Organ and Tissue Regeneration: Research Perspectives* ed F J Seil (New York: Academic), pp 51–88

Mescher A L 2013 Junquiera's Basic Histology 13th edn (New York: McGraw-Hill)

Messina E *et al* 2004 Isolation and expansion of adult cardiac stem cells from human and murine heart *Circ. Res.* **95** 911–21

Mitashov V I 1996 Mechanisms of retina regeneration in urodeles *Int. J. Dev. Biol.* **40** 833–44

Miyaoka Y and Miyajima A 2013 To divide or not to divide: revisiting liver regeneration *Cell Div.* **8** 8

Miyazawa K, Shimomura T and Kitamura N 1996 Activation of hepatocyte growth factor in injured tissues is mediated by hepatocyte growth factor activator *J. Biol. Chem.* **271** 3615–8

Miyazawa K 2010 Hepatocyte growth factor activator (HGFA): a serine protease that links tissue injury to activation of hepatocyte growth factor *FEBS J.* **277** 2208–14

Michalopoulos G K and DeFrances M C 1997 Liver regeneration *Science* **276** 60–6

Mokalled M H *et al* 2016 Injury-induced *ctgfa* directs glial bridging and spinal cord regeneration in zebrafish *Science* **354** 630–4

Nag A C, Healy C J and Cheng M 1979 DNA synthesis and mitosis in adult amphibian cardiac muscle cells *in vitro Science* **205** 1281–2

Nakamura K and Chiba C 2007 Evidence for notch signaling involvement in retinal regeneration of adult newt *Brain Res.* **1136** 28–42

Oberpriller J O and Oberpriller J C 1971 Mitosis in adult newt ventricle *J. Cell Biol.* **49** 560–3

Oberpriller J O and Oberpriller J C 1974 Response of the adult newt to ventricle injury *J. Exp. Zool.* **187** 249–60

Oberpriller J O, Ferrans V J and Carroll R J 1983 DNA synthesis in rat atrial myocytes as a response to left ventricular infarction: an autoradiographic study of enzymatically dissociated myocytes *J. Mol. Cell Cardiol.* **15** 31–42

Oh H *et al* 2003 Cardiac progenitor cells from adult myocardium: homing, differentiation, and fusion after infarction *Proc. Natl Acad. Sci. USA* **100** 12313–8

O'Neill P D, Whalley K and Feretti P 2004 Nogo and Nogo-66 receptor in human and chick: implications for development and regeneration *Dev. Dyn.* **231** 109–21

Otu H H *et al* 2007 Restoration of liver mass after injury requires proliferative and not embryonic transcriptional patterns *J. Biol. Chem.* **282** 11197–204

Overturf K *et al* 1997 Serial transplantation reveals the stem-cell-like regenerative potential of adult mouse hepatocytes *Am. J. Pathol.* **151** 1273–80

Parrinello S *et al* 2010 EphB signaling directs peripheral nerve regeneration through Sox2-dependent Schwann cell sorting *Cell* **143** 145–55

Porrello E R *et al* 2011 Transient regenerative potential of the neonatal mouse heart *Science* **331** 1078–90

Poss K D, Wilson L G and Keating M T 2002 Heart regeneration in zebrafish *Science* **298** 2188–90

Power C and Rasko J 2008 Wither Prometheus' liver? Greek myth and the science of regeneration *Ann. Intern. Med.* **149** 421–6

Qadir M M F *et al* 2018 P2RY1/ALK3-expressing cells within the adult human exocrine pancreas are BMP-7 expandable and exhibit progenitor-like characteristics *Cell Rep.* **22** 2408–20

Raven A *et al* 2017 Cholangiocytes act as facultative liver stem cells during impaired hepatocyte regeneration *Nature* **547** 350–4

Raya A *et al* 2003 Activation of Notch signaling pathway precedes heart regeneration in zebrafish *Proc. Natl Acad. Sci. USA* **100** 11889–95

Reimer M M *et al* 2009 Sonic hedgehog is a polarized signal for motor neuron regeneration in adult zebrafish *J. Neurosci.* **29** 15073–82

Reyer R W 1977 The amphibian eye: development and regeneration *Handbook of Sensory Physiology, Vol VII/5 The Visual System in Vertebrates* ed F Crescitelli (Berlin: Springer), pp 309–90

Reyer R W 1990a Macrophage invasion and phagocytic activity during lens regeneration from the iris epithelium in newts *Am. J. Anat.* **188** 329–44

Reyer R W 1990b Macrophage mobilization and morphology during lens regeneration from the iris epithelium in newts: studies with correlated scanning and transmission electron microscopy *Am. J. Anat.* **188** 345–65

Robey T E and Murry C E 2008 Absence of regeneration in the MRL/Mpj mouse heart following infarction or cryoinjury *Cardiovasc. Pathol.* **17** 6–13

Rumyantsev P P 1992 Reproduction of cardiac myoctyes developing in vivo and its relationship to processes of differentiation *Growth and Hyperplasia of Cardiac Muscle Cells* ed P P Rumyantsev (New York: Harwood), pp 70–159

Schweitzer J, Becker T, Becker C G and Schachner M 2003 Expression of protein zero is increased in lesioned axon pathways in the central nervous system of adult zebrafish *Glia* **41** 301–17

Schweitzer J *et al* 2007 Contactin 1a expression is associated with oligodendrocyte differentiation and axonal regeneration in the central nervous system of zebrafish *Mol. Cell Neurosci.* **35** 194–207

Sincropi D V and McIlwain D L 1983 Changes in the amounts of cytoskeletal proteins within the perikarya and axons of regenerating frog motoneurons *J. Cell Biol.* **96** 240–7

Sirbulescu R F and Zupanc G K 2010 Spinal cord and repair in regeneration-competent vertebrates: adult teleost fish as a model system *Brain Res. Rev.* **67** 73–93

Slack M 1995 Developmental biology of the pancreas *Development* **121** 1569–80

Smart N *et al* 2007 Thymosin *β*4 faciltates epicardial neovascularization of the injured adult heart *Ann. N. Y. Acad. Sci.* **1194** 97–104

Smart N *et al* 2011 *De novo* cardiomyocytes from within the activated adult heart after injury *Nature* **474** 640–4

Smith R R *et al* 2007 Regenerative potential of cardiosphere-derived cells expanded from percutaneous endomyocardial biopsy specimens *Circulation* **115** 896–908

Stocum D 2012a Regeneration of cardiac muscle and hematopoietic tissues Regenerative Biology and Medicine 2nd edn (New York: Academic) chapter 7

Stocum D 2012b Regeneration of epidermal structures Regenerative Biology and Medicine 2nd edn (New York: Academic) chapter 3

Tateno C *et al* 2000 Heterogeneity of growth potential of adult rat hepatocytes *in vitro Hepatology* **31** 65–74

Teta M *et al* 2007 Growth and regeneration of adult beta cells does not involve specialized progenitors *Dev. Cell* **12** 817–26

Taub H 1996 Transcriptional control of liver regeneration *FASEB J.* **10** 413–27

Trembly J H and Steer C J 1998 Perspectives on liver regeneration *J. Minn. Acad. Sci.* **63** 37–46

Tsiotos G G, Barry M K, Johnson C D and Sarr M G 1999 Pancreas regeneration after resection: does it occur in humans *Pancreas* **19** 310–3

Tsonis P A, Jang W, Del-Rio Tsonis K and Eguchi G 2001 A unique aged human retinal pigmented epithelial cell line useful for studying lens differentiation *in vitro Int. J. Dev. Biol.* **45** 753–8

Tsonis P A *et al* 2000 Role of retinoic acid in lens regeneration *Dev. Dyn.* **219** 588–93

Tsonis P A *et al* 2000 Role of retinoic acid in lens regeneration *Dev. Dynam.* **219** 588–93

Tsonis P A *et al* 2004 A novel role of the hedgehog pathway in lens regeneration *Dev. Biol.* **267** 450–61

Turner J T and Singer M 1974 The ultrastructure of regeneration in the severed newt optic nerve *J. Exp. Zool.* **190** 249–88

Van Amerongen M J *et al* 2008 Cryoinjury: a model of myocardial regeneration *Cardiovasc. Pathol.* **17** 23–31

Vergara M N and Del-Rio Tsonis K 2009 Retinal regeneration in the *Xenopus laevis* tadpole: a new model system *Mol. Vis.* **15** 1000–13

Vitalo A G, Sirbulescu R F, Ilies J and Zupanc G K H 2016 Absence of gliosis in a teleost model of spinal cord regeneration *J. Comput. Physiol.* A **202** 445–56

Wang H *et al* 2001 C/EPB*α* arrests cell proliferation through direct inhibition of cdk2 and cdk4 *Mol. Cell* **8** 817–28

Wang H *et al* 2002 C/EBPα triggers proteasome-dependent degradation of cdk4 during growth arrest *EMBO J.* **21** 930–41

Wang B *et al* 2015 Self-renewing diploid axin2+ cells fuel homeostatic renewal of the liver *Nature* **524** 180–5

Wang J, Cao J, Dickson A L and Poss K D 2015 Epicardial regeneration is guided by cardiac outflow tract and Hedgehog signaling *Nature* **522** 226–30

Wu S M, Chien K R and Mummery C 2008 Origins and fates of cardiovascular progenitor cells *Cell* **132** 537–42

Xu X *et al* 2008 β cells can be generated from endogenous progenitors in injured mouse pancreas *Cell* **132** 197–207

Yamada T, Reese D H and McDevitt D S 1973 Transformation of iris into lens in vitro and its dependency on neural retina *Differentiation* **1** 65–82

Yanger K *et al* 2014 Adult hepatocytes are generated by self-duplication rather than stem cell differentiation *Cell Stem Cell* **15** 340–9

Yannas I V 2001 *Tissue and Organ Regeneration in Adults* (New York: Springer)

Yimiamai D *et al* 2014 Hippo pathway activity influences liver cell fate *Cell* **157** 1324–38

Zalik S E and Scott V 1972 Cell surface changes during dedifferentiation in the metaplastic transformation or iris into lens *J. Cell Biol.* **55** 134–46

Zalik S E and Scott V 1973 Sequential disappearance of cell surface components during dedifferentiation in lens regeneration *Nature (New Biol.)* **244** 212–4

Zaret K S and Grompe M 2008 Generation and regeneration of cells of the liver and pancreas *Science* **322** 1490–4

Chapter 5

Appendage regeneration

Summary

The urodele amphibians (salamanders and newts) can regenerate their limbs and tails, a fascinating ability that we would like to understand and acquire. The urodeles regenerate amputated limbs by a combination of activating resident progenitor cells and creation of progenitor cells by dedifferentiation, to form a blastema that grows and regenerates the missing parts. The formation, release and aggregation of these progenitor cells is regulated by proteases, macrophages, and a crucial signaling center, the apical epidermal cap (AEC) that bounds the apex of the blastema. The cells of the accumulation blastema synthesize DNA without dividing, and express a more limb bud-like gene activity profile and ECM. Growth and signaling factors provided by nerves regenerating into the blastema and by the apical epidermal cap (AEC) are essential for the mitotic growth of the blastema. All of the limb tissues except keratinocytes and chondrocytes contribute cells to the blastema. Blastema cells remember their phenotype of origin and redifferentiate according to this memory; fibroblast-derived blastema cells are also able to transdifferentiate into chondrocytes and perhaps tendon cells. Blastema cells derived from fibroblasts also retain a memory of their position on the PD axis that prevents them from forming structure proximal to the level of amputation and gives them self-organizational capacity; this positional memory is encoded in cell surface molecules. Blastema cells derived from other cell phenotypes lack a positional memory and their redifferentiation is regulated by the fibroblast-derived cells. A number of models have been proposed to explain how the pattern of redifferentiation is regulated, based on the results of axial reversal experiments, experiments on the regeneration of half and double half limbs, and experiments using retinoic acid to alter positional identity of blastema cells. Finally, we explore the regeneration of other appendages: the poorly regenerating *Xenopus* limb and mouse digits, and the regeneration-competent ear tissue of some mouse species, and deer antlers. These models provide opportunities to compare what regeneration-competent species and tissues have in common, and how they differ from regeneration-deficient species.

5.1 Limb regeneration in urodele amphibians

Appendages are used for a variety of functions such as locomotion and grasping (fins and limbs), gathering sound waves and thermal regulation (external ear tissue), and fighting for mates (antlers of male Cervidae, the deer family). The ability to regenerate external ear tissue is restricted to certain mice and rabbits, and that of antlers to the deer family. Perhaps surprisingly, fin and limb regeneration is seen throughout the animal kingdom. The arms of sea stars, octopi and squids, insect legs, the fins of teleost fish, the arms, legs and tails of the urodele amphibians (newts and salamanders), lizard tails, and even the tips of mammalian digits all can regenerate. Some gene expression profiles, progenitor cells and tissue interactions in amputated mouse digit tips are similar to those of salamander limb regeneration (Simkin *et al* 2015, Zielins *et al* 2016). These similarities have encouraged the idea that mammals have a latent capacity for appendage regeneration that can be activated to achieve the goal of regenerating a complete human limb. This chapter will focus on limb regeneration in urodeles, where it has been best studied, and also touch on the regeneration of mouse digits, mouse external ear tissue, and deer antlers. Much of the text on urodele limb regeneration is drawn from a more extensive review (Stocum 2017) and on mammalian epimorphic regeneration from reviews by Simkin *et al* (2015) and Seifert and Muneoka (2017).

Evidence from the fossil record indicates that ancestral urodeles of the Permian period (the last period of the Paleozoic era, ~300 MYA) were capable of limb regeneration (Frobisch *et al* 2014). How the urodeles evolved to be the only modern vertebrate capable of limb regeneration throughout life is a matter of speculation (Brockes 2015). Since their regenerative powers were first reported by Spallanzani in 1768, the urodeles have been the primary model experimental organism by which to probe the secrets of limb regeneration.

5.1.1 Phases and stages of limb regeneration

The tetrapod limb has three axes: proximal to distal (PD, girdle to digit tip), anterior to posterior (AP, thumb/big toe to little finger/toe), and dorsal to ventral (DV, extensor muscle side to flexor muscle side). From proximal to distal, the limb skeleton consists of a girdle (shoulder or pelvic), stylopodium (humerus or femur), zeugopodium (radius/ulna or tibia/fibula), and autopodium (carpals and fingers or tarsals and toes), along with associated muscles and tendons. Depending on their phylogenetic position, the skeletal elements and associated tissues in limbs of different classes of tetrapods vary in shape, and number, and organization that reflect the function of the limb in swimming (turtle or seal forelimb), flying (birds, bats), locomotion on land (horse, human), or grasping (primates). These differences are most prominent in the autopodium. Urodele limbs are similar to human limbs, except for having only four digits in the forelimb instead of five (figure 5.1).

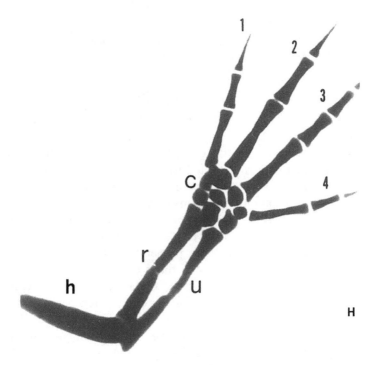

Figure 5.1. The right forelimb skeleton of a urodele larva. H = humerus; r = radius; u = ulna; C = eight carpals; 1–4 = digits from anterior to posterior. The limb skeleton looks very much like a human arm skeleton, with the exception that it has only four digits. The autopodial skeleton of the urodele hind limb has five digits, just like the human foot.

The events of urodele limb regeneration after amputation at any PD level can be divided into two phases (figure 5.2). The first phase creates a collection of progenitor cells at the level of amputation called the accumulation or early bud blastema that is similar to the early embryonic limb bud. The ability to create a blastema from differentiated limb tissues is a unique feature separating the urodeles from other tetrapod species. In the second phase, the accumulation blastema passes through a series of developmental stages involving growth, morphogenesis and differentiation to reproduce the amputated structures. Each stage is characterized by its own morphological and histological features: a conical medium bud of proliferating cells, a larger late bud or palette during which differentiation and morphogenesis of the amputated segments takes place in a PD and AP order, culminating in the appearance of digits. Amputated limbs follow the 'rule of distal transformation'; i.e. they regenerate only those parts distal to the level of amputation even when the PD polarity of the limb is reversed by implanting its distal end in a pocket made in the flank and amputating through the upper arm. The blastemas that form on each cut end of the stylopodium regenerate only those parts distal to the level of the cut.

5.1.2 Mechanism of blastema formation

Emergence of the accumulation blastema is a result of three processes: (1) wound closure by epidermal migration from the cut edges of the skin; (2) generation of

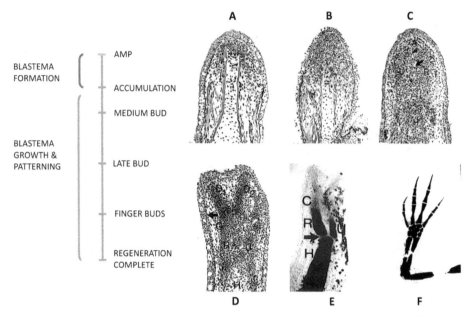

Figure 5.2. Left: phases and stages of urodele forelimb regeneration after amputation through the mid-humerus. Right: longitudinal sections of regenerating *Amblyomma maculatum* fore limbs. (A) Accumulation blastema. (B) Medium bud blastema. (C) Late bud blastema. The cartilage condensations of the radius and ulna (R, U) are becoming visible. H = humerus. Arrow indicates blood vessel. (D) Notch (two fingerbud) stage. Condensations of digits 1 and 2 are clearly visible; the faint condensations of digits 3 and 4 are visible on the posterior (right) side of the section. C = carpal condensation; R, U = radial and ulnar condensations, H = humerus. (E) Two-digit stage whole mount stained with methylene blue for cartilage matrix. Arrow indicates elbow joint. (F) The fully regenerated limb. Reproduced from (Stocum 2017). Wiley, Open Access.

blastema cells by histolysis and dedifferentiation of stump tissues; and (3) blastema cell migration and accumulation under an apical thickening of the wound epidermis, the apical epidermal cap (AEC). The AEC is a distal signaling center for promoting blastema cell mitosis that is analogous to the apical ectodermal ridge (AER) of tetrapod limb buds (Thornton 1968, for a review).

5.1.2.1 Epidermal wound closure
Immediately after amputation the wound is sealed by a thrombin-catalyzed blood clot. The epidermal basal cells at the wound margin lose their intercellular junctions and the hemidesmosomal junctions that anchor them to the basement membrane, and migrate under the clot to close the wound within a few hours. The migrating cells do not divide, but a zone of dividing epidermal cells proximal to the wound edge supplies a continual stream of migrating cells. The cells of this wound epidermis express *Sp9*, a characteristic marker gene that may be involved in formation of the AEC (Satoh *et al* 2010). Ion channels in the wound epidermis generate early signals obligatory for blastema formation, including Na^+ influx/H^+ efflux (Jenkins *et al* 1996, Adams *et al* 2007). The basal cells of the wound epidermis also have secretory functions important for blastema growth and patterning. To carry out these

functions, the wound epidermis and the AEC must be in close contact with the underlying blastema cells and no basement membrane is re-formed under the wound epidermis until the PD axis is complete.

5.1.2.2 Histolysis and the generation of blastema cells

Histolysis is the degradation of the extracellular matrix (ECM) of tissues local to the amputation surface by proteolytic enzymes, particularly lysosomal acid hydrolases and matrix metalloproteinases (MMPs), without which a blastema cannot not form (Schmidt, 1968, Santosh *et al* 2011, Vinarsky *et al* 2005). Histolysis liberates fibroblasts of the dermis, interstitial connective tissue of muscle, periosteum and nerve sheath, as well as Schwann cells of the peripheral nerves. Myofibers fragment at their cut ends and break up into mononucleate cells while simultaneously releasing Pax7[+] satellite cells. The basal layer of the wound epidermis and tissue-resident macrophages are major sources of MMPs, which prevent the reassembly of a basement membrane, thereby insuring contact between the wound epidermis and the underlying tissues (Godwin *et al* 2013). Histolysis of stump tissue continues until the medium bud stage, when it ceases due to the activity of tissue inhibitors of metalloproteinases (TIMPS) (Stevenson *et al* 2006).

Mononucleate cells derived by histolysis undergo dedifferentiation to progenitor blastema cells with large nuclei and sparse cytoplasm that exhibit intense RNA and protein synthesis (Thornton 1968, for a review). Blastema cells appear within 2–3 days post-amputation in larval urodeles and within 4–5 days in adult newts. As blastema cells accumulate, new capillaries and nerve axons regenerate from their cut ends into the accumulation and the wound epidermis thickens into the AEC.

Dedifferentiation involves epigenetic nuclear reprogramming to suppress the transcription of differentiation genes, while activating transcription of genes associated with progenitor status, reduction of cell stress and remodeling of internal structure (Stewart *et al* 2013, Voss *et al* 2015, Bryant *et al* 2017). Inhibition of these transcriptional changes by actinomycin D does not affect histolysis, but does prevent or retard dedifferentiation, leading to regenerative failure or delay (Carlson 1969). Dedifferentiated cells express a more limb bud-like ECM in which type I collagen synthesis and accumulation are reduced, and fibronectin, tenascin and hyaluronate accumulate (Mescher 1996, for a review). The molecular details of transcriptional regulation during dedifferentiation are only partly known. Genes characteristic of stem cells such as *msx1*, *nrad*, *rfrng* and *notch* are up regulated during blastema formation (Geraudie and Ferretti 1998, for a review). *Msx1* inhibits myogenesis and its forced expression in mouse myotubes causes cellularization and reduced expression of muscle regulatory proteins (Odelberg *et al* 2001). Inhibiting *msx1* expression with anti-*msx* morpholinos in cultured newt myofibers prevents their cellularization, and reduces their expression of muscle regulatory proteins (Kumar *et al* 2004). *Nrad* expression is correlated with muscle dedifferentiation (Shimizu-Nishikawa *et al* 2001), and *Notch* is a major mediator of stem cell self-renewal (Lundkvist and Lendhal 2001). Global analysis of transcript expression by micro-array and RNA-Seq has identified suites of genes for progenitor cell markers and specific stages of regeneration, as well as genes regulated by neural signals

(Monaghan and Maden 2012, Mercer *et al* 2012, Knapp *et al* 2013, Looso *et al* 2013). Since not all transcripts may be translated into proteins, proteomic analyses of blastema formation have also been carried out (King *et al* 2009, Rao *et al* 2009, 2014). These studies have identified highly up regulated and down regulated genes and proteins that can now be the focus of specific analysis (Jhamb *et al* 2011).

Three of the six Yamanaka factors (*klf4, sox2, c-myc*) used to reprogram mammalian adult somatic cells to induced pluripotent stem cells (iPSCs) (Takahashi *et al* 2007, Yu *et al* 2007) are up regulated during blastema formation in regenerating newt limbs. The Lin 28 protein, the product of a fourth transcription factor gene used to derive iPSCs, also is up regulated during blastema formation in regenerating axolotl limbs (Rao *et al* 2009). Blastema cells, however, are not pluripotent (Christen *et al* 2010), indicating the existence of regulatory factors that prevent reprogramming to this extreme. The further molecular characterization of transcription factors and changes in epigenetic marks via chromatin-modifying enzymes, will be crucial for understanding the mechanism of reprogramming in regenerating amphibian limbs.

The molecular details of internal structural remodeling in dedifferentiating cells are poorly understood. Dismantling of phenotypic structure and function is most visible in myofibers, but the molecular details of the process are largely uninvesti- gated for any limb cell type. Two small purine molecules dubbed myoseverin and reversine that cause cellularization of C2C12 mouse myofibers have been screened from combinatorial chemical libraries. (Rosania *et al* 2000, Chen *et al* 2004). Myoseverin disrupts microtubules and up regulates genes for growth factors, immunomodulatory molecules, ECM remodeling proteases, and stress-response genes, consistent with the activation of pathways involved in wound healing and regeneration, but does not activate the whole program of myogenic dedifferentiation in newt limbs (Duckmanton *et al* 2005). Reversine treatment of C2C12 myotubes resulted in mononucleate cells that mimicked MSCs in their ability to differentiate *in vitro* into osteoblasts and adipocytes, as well as muscle cells (Anastasia *et al* 2006). Myoseverin and reversine might thus be useful in analyzing the events of structural remodeling, and may have natural counterparts that can be isolated.

Canonical and non-canonical Wnt signaling appears to be important in axolotl limb and zebrafish fin regeneration. Genes and proteins for both pathways are expressed in the regenerating axolotl limb (Rao *et al* 2009). The canonical pathway (via Wnt8) promoted zebrafish fin regeneration whereas the non-canonical pathway inhibited it (Ghosh *et al* 2008, Stoick-Cooper *et al* 2007). The canonical Wnt pathway has also been implicated in deer antler regeneration (Mount *et al* 2006) and *Xenopus* tadpole tail regeneration (Lin and Slack 2008). Further studies will be required to understand the details of how Wnt signaling pathways regulate appendage regeneration in different species.

5.1.2.3 Cell cycling during blastema formation
[^{3}H]-thymidine labeling studies have shown that cells of amputated urodele limbs initiate cell-cycle entry coincident with their histolysis and dedifferentiation. The pulse-labeling index reaches 10%–30% during formation of the accumulation

blastema (Mescher and Tassava 1976, Loyd and Tassava 1980). By contrast, the mitotic index is very low, between 0.1 and 0.7% (average ~0.4%, or 4/1000 cells) in both *Ambystoma* larvae (Kelly and Tassava 1973) and adult newts (Mescher 1976). The fact that blastema cells synthesize DNA but divide only infrequently during formation of the accumulation blastema suggests that a large proportion of dedifferentiating cells arrest in G_2. What prevents them from undergoing mitosis is unknown, but might involve the ecotropic viral integration factor 5 (Evi5), a centrosomal protein that prevents mammalian cells from prematurely entering mitosis (Eldridge *et al* 2006) and which is strongly up regulated in the axolotl accumulation blastema (Rao *et al* 2009).

The signals that drive liberated cells to enter the cell cycle have been studied in myotubes derived from the newt A1 cell line of myogenic precursors (Ferretti and Brockes 1988) and involve multiple players. Serum stimulation of A1 myotubes induces their partial dedifferentiation, as manifested by down regulation of the *Myf5* gene (Imokawa *et al* 2004). An as yet unidentified thrombin-activated factor present in the serum of all vertebrates tested thus far promotes progression through G_1 and S in cultured newt myotubes (Tanaka *et al* 1997) by activating a sustained extrac-ellular signal-regulated kinase (ERK1/2) pathway that down regulates the Sox6 and p53 proteins, facilitating phosphorylation and inactivation of the retinoblastoma protein (pRb) to block entry into S (Yun *et al* 2013, 2014). A MARCKS (myristoylated alanine-rich C-kinase substrate)-like protein called the muscle LIM protein (MLP) initiates entry into the cell cycle of muscle-derived blastema cells (Sugiura *et al* 2016). MLP is secreted within 12 h after amputation of a newt limb and its activity is essential for entry into the cell cycle. Whether MLP is a general initiator of cell-cycle entry for blastema cells derived from other limb tissues is unknown, as is its relation to the thrombin-activated factor.

Mouse myonuclei do not synthesize DNA in response to serum stimulation (Tanaka *et al* 1997). They briefly activate the ERK1/2 pathway, but do not sustain the activity, and fail to deactivate pRb (Yun *et al* 2014). In addition, mammalian myotubes must overcome an additional block to DNA synthesis by the ARF tumor suppressor protein encoded by the *ink4a* locus, which is expressed only in taxa above the urodeles (Pajcini *et al* 2010). Newt blastema extract promotes DNA synthesis in both newt and mouse myotubes *in vitro* (McGann *et al* 2001), and mouse myonuclei will synthesize DNA if they are part of a mouse/newt heterokeryon (Velloso *et al* 2001), indicating that blastema extract provides factors for DNA synthesis that are not present in serum. Although cell-cycle factors are necessary and sufficient to stimulate newt myonuclei to enter the cell cycle, they are not sufficient to drive them into mitosis, and the myonuclei arrest in G_2. Cell-cycle entry is independent of myofiber cellularization, since cell-cycle-inhibited myofibers implanted into newt limb blastemas break up into mononucleate cells, but mitosis requires mononucleate cell status (Velloso *et al* 2001).

5.1.2.4 Macrophages are necessary for blastema formation
Macrophages of the innate immune system are important mediators of wound repair in mammals via their bactericidal and phagocytic activities and their secretion of

growth factors and cytokines that modulate inflammation and initiate structural repair by fibroblasts (chapter 2). They are essential for successful blastema formation during urodele limb regeneration (Godwin and Rosenthal 2014, Mescher 2017, Mescher et al 2017). Macrophages make the switch from inflammatory to anti-inflammatory as in mammalian wound repair, and make MMPs that help maintain a basement membrane-free zone between the wound epithelium and underlying mesenchyme of the forming and growing blastema. Macrophage depletion by liposome-encapsulated clodronate during blastema formation results in regenerative failure and scarring of the limb stump (Godwin et al 2013). The epidermis closes the wound, but does not develop an AEC and dermal scar tissue is interposed between the wound epidermis and underlying tissues. By contrast, depletion after a blastema enters the growth phase only delays regeneration. These results suggest a central role for macrophages in limb regeneration by shifting cytokine ratios in favor of the anti-inflammatory subset to resolve inflammation, and by the degradation of ECM, including the basement membrane. Macrophages are also necessary for the regeneration of ear tissue removed by punch holes in the African spiny mouse *Acomys*, which occurs by the formation of a blastema around the rim of the wound (Simkin et al 2017, Seifert and Muneoka 2017). Pro-inflammatory macrophages fail to penetrate the blastema tissue, but clodronate treatment eliminating both pro- and anti-inflammatory macrophages results in scarring.

An important function of macrophages on limb regeneration appears to be their removal of senescent cells. In mammals, senescent cells accumulate in aging tissues. Very few cells undergoing apoptosis are detected in regenerating urodele limbs (Mescher et al 2000). Yun et al (2015) demonstrated that cell senescence is induced during blastema formation in regenerating urodele limbs, but that senescent cells are not accumulated because they are cleared by macrophages. Furthermore, this clearance is obligatory for blastema formation. There is evidence for a reciprocal relationship between mammalian tissue damage, cell senescence and ease of reprogramming somatic cells by Yamanaka transcription factors (Mosteiro et al 2016), raising the question of whether senescent cells release factors that facilitate reprogramming of differentiated limb cells to blastema cells prior to being eliminated by macrophages?

5.1.2.5 Tissue contributions to the blastema

Experiments in which transgenic GFP neurula stage tissues were grafted in place of their counterparts in non-GFP showed that dermal fibroblasts, Schwann cells, skeletal cells and myogenic cells all contribute to the blastema (Kragl et al 2009) (figure 5.3). Dermal fibroblasts make the highest contribution (Muneoka et al 1986). Other sources of fibroblasts—periosteum, interstitial muscle tissue, and nerve sheath—have not been traced, but most likely contribute as well. The myogenic contribution, however, varies with species and phase of the life cycle (Sandoval-Guzman et al 2014, Tanaka et al 2016). Satellite cells are the source of regenerated muscle in regenerating limbs of larval and metamorphosed axolotls. The larval newt limb also mobilizes satellite cells to regenerate muscle during limb regeneration, but the adult newt limb switches to dedifferentiation of mononucleate myofiber fragments as the primary source of muscle progenitors. Triploid and GFP tracing of

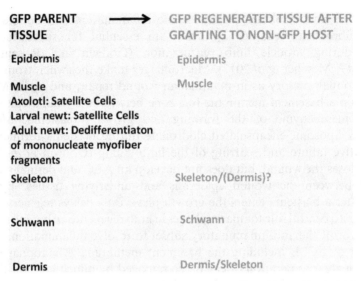

Figure 5.3. Contributions of axolotl limb tissues to the blastema, as assessed by grafting individual limb tissues from GFP animals to non-GFP limbs. The contributions are largely lineage-specific. The axolotl regenerates muscle via satellite cells throughout life. The larval newt also regenerates muscle via satellite cells, but the adult newt regenerates muscle via the dedifferentiation and proliferation of mononucleate myofiber fragments. Reproduced from (Stocum 2017). Wiley, Open Access.

chondrocytes of has failed to reveal any contribution to the blastema (McCusker *et al* 2016)

There are many questions yet to be answered about the origin of the blastema. For example, what percentage of the blastema cells is contributed by non-dermal fibroblasts and Schwann cells of the nerve? Might specific subpopulations of stem cells exist in dermal and other fibroblast populations that contribute to the blastema, as opposed to dedifferentiation? Is the switch from satellite cells to myofiber dedifferentiation in adult muscle regeneration an all or none event, or is it gradual, and what regulates this switch?

5.1.2.6 Blastema cells have lineage-specific and positional memories
GFP labeling experiments have shown that blastema cells have two types of memory. The first is a memory of limb and parent cell phenotype which dictates that blastema cells derived from muscle and Schwann cells redifferentiate in a lineage-specific manner as myogenic cells and Schwann cells (Kragl *et al* 2009) (figure 5.3). Blastema cells derived from fibroblasts, on the other hand, in addition to redifferentiating into fibroblasts, transdifferentiate into chondrocytes and tendon cells. In fact, a complete skeleton can regenerate distal to the plane of amputation from dermal cells of the skin, as shown in experiments grafting normal skin in place of the skin of irradiated limbs (Namenwirth 1974).

The second type of memory is positional identity, a memory of where parent cells reside in relation to their neighbors. Positional identity is restricted to fibroblast-derived blastema cells, and insures that blastemal cells derived from these fibroblasts at a given amputation level can regenerate only the positional identities of those structures distal to that level (Nacu *et al* 2013). Blastema cells derived from muscle and Schwann cells lack a memory of their position of origin, and the position they come to occupy during pattern formation is regulated by fibroblast-derived cells. Phenotypic and positional memory is likely due to retention of the original epigenetic codes imposed during limb bud development, as reflected in a stably maintained histone methylation pattern of blastema cell DNA (Hayashi *et al* 2015).

In vitro and *in vivo* cell sorting assays (figure 5.4) have demonstrated that positional identity is manifested in a gradient of cell surface adhesive strength of newt and axolotl blastema cells derived from different PD levels of the limb (high distal, low proximal) (Nardi and Stocum 1983, Crawford and Stocum 1988a, Egar 1993, Echeverri and Tanaka 2005). Retinoic acid (RA), a small lipid molecule involved in numerous metabolic reactions, proximalizes the positional identity of blastema cells (Crawford and Stocum 1988b, see section 5.1.4 below). Genetic marking experiments showed that the PD adhesive differentials exist at the single cell level (Kragl *et al* 2009). Position-dependent adhesive differentials of cells in the

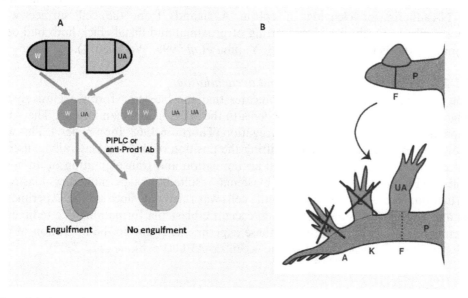

Figure 5.4. Assays demonstrating proximodistal (PD) differences in blastema cell adhesion. (A) *In vitro* assay in which blastemas from different PD levels are cultured base-to-base in pairs. W = wrist; UA = upper arm. The more proximal blastemal of the pair surrounds the more distal one, whereas simple fusion occurs between two blastemas from the same level. Engulfment in a PD pair is disrupted by phosphatidylinositol phospholipase C (PIPLC) or antibody to the blastema cell surface protein Prod1. (B) *In vivo* 'affinophoresis' assay. Forelimb blastemas (red) derived from the mid-upper arm (UA), elbow (E) or wrist (W) were grafted onto a wound bed made at the blastema-stump junction of a hind limb amputated through the mid-femur (F). P = posterior. Subsequently, the wrist and elbow blastemas sort to their corresponding levels on the regenerating hind limb. The X indicates that proximalization of blastemal PD positional identity to the level of the upper arm causes wrist and elbow blastemas to develop as if they were derived from the upper arm, and to remain at the mid-femur grafting level, showing that positional identity and adhesion are causally related. From (Stocum 2017). Wiley, Open Access.

developing and regenerating limb bud of the early *Xenopus* tadpole have also been demonstrated by *in vitro* cell sorting assays (Ohgo *et al* 2010).

Several cell surface molecules have been implicated in establishing PD adhesive differentials that reflect positional identity. Antibody blocking of Prod1, a member of the Ly6 family of three-finger proteins anchored to the cell surface by a glycosylphosphatidyl inositol (GPI) linkage, or its removal from the blastema cell surface by phosphatidylinositol-specific phospholipase C (PIPLC), inhibits the recognition of adhesive differentials between distal and proximal blastemas (Morais da Silva *et al* 2002), whereas over-expression of Prod1 in distal blastema cells causes them to sort to a more proximal (less adhesive) position when grafted into proximal blastemas (Echeverri and Tanaka 2005). These results suggest a proximal (high) to distal (low) gradient of Prod1 that establishes adhesive differentials reflecting blastema cell positional identity. Other surface molecules that may be involved in position-dependent adhesion of blastema cells are CD-59, ephrins and cadherins. Antibodies to CD-59, which is expressed in a high to low gradient along the PD axis of the gecko tail abolished the sorting *in vitro* of distal tail blastemas from proximal blastemas (Wang *et al* 2011). Antibodies to the EphA4 receptor and to N-cadherin, or cleaving of ephrin A ligands from the cell surface with phospholipase C, abolished the sorting of proximal and distal chick limb bud cells from one another (Wada *et al* 1998, Yajima *et al* 1999, Wada 2011).

5.1.2.7 Blastema cell migration and accumulation
The G_2 arrest of blastema cells indicates that the blastema forms exclusively by migration and aggregation of cells beneath the AEC rather than mitosis. The AEC appears to direct migration and aggregation (Thornton 1968, for a review). This was shown by experiments in which shifting the position of the AEC laterally caused a corresponding shift in blastema cell accumulation and transplantation of an additional AEC to the base of the blastema resulted in supernumerary blastema formation. Nerve guidance of blastema cells was ruled out, since similar experiments on aneurogenic limbs also resulted in eccentric blastema formation. The redirected accumulation of blastema cells in these experiments is due to the migration of the cells on fibronectin produced by the eccentric AEC (Levesque *et al* 2007).

5.1.3 Blastema growth

Blastema growth requires two synergistic conditions (1) the expression of mitogenic factors by the AEC and nerves, and (2) interaction between cells with non-neighboring positional identities. Unless these two conditions, along with re-vascularization are met, cells of the accumulation blastema do not divide and are removed.

5.1.3.1 Nerve:AEC interaction
Neither denervation of an amputated limb nor preventing the wound from being covered by a wound epidermis that forms an AEC prevents histolysis, dedifferentiation and entry of blastema cells into the cell cycle, but the absence of either

results in failure of the cells to proliferate (figure 5.5). The AEC forms in the absence of nerves, but is not maintained. Coincident with innervation of the AEC, [³H]-thymidine labeling and mitotic indices of the accumulation blastema rise as much as ten-fold, but these increases do not take place in limbs that are either denervated or deprived of wound epidermis.

When the growing blastema achieves a critical mass of cells (by the medium bud stage) it becomes independent of the nerve for differentiation and morphogenesis, but its cells still require nerve axons for proliferation, and so forms a morphologically normal, but miniature regenerate when denervated (Singer and Craven 1948). A medium bud blastema deprived of both innervation and AEC forms a distally truncated miniature regenerate when cultured in a dorsal fin tunnel, but if its distal end is positioned to stick out of the fin tunnel it is re-covered by fin epidermis and forms a morphologically normal but miniature regenerate like the denervated limb (Stocum and Dearlove 1972). These results suggest that the AEC has a role in proliferation and in proximodistal patterning.

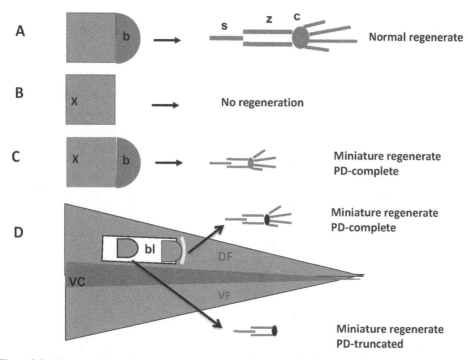

Figure 5.5. Nerves and the apical cap are essential for blastema cell mitosis. (A) Normal development of a forelimb medium bud blastema derived from the mid-stylopodium. S = stylopodium; Z = zeugopodium; C = carpals. (B) Limbs denervated by transection at the brachial plexus (X) at the time of amputation fail to regenerate. (C) Regeneration of a forelimb medium bud blastema after denervation at the brachial plexus. Morphogenesis proceeds, but the result is a miniature regenerate due to lack of mitosis. (D) Result of implanting a medium bud blastema denuded of epidermis to a fin tunnel so that it either cannot contact fin epidermis, or can contact fin epidermis (yellow line). The result is a miniature regenerate truncated at the level of the carpals in the former case, and a miniature regenerate complete in the PD axis in the latter case. VC = vertebral column; DF = dorsal fin; VF = ventral fin.

In an extensive series of experiments done in the 1940s, Singer developed the neurotrophic hypothesis of limb regeneration (summarized in Singer 1952). He found that a threshold level of innervation is required for blastema cell proliferation. Sensory axons alone, which innervate the blastema epidermis, are sufficient to meet this threshold, whereas motor axons alone are not, unless they are augmented by eliminating sensory innervation and allowing regenerating motor axons to sprout into empty sensory endoneurial tubes. This was proof that the type of innervation was irrelevant, and that the nerve requirement was quantitative. The limb bud does not become fully innervated until late in development, so its growth does not require nerves, but it does require signals from the apical epithelium (Tschumi 1957). In amniote limb buds, the apical epithelium is configured into a ridge, the apical ectodermal ridge (AER), whereas in amphibian limb buds there is no ridge, but the apical ectoderm is functionally equivalent to the AER (Tarin and Sturdee 1971). Once fully developed, however, the amphibian limb is nerve-dependent for regeneration (Fekete and Brockes 1987) and the wound epidermis formed after amputation is configured into the AEC following amputation. Urodele limb buds rendered aneurogenic during embryogenesis never acquire nerve dependence for regeneration, but remain dependent on the wound epidermis for blastema cell mitosis (Yntema 1959a, 1959b). Nerve dependence of aneurogenic limbs can be instituted by allowing them to become innervated (Thornton and Thornton 1970).

There are three hypotheses regarding how the nerves and AEC function together to promote mitosis of blastema cells (figure 5.6). The first postulates that the nerve and the AEC have different functions in the cell cycle, with the AEC maintaining blastema cells in a dedifferentiated state, and the nerve providing mitogenic factors (Tassava and Mescher 1975). Evidence consistent with this hypothesis is that adult newt blastemas cultured transfilter to dorsal root ganglia (DRG) or brain cells fail to undergo mitosis in the absence of the AEC, withdraw from the cell cycle, and differentiate prematurely as cartilage, whereas in the presence of the AEC they are

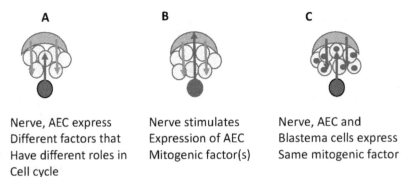

A	B	C
Nerve, AEC express Different factors that Have different roles in Cell cycle	Nerve stimulates Expression of AEC Mitogenic factor(s)	Nerve, AEC and Blastema cells express Same mitogenic factor

Figure 5.6. Three hypotheses for the role of nerve and apical epidermal cap (AEC) factors in mitosis of blastema cells. (A) Nerve (blue arrow) and AEC (red arrow) express different factors that affect different aspects of the cell cycle without interacting. (B) Nerve factor stimulates the AEC to express the mitogenic factors. The nerve factor does not directly affect the cell cycle. (C) The nerve and the AEC both affect the cell cycle directly by expressing the same factor, one that is also expressed by the blastema cells (blue nuclei).

maintained in an undifferentiated state and proliferate (Globus *et al* 1980, Smith and Globus 1989).

This same evidence is also consistent with the second hypothesis, which is that the AEC provides mitogenic factors to the blastema cells, but requires the nerve to express them (Stocum 2011). Additional evidence for this idea is that two growth factors expressed by the AEC, the anterior gradient protein (AG) and Fgf2, stimulate blastema cell proliferation *in vitro* and *in vivo*, and are down regulated by denervation (Mullen *et al* 1996, Kumar and Brockes 2007). Blastema cells express the receptors for Fgf2 (FGFR2) and AG (Prod1). Importantly, the gene for AG supports regeneration of denervated limbs from dedifferentiation to digit stages when electroporated into the limb five days post-amputation (Kumar and Brockes 2007). Fgf2 was also reported to support regeneration to digits when delivered in beads to late bud blastemas (Mullen *et al* 1996).

Conditioned medium of Cos7 cells transfected with the AG gene stimulates BrdU incorporation into cultured blastema cells; this incorporation is blocked by anti-bodies to Prod1, indicating that AG acts directly on blastema cells through Prod1 to stimulate proliferation. AG is expressed in the AEC of regenerating aneurogenic limbs as would be predicted in this model (Kumar *et al* 2011); whether this is true for Fgf2 as well is unknown. DRG neurons express several other factors that promote blastema cell proliferation *in vitro*, including transferrin and substance P. Fgf-8 and BMP are also expressed in DRG neurons and are detectable in peripheral limb nerve axons (Satoh *et al* 2016). Furthermore, in combination they can substitute for the nerve in inducing a supernumerary limb (Makanae *et al* 2014).

The third idea is that a single mitogen expressed by nerve, AEC, and blastemal cells themselves is required to promote blastema cell mitosis. Glial growth factor 2 (Ggf2, neuregulin 1) was suggested as a neurotrophic factor 35 years ago (Brockes 1984, Brockes and Kintner 1986) and was briefly mentioned to rescue regeneration to digit stages in denervated axolotl limbs when injected intraperitoneally during blastema formation (Wang *et al* 2000). Blastema cells, DRG neurons, and basal epidermal cells of the AEC all express NRG1 (Farkas *et al* 2016). Denervation reduced NRG1 expression by medium bud blastema cells by 25%, indicating that blastema cells have a basal level of NRG1 expression that is augmented by innervation. Inhibition of NRG1 signaling by mubritinib abolished blastema formation in innervated limbs and produced miniature regenerates from 16 day blastemas, equivalent to the regenerates obtained by denervation of medium bud blastemas. NRG1-soaked beads implanted from 19–36 days post-amputation in denervated limbs supported regeneration to digit stages, though not to the same degree as innervated controls.

While further experiments are required to verify and extend these results, the work of Farkas *et al* (2016) suggests a synergistic relationship between nerve, AEC and blastema cells, in which blastema cells autonomously express NRG1 in the absence of nerves, but at a level insufficient for mitosis. The sum of NRG1 from motor axons, the AEC (induced by sensory axon and perhaps blastema cell NRG1) would provide sufficient NRG1 to reach the threshold for blastema cell mitosis. This kind of synergism would easily explain why augmenting motor innervation in the absence of sensory innervation enables complete regeneration, because the required threshold level of NRG1 could be reached in the absence of sensory nerves. The addiction to nerve for regeneration that arises during limb development could thus be interpreted as a quantitative increase in the requirement by blastema cells for NRG1.

Which of these hypotheses will prove to be the most explanatory for how the nerve and AEC promote limb regeneration remains to be seen. It may be that elements of each will be required to construct a comprehensive picture of the role of nerve and AEC in blastema growth.

5.1.3.2 Interaction between cells of different AP position

Blastema cells migrate centripetally from the dermis of the skin as the accumulation blastemal forms (Gardiner et al 1986). Even in the presence of nerves and AEC, blastema cells fail to undergo mitosis and persist unless their APDV positional identities are sufficiently different when confronted during centripetal migration. This was shown by rotating a longitudinal strip of unirradiated anterior or posterior skin 90° and grafting it around the limb circumference. The limbs failed to regenerate after amputation, but normal regeneration ensued when shorter longitudinal skin strips representing two or more quadrants were rotated and grafted (Lheureux 1975).

Lheureux (1977) devised an experimental system that showed the synergistic effect on blastema cell mitosis of the nerve, AEC and differing circumferential positional interactions (figure 5.7(A)). After making a wound on the A or P side of a limb, a nerve was deviated to the wound site, and a piece of skin from the opposite side of the limb grafted to the wound. When carried out separately, nerve deviation or grafting opposite side skin to the wound site resulted in the formation of an AEC and blastema cells, neither of which persisted. Together, however, these operations stimulated the formation of a supernumerary blastema that grew and regenerated a complete limb. Nacu et al (2016) used this system to show that Shh could substitute for posterior skin in an anterior wound and that Fgf-8 could substitute for anterior skin in *a posterior* wound (figure 5.7(B)). These results indicate that Shh and Fgf-8 cooperate, along with neural and AEC factors, to promote blastema growth, but how these elements are integrated is unclear.

5.1.4 Pattern formation in the blastema

Pattern formation is the process used by the blastema to restore a complete three-dimensional map of positional identities from the progeny of cells derived from one particular PD level of the limb. This must involve a linkage between mitosis, APDV positional identities, and the progressive distalization of the PD positional identity of blastema cells.

5.1.4.1 Patterning in the AP axis

Patterning in the AP axis of limb buds and regeneration blastemas has been attributed to a posterior to anterior morphogen gradient of Shh emanating from a

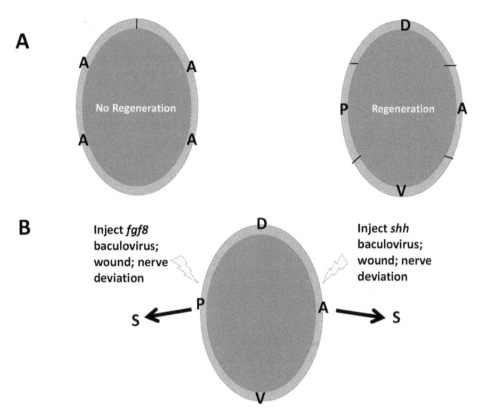

Figure 5.7. (A) Lheureux experiment showing that skin dermal cells possessing the same positional identity cannot form a blastema after being grafted to an amputated irradiated limb, but if skin from different quadrants of the limb is grafted in the same way, a blastema will form and the limb regenerates. (B) Experiment showing that Fgf-8 can substitute for anterior skin when injected posteriorly, and Shh can substitute for posterior skin when injected anteriorly, to evoke supernumerary limbs (S) after wounding and nerve deviation. After (Stocum 2017). Copyright 2017, the Author.

posterior patch of mesenchyme called the zone of polarizing activity (ZPA) (MacCabe *et al* 1973, Fallon and Crosby 1975, Cameron and Fallon 1977, Riddle *et al* 1993, Endo *et al* 1997, Imokawa and Yoshizato 1997). Transfection of *shh* into anterior blastema tissue of the axolotl limb by vaccinia virus results in the regeneration of supernumerary digits, but not more proximal structures (Roy *et al* 2000). The *shh* gene in the axolotl blastema requires a hypomethylated enhancer gene for expression. The enhancer is hypermethylated in the fibroblastema formed by *Xenopus* amputated late tadpole limbs, resulting in regeneration of only a cartilage spike (Yakushiji *et al* 2007). Treatment of amputated froglet limbs with a synthetic agonist of Hedgehog signaling induced activation of Shh target genes and the formation of multiple cartilaginous structures instead of a single spike, but nothing resembling a normal limb skeletal pattern (Yakushiji-Kaminatsui *et al* 2009).

However, other evidence argues against Shh as an AP morphogen. The skeletal elements of urodele limb buds and blastemas differentiate in an anterior to posterior order, not posterior to anterior as in other tetrapod limb buds. Shh is not required for AP patterning of elements proximal to the digits, either in regenerating limbs or chick limb buds (Roy and Gardiner 2000, Leitingtung *et al* 2002), but has been implicated late in limb bud development in defining digit number and identity by regulating BMP expression in the prospective chick autopodium (Dahn and Fallon 2000, Drossopoulou *et al* 2000). However, BMP expression was found to be independent of Shh signaling in regenerating axolotl limbs (Guimond *et al* 2010). Shh in regenerating limbs would therefore appear to be more essential for blastema cell mitosis and distalization of PD positional identity, as indicated above. Distalization and APDV patterning in segments proximal to the digits are interdependent, as shown by the fact that distalization will not occur in the absence of interaction between disparate APDV positional identities.

5.1.4.2 Models of distalization for terminal regeneration

Several models of distalization have been proposed to explain the terminal regeneration that occurs after simple amputation: the polar coordinate model (French *et al* 1976, Bryant *et al* 1981); the boundary model (Meinhardt 1983a, 1983b); an autonomous timing model (Saiz-Lopez *et al* 2015); and an intercalary averaging model (Maden 1977).

The polar coordinate model assigns each cell at a given PD level of the limb angular and radial positional values (figure 5.8). The angular value represents the position of a cell on the circumference (most often illustrated as a series of 'clock-face' numbers) and the radial value represents its position on the PD axis, with the level of amputation being the most proximal and the most distal structure represented at the center. Following histolysis, blastema cells with differing angular (circumferential) identities and the same PD identity migrate centripetally (Gardiner *et al* 1986) and engage in short range interactions between the different angular identities to intercalate a complete set of circumferential/radial identities that can adopt the next most distal (radial) identity. This interactive process is recursive, and continues until the PD pattern is restored. Interfering with APDV interactions by treating amputated limbs with beryllium induces pattern abnormalities by interfering with the migration and interaction of blastema cells derived from different circumferential positions on the skin (Cook and Seifert 2016). It is important to understand that the radial positions in the planar scheme of the model represent the successive regeneration of PD identities, not a literal map of these identities on the amputation surface. From the results of a variety of experiments on the regeneration of surgically constructed half and double half limbs, the model assigns the majority of the circumferential positional identities to the posterior and dorsal halves of the limb.

The boundary model (Meinhardt 1983a) postulates four structural domains differing in positional identity within the cross-section of the amputation surface. Large anterodorsal (ADOR) and anteroventral (AVEN) domains confront smaller posterodorsal (PDOR), and posteroventral (PVEN) domains (figure 5.9(A)). The DV boundary evenly divides dorsal and ventral, whereas the AP boundary is located more posteriorly. Where these boundaries intersect, interactions take place that induce the expression of a diffusable morphogen (s). It might be presumed that this is

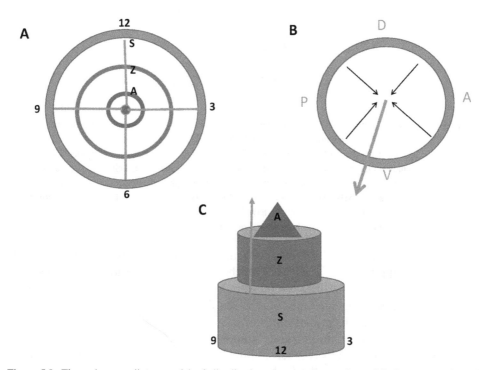

Figure 5.8. The polar coordinate model of distalization. Amputation surface of limb amputated at the proximal extreme of the stylopodium. (A) A map of prospective PD positional identities. Positional identities are represented by two values, position on the radius and position on a circumference. Proximodistal positional identities for stylopodial, zeugopodial and autopodial segments (S, Z, A) are represented on the radii of the circumference of the amputation surface (red lines) and circumferential position is represented by clock-face numbers on the concentric green, blue and purple circles. The distal tip of the regenerate is represented by the purple dot. This planar representation does not indicate literal positional identities on the amputation surface. (B) Diagram indicating that blastema cells from different positions on the limb circumference (dermis of the skin) at the level of amputation must migrate centripetally and interact to generate the next distal positional identity in the sequence, the blue circle, etc. (C) Recursive interactions of circumferential cells at the tip of the growing blastema specify progressively more distal positional identities, telescoping the regenerating pattern out from the amputation surface (red arrow).

Shh and Fgf-8, but as argued earlier, these signals more likely mediate mitosis-driven distalization, with a direct patterning effect of Shh only on the autopodium.

Meinhardt's boundary model proposed a 'bootstrap' mechanism of distalization driven by the production of an AEC morphogen (Meinhardt 1983b) (figure 5.9(B)). Higher concentrations of the morphogen specify more distal positional identities. The blastema cells produce an AEC maintenance factor (AECMF) that controls production of the AEC morphogen. The concentration of morphogen is lowest at the earliest stages of regeneration and will specify the most proximal structure to be regenerated. Once specified, these cells now ramp up their production of AECMF, resulting in production of a higher concentration of morphogen by the AEC that specifies the next more distal structure, and so on. Progressively more distal PD positional identities are thus serially 'bootstrapped' into existence by stepwise

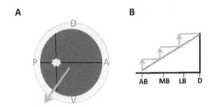

Figure 5.9. The boundary model of distalization. (A) A morphogen (star) is produced at the intersection of the AP and DV boundaries that generates an AP and DV pattern and allows distalization. (B) Another morphogen is produced by the AEC (green steps). The strength of expression is determined by an apical epidermal maintenance factor (AEMF) (red line) produced by the blastemal cells. At the accumulation blastema stage, the concentration of AEMF is low, and the AEC morphogen specifies the next most proximal part. By medium bud stage (MB) the concentration of AEMF has increased so that the AEC morphogen increases, etc, until all the PD positional identities have been 'bootstrapped' into existence. AB = accumulation blastema; MB = medium bud; LB = late bud; D = digit stage.

increases in AECMF and AEC morphogen. Combined with the polar coordinate model, bootstrapping gives a physical mechanism for assignation of progressively more distal positional identities. Nerves are not involved in establishing PD axial patterning, since aneurogenic limbs of reversed PD polarity regenerate distally in the same way as reversed innervated limbs (Wallace, 1980).

The polar coordinate model and boundary models were initially devised to explain the formation of supernumerary limbs after reversal of the AP or DV axes of the blastema with respect to the limb stump (figure 5.10). Confrontation of cells at A:P or DV junctions after axial reversal evokes short arc intercalation of a supernumerary set of circumferential values that drive the formation of an accessory blastema (French *et al* 1976, Bryant *et al* 1981). The number (two), location (where anterior and posterior or dorsal and ventral tissues are maximally confronted) and handedness (host limb handedness) of the supernumerary limbs that arise after axial reversal of the blastema is predicted by the model (Bryant *et al* 1981). Meinhardt's boundary model makes exactly the same predictions as the polar coordinate model with regard to supernumeraries generated by AP or DV axial reversal, because reversal of either the AP or DV axis creates new sets of intersecting AP and DV boundaries. Following reversal of the blastema AP axis, two zones of intersecting boundaries are established on anterior and posterior sides of the limb in addition to the original (primary) set; supernumeraries can now arise from both locations via mitosis and distalization.

The outcome of reversing both the AP and DV axes simultaneously by 180° inversion of the blastema on its limb stump is more complex. APDV reversal evokes the formation of up to three supernumeraries whose loci and handedness are variable. Anatomically normal right or left limbs are only one structural class of supernumerary seen after this operation. The musculoskeletal structure of supernumeraries at the carpal/metacarpal levels revealed three additional structural classes: mirror-imaged (double dorsal or double ventral), part normal/part mirror-imaged, and part normal/part inverted (mixed handed) limbs. The boundary model proved better than the polar coordinate model at predicting the handedness, location and number of these additional classes (Maden 1983).

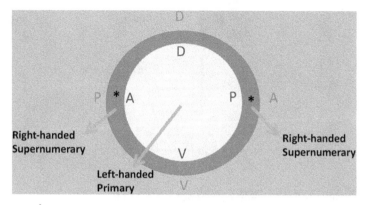

Figure 5.10. Reversal of the AP axis of the left-handed blastema (yellow) with respect to the AP axis of the right-handed stump (red) by contralateral grafting evokes the formation of supernumerary limbs where the posterior half of the blastemal meets the anterior half of the stump and vice versa (asterisks). The two supernumerary limbs have stump-handedness (right) and the primary regenerate is left-handed.

Supernumerary limbs of high complexity also arise after 180° rotation of skin and muscles, or cross-transplant of muscles and amputation through the grafted region, or after implanting carcinogens (Carlson 1974, 1975a, 1975b). Presumably the mechanisms underlying the development of these supernumeraries involve the same kinds of interactions as those operating after blastema axial reversals.

The regenerative capacity of half limbs and symmetrical double half limbs has provided insights into regional cellular contributions to the blastemal and into cellular interactions after amputation (figure 5.11). Double anterior half and double posterior half limbs can be made by exchanging anterior and posterior muscle and skin in the stylopodium or exchanging anterior and posterior halves of the zeugopodium and autopodium. Surgically constructed double anterior stylopodia regenerate only a short cone of cartilage, whereas double posterior stylopodia regenerate double posterior limbs with a symmetrical stylopodial bone, two symmetrical zeugopodial bones and two symmetrical sets of digits. The polar coordinate model explains these results by postulating a smaller contribution of angular positional identities in the anterior versus posterior half of the limb. Centripetally migrating cells of double anterior limbs have little or no positional disparity and their interaction would rapidly result in their convergence to a uniform anterior identity, halting distalization at a stylopodial level. Surgically constructed double anterior zeugopodia regenerate two symmetrical zeugopodial bones and one symmetrical digit, whereas double posterior zeugopodia regenerate two symmetrical zeugopodial bones and two mirror-image sets of digits minus the anterior ones. This suggests that there is an asymmetry in digit-forming potential of the blastema, so that when regeneration reaches the level of the autopodium, the posterior cells are contributing to most of the digits.

The timing model
The timing model for distalization (figure 5.12) postulates that positional identities representing the segment of amputation are generated by RA, whereas more distal

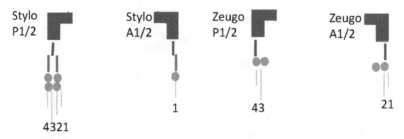

A Right posterior and anterior half fore limbs , dorsal view

B Double anterior half and double posterior half fore limbs

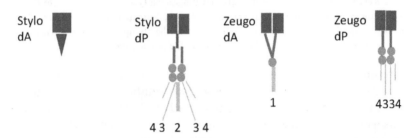

Figure 5.11. Results of amputating anterior and posterior half fore limbs (A1/2 and P1/2) and double anterior and double posterior half fore limbs (dA and dP) made in the stylopodium and zeugopodium. Blue lines represent regeneration of stylopodium and/or zeugopodium; red circles represent carpals; green lines indicate digits. Reproduced from (Stocum 2017). Wiley, Open Access.

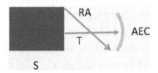

Figure 5.12. The timing model of distalization. A retinoic acid gradient diminishing distally (red arrow) interacts with a mitotic counting mechanism (T, green arrow) generated by an environment created by the AEC. Reproduced from (Stocum 2017). Wiley, Open Access.

PD identities are specified by an autonomous timing mechanism linked to mitosis, promoted by a blastema environment created by the AEC (Summerbell *et al* 1973, Rosello-Diez and Torres 2014, Saiz-Lopez *et al* 2015). This model would explain why the stylopodium of an axolotl limb, but nothing more distal, can regenerate after grafting a mature hand to the mid-stylopodium (Pescitelli and Stocum 1981) and why BMP-2 or 7 can stimulate regeneration of the amputated second phalange and neonatal zeugopodium to complete themselves, even though they cannot regenerate the terminal phalange or autopodium, respectively (Yu *et al* 2010, 2012, Masake and Ide 2007, Ide 2012).

The averaging model
The polar coordinate, boundary and timing models view the specification of positional identities as taking place serially from proximal to distal, which fits the spatial and temporal expression of HoxA-9–13 genes associated with the specification of PD pattern (Yokoyuchi *et al* 1991, Izpisua-Belmonte *et al* 1992, Yakushiji-Kaminatsui *et al* 2009, Tamura *et al* 2009, Ohgo *et al* 2010, Roensch *et al* 2013). By contrast, Maden (1977) proposed a non-serial model of distalization based on the successive intercalary regeneration of positional identities between proximal and distal boundaries by an averaging mechanism (figure 5.13). The proximal boundary is the fibroblast positional identity of the amputation level. The distal boundary is unknown, but might be conferred on the initial fibroblast-derived blastema cells autonomously or by virtue of their contact with the AEC (Nye *et al* 2003). Confrontation of the two boundaries initiates an averaging cascade of intercalation. The first averaging event leads to intercalation of the positional identity halfway between the autopodium and the level of amputation. There are now three positional identities dividing the blastema into compartments, which could be segmental boundaries. Progressive mitosis and intercalary averaging would continue to fill in the nearest neighbor map within each compartment. There is evidence from mapping experiments that compartments representing the different limb segments are already present in the very early chick limb bud (Stark and Searls 1973, Dudley *et al* 2002) and in the urodele limb regeneration blastema (Echeverri and Tanaka 2005). The problem is how to reconcile this model with the proximal to distal sequence of expression of the HoxA-9–13 genes. The assumption would have to be made that serial expression of these genes does not adequately reflect the actual patterning mechanism itself, raising the speculation that serial Hox gene expression might be the result rather than the cause of the patterning mechanism.

5.1.4.3 Genes associated with pattern specification
A number of genes associated with blastema pattern formation have been identified. Several homeobox genes encoding transcription factors are expressed in blastemas

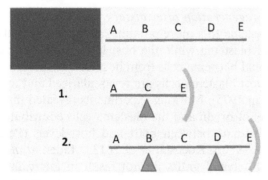

Figure 5.13. The averaging model of distalization. Blue, stylopodium; A-E positional identities removed by amputation; green = AEC. Red triangles indicate intercalations. A and E are boundary positional identities. 1. Intercalation halfway between A and E. 2. Intercalations between A and C, and C and E. Reproduced from (Stocum 2017). Wiley, Open Access.

formed after amputation through the proximal stylopodium. (Gardiner and Bryant 1996, Geraudie and Ferretti 1998). Forelimb identity is associated with the *tbx5* gene and hind limb identity with the *tbx4* gene (Simon *et al* 1997). The spatiotemporal expression pattern of *Hoxa* gene combinations (*Hoxa* codes) is considered to reflect the order of PD specification of regenerating limb segments. *Hoxa*-9, *Hoxa*-11 and *Hoxa*-13 are expressed serially in proximal to distal temporal and spatial order in the mesenchyme of the blastema (Roensch *et al* 2013). *Hoxa*-9 is expressed first throughout the blastema, *Hoxa*-11 next in the prospective zeugopodial region, and *Hoxa*-13 last in the prospective autopodial region. *Meis*-2 is another gene expressed preferentially in the prospective stylopodial and zeugopodial regions (Mercader *et al* 2005). These observations implicate *Hoxa*-9/*Meis*-2 in specifying the PD pattern of the stylopodium, *Hoxa*-9/-11/*Meis*-2 the zeugopodium and *Hoxa*-9/-13 the autopodium. Along with *shh* and Fgf-8, *Hoxd*-8, -10 and -11 are associated with anterior–posterior patterning in limb buds, and are also expressed in the regeneration blastema (Torok *et al* 1998).

5.1.4.4 Effects of retinoic acid on blastema pattern

Figure 5.14 shows that retinoic acid proximalizes, posteriorizes and ventralizes the positional identities of blastema cells in a dose-dependent fashion, resulting at the higher doses in complete limbs extending from the wrist (Niazi 1996, Maden 1998, Stocum 2017, for reviews). Genes associated with stylopodial development are up regulated in the autopodial blastema by retinoic acid and genes associated with autopodial development are down regulated. The fact that RA can alter positional identities of blastema cells suggests that it might be an important component of the molecular mechanism that patterns the blastema. This idea is supported by the fact that RA is present in a posterior–anterior gradient across the blastema and that proximal blastemas contain 3.5-fold higher levels of RA than distal blastemas. Five RA receptor isoforms (RARs) have been detected in the blastema: α1, 2, δ1a, b and δ2. The δ2 receptor is responsible for change in positional identity. Finally, inhibitors of RA synthesis, such as disulfiram, inhibit limb and tail regeneration.

5.1.4.5 Intercalary regeneration after distal to proximal grafting

Grafting a distal blastema to a more proximal stump level results in developmental delay of the grafted blastema while the host level tissue undergoes histolysis and liberation of additional blastema cells from host level tissues. Development resumes once the graft and host blastema cells are revascularized and reinnervated (Stocum 1975, Iten and Bryant 1975). Marking experiments revealed that the graft develops according to its level of origin and the blastema cells contributed by the host form the structures intermediate between graft and host levels (Pescitelli and Stocum 1980, Maden 1980a, 1980b, Roensch *et al* 2013, Maden *et al* 2015) (figure 5.15). Proximal to distal blastemal grafts do not result in intercalation, but rather the formation of serially duplicated limb segments. Furthermore, when either a normal or double anterior distal zeugopodial blastema was grafted to a double anterior stylopodial limb stump, the graft developed according to origin and a single symmetrical stylopodial and zeugopodial skeletal element was regenerated by cells

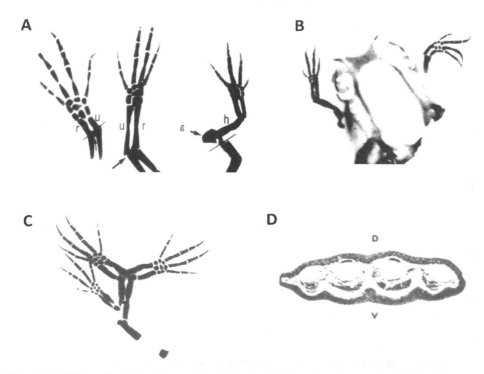

Figure 5.14. Effects of retinoic acid on positional identity. (A) Dose-dependent proximalization of positional identity after amputation through the very distal zeugopodium of a forelimb results in serial duplications of structure. R = radius; u = ulna; h = humerus; g = shoulder girdle; arrow and lines indicate amputation level. (B) Regeneration of RA-treated anterior half zeugopodia grafted to the orbit of the eye. AP/DV complete, proximalized regenerates were formed. (C) Regeneration of RA-treated double anterior half zeugopodia. Mirror-image proximalized limbs were regenerated, along with a proximalized supernumerary limb (S) where anterior tissue was juxtaposed to posterior tissue on one side of the limb. The result in both (B) and (D) indicates that RA posteriorizes positional identity. (D) Cross-section of RA-treated dorsal half forelimb at the level of the metacarpal muscles. The regenerate was proximalized and the muscle patterns were normal, indicating that RA ventralizes positional identity. In (B) and (C), the results are due to the confrontation of posteriorized blastema cells with unaffected anterior stump cells, and in (D) to the confrontation of ventralized blastema cells with unaffected dorsal stump cells. Reproduced from (Stocum 2017). Wiley, Open Access.

contributed by the stump (Stocum 1980, 1981). These results imply that interactions between anterior and posterior cells (and by implication, Shh and Fgf-8) are not required to generate intermediate positional identities. Intercalary regeneration in the PD axis thus appears to require only an interaction between distal and proximal positional identities without regard to any interactions in the cross-section of the limb. This type of regeneration is best explained by the averaging model (Maden 1977).

5.1.4.6 Is blastema patterning autonomous or induced?
A long-standing question of limb regeneration is whether the mechanism of distalization is induced in the blastema by signals from adjacent stump tissues or

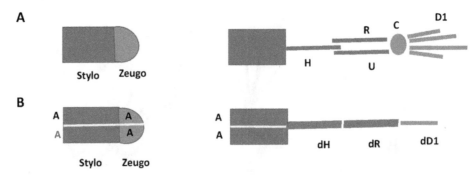

Figure 5.15. (A) Intercalary regeneration of the distal half of the stylopodium and zeugopodium after grafting a blastema derived from the very distal zeugopodium to the mid-stylopodial stump of a forelimb. H = humerus; R = radius; U = ulna; C = carpals; D1 = anteriormost digit. (B) Intercalary regeneration after grafting the blastema from the distal level of a double half anterior zeugopodium to the mid-stylopodial level of a double half stylopodial forelimb stump. The humerus and a radius and digit are regenerated, all of which are thicker than normal and are therefore probably fused double structures (dH, dR, dD1).

is this mechanism inherent to the blastema? This question was tested from the 1920s to the 1940s by the reciprocal grafting of early blastemas between limb and tail stumps to see if they developed according to graft origin or host origin, and by grafting blastemas to ectopic sites to determine their ability to self-differentiate (Stocum 1984, for a review). The results suggested that young blastemas had no power to self-differentiate (were 'nullipotent'). Rather, their differentiation and morphogenesis was directed by the tissues with which they were in contact. Thus, for example, a young limb blastema would form a tail if grafted to a tail stump, or a lens if grafted into the pupil of the lentectomized eye. Once they had been sufficiently programmed by differentiated tissues at their site of origin, the blastemas could develop autonomously when grafted to other locations. However, as pointed out by several investigators, these results were suspect because the grafts were not marked in a way that could distinguish their conversion to other tissues from cell death and replacement by cells from the host graft site.

Later experiments from the 1960s to 1980s using markers such as color, triploidy or immune rejection suggested that transplanted early blastemas can self-organize if they receive adequate vascularization and innervation. Young limb blastemas of larval and adult urodeles grafted ectopically to a wound bed on the dorsal fin, grafted proximal or distal to their level of origin, or exchanged between forelimb and hind limb could self-organize according to their level of origin (Stocum 1984, 2017, Stocum and Cameron 2011, for reviews). Such experiments have been repeated with genetically marked blastemas. These experiments showed that when grafted to a more proximal level of the limb stump, blastema cells derived from muscle or Schwann cells, which carry no positional identity, can be incorporated into positions proximal to their level of origin in the intercalary regenerate, but that all other tissues of the graft develop according to origin (Roensch et al 2013, Maden et al 2015).

Other experiments also provide evidence that the early blastema is a self-organizing system. Young blastemas derived from a normal tissue organization develop autonomously when grafted to a wound bed made on the dorsal fin and develop according to origin when grafted to a double anterior stump and vice versa (Stocum 1980, 1981). Grafted blastema cells forced to dedifferentiate when recovered by wound epidermis, as when several blastemal mesenchymes are massed and grafted to the back (Polezhaev 1937, DeBoth 1970), or proximal halves of stylopodial forelimb blastemas are grafted to the ankle stump of the hind limb (Stocum and Melton 1977), develop according to origin.

5.2 Appendage regeneration in *Xenopus*

The early limb buds of anuran tadpoles regenerate perfectly, but lose regenerative capacity in a proximal to distal direction coincident with the onset of differentiation of the limb tissues. With some exceptions, the amputated limbs of most adult anurans end in a cartilage callus and scar tissue under the skin (Stocum 1995, for a review). One of these exceptions is the South African Pipid clawed frog *Xenopus laevis*, which regenerates hypomorphically as a long symmetrical spike of cartilage lacking muscle.

At stage 51, the *Xenopus* hind limb bud is conical in shape with no evidence of segmentation or differentiation. By stage 52, it has doubled its length and a constriction distally separates the prospective footplate from the more proximal prospective zeugopodial and stylopodial segments of the bud. By stage 53, the cells of the limb bud have begun to differentiate in the stylopodial part of the limb bud and by stage 54, differentiation is ongoing in the zeugopodium and autopodium. The hind limb is fully differentiated by stage 57; by stage 59 the differentiating forelimb bud emerges from beneath the operculum. By stage 66 the tadpole has resorbed its tail to become a froglet. The limb bud is competent to form a mesenchymal blastema and regenerates normally from any level of amputation at stages 51 and 52, but progressively loses regenerative capacity from proximal to distal, coincident with differentiation of the limb segments (figure 5.16). By stage 57 and thereafter, the hind limb regenerates a symmetrical cartilage spike lacking muscle after amputation at any level of amputation. The amputated hind limb forms a blastema-like structure, but it is fibroblastic in histological appearance and undergoes premature differentiation into cartilage (Dent 1962). The forelimb follows a similar pattern of regenerative loss.

The proximal to distal loss of regenerative capacity is associated with a loss of interaction between the AEC and underlying fibroblastema cells. Growth of the stage 52 limb bud and blastema mesenchyme is dependent on Fgf-8 secreted by the apical epidermis that is induced by Fgf-10 from the subjacent mesenchyme (Yokoyama *et al* 2000). Fibroblastema cells fail to express Fgf-10 after amputation through differentiated regions, and thus Ffg8 expression fails as well. To test whether innervation during limb development might somehow render limb cells less responsive to AEC signals after amputation, stage 52 limb buds (which are not yet innervated) were grafted to denervated stage 57 limbs and allowed to develop (the equivalent of making aneurogenic limbs in urodeles). When amputated, these limbs regenerated hypomorphically, indicating that intrinsic changes in limb cells, not

Xenopus Tadpole vs Xenopus Froglet

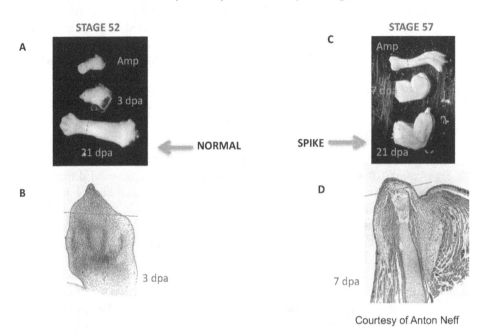

Courtesy of Anton Neff

Figure 5.16. Regeneration of the *Xenopus* tadpole hind limb at stage 52 versus stage 57. (A) Amputation of stage 52 limb bud through the prospective tarsal region (dashed line). A blastema forms within three days post-amputation and by 21 days limb morphology is restored. (B) Longitudinal section of the blastema at three days post-amputation. The blastema is composed of a mass of undifferentiated cells similar to the medium bud blastema of an amputated axolotl limb. (C) Amputation of a stage 57 limb through the level of the tibia-fibula (dashed line). A fibroblastema has formed by 7 days post-amputation, but fails to grow normally. By 21 days post-amputation, the blastemal is forming a spike of cartilage. (D) Longitudinal section of the blastema at 7 days post-amputation. The blastema is of poor histological quality with few proliferating cells. Courtesy of Dr Anton Neff, Indiana University School of Medicine. Reproduced with permission from (Stocum 2012). Copyright 2012 Elsevier.

innervation, were responsible for the change in regenerative capacity (Filoni *et al* 1991). Furthermore, reciprocal exchange of stage 52 blastemas and stage 57 fibroblastemas did not alter the differential in regenerative capacity between these two stages (Sessions and Bryant 1988). These results strongly suggest that intrinsic changes related to the differentiation of limb cells render fibroblastema cells unable to produce Fgf-10. Even hypomorphic regeneration, however, is dependent on the apical epidermis and nerves, as in the urodele limb bud (Goss and Holt 1992, Filoni *et al* 1995).

The effects of these intrinsic changes on blastema formation and development have been studied in comparisons of amputated stage 52 limb buds and stage 57 or froglet fully differentiated limbs. Fibroblastema formation is associated with impaired histolysis and dedifferentiation and a thinner AEC (Wolfe *et al* 2000). Although fibroblastema cells up regulate stem genes such as *sall4*, they do so at levels much lower level than blastema cells derived from the stage 52 limb bud, and they do not up regulate any Yamanaka factor genes (Pearl *et al* 2008). These observations suggest an inability to degrade ECM and to undergo dedifferentiation (Santosh *et al* 2011). Induction of stress-response pathways is requisite for successful regeneration in stage 52 hind limbs, but whether these pathways are up or down regulated after amputation of froglet limbs is unknown. Genes for cartilage differentiation are strongly up regulated during fibroblastema formation, whereas the expression of such genes is delayed until much later in the growing axolotl limb blastema (Rao *et al* 2014). The development of the immune system in *Xenopus*, which is much stronger than in urodeles, is also hypothesized to be a major factor in deficient limb regeneration in *Xenopus* and mammals (Mescher *et al* 2017, for a review).

A major departure of the fibroblastema from the stage 52 mesenchyme blastema and the axolotl mesenchyme blastema is its failure to activate *Hoxa* genes associated with PD patterning in the correct spatial and temporal order. In contrast to its expression on the posterior side of the unamputated stage 52 limb bud and the blastema formed after amputation of a stage 52 limb bud, *Shh* is not expressed on the posterior edge of the fibroblastema. This conceivably might be due to a lack of interaction between expressing Shh and Fgf-8, and thus the dedifferentiation and mitotic activity required to form a normal blastema.

Loss of *Shh* expression is correlated with hypermethylation of its enhancer sequence during limb bud differentiation, in contrast to much lower levels of methylation in the stage 52 limb bud/blastemal and axolotl blastema. Expression of genes promoting asymmetry in the DV axis such as *Lmx*-1, and *Bmp*, as well as genes involved in coordinating pattern in the PD, AP, and DV axes of embryonic limb buds, are down regulated in the fibroblastema (Matsuda *et al* 2001, Rao *et al* 2014). A comparison of gene activity between regeneration-competent and deficient *Xenopus* limb buds indicates that they share certain gene subsets, while expressing distinct subsets of other genes (King *et al* 2003).

5.3 Mouse digit tip regeneration

Figure 5.17 shows the histological anatomy of the terminal phalange of the unamputated mouse digit and its regeneration. This phalange regenerates the missing distal structures after amputation of up to its distal two-thirds (Borgens 1982, Simkin *et al* 2015, Seifert and Muneoka 2017, for reviews). The wound epidermis constricts around the exposed bone and cuts it off a bit further proximally, opening up the marrow cavity. A mesenchymal blastema forms at this new level. The origin of the blastema cells is a cylinder of connective tissue surrounding the bone, cells of the nail bed, Schwann cells and possibly the marrow cavity itself (Lehcozky *et al* 2011). Blastema formation and growth requires BMP and Wnt signaling; inhibiting this signaling inhibits both nail growth and the regenerative response (Zhao and Neufeld 1995, Han *et al* 2003, Takeo *et al* 2013, Rinkevich *et al* 2013). Recently, nerve-associated Schwann cell precursors were shown to dedifferentiate, become part of the blastema, and secrete the paracrine factors PDGF-AA and oncostatin M, which promote blastema cell proliferation (Johnston *et al* 2016).

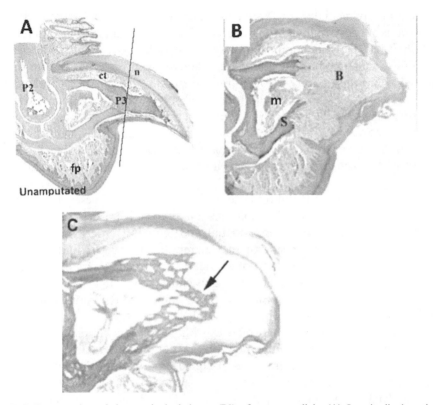

Figure 5.17. Regeneration of the terminal phalange (P3) of a mouse digit. (A) Longitudinal section of unamputated phalange showing midpoint amputation level (line), just proximal to the fat pad (fp). N = nail; ct = connective tissue. P2 = second phalange. (B) Blastema (B) that arises from stump tissue (S). The bone proximal to the amputation level has been removed by constriction, moving the amputation proximally into the fat pad and opening the marrow cavity (m). (C) Bone (arrow) differentiating from the blastema cells to restore the phalange. Reproduced from (Simkin *et al* 2015). Wiley, Open Access.

The blastema cells differentiate directly into the osteoblasts of the missing phalangeal bone. Amputation through more proximal levels of the terminal phalange and beyond elicits only capping of the bone and the deposition of a thick pad of scar tissue.

5.4 Regeneration of mouse ear tissue

The mouse ear is essentially a thin plate of cartilage covered by skin. These tissues can regenerate to fill in a punch hole made in the ears of the MRL/lpj mouse strain and several species of *Acomys* (African spiny mouse), as well as the rabbit (Williams-Boyce and Daniel 1980, Clark *et al* 1998, Gawriluk *et al* 2016). *Acomys* can regenerate across a hole of up to 8 mm, whereas the regenerative capacity of the MRL/mpj ear is 2 mm (Gawriluk *et al* 2016). By contrast, *Mus* species such as B6, BALB/c and CD1 exhibit only partial closure of ear holes. Regeneration is faster in females than in males.

Ear hole closure in these regenerative-competent species is accomplished by proliferating mesenchymal cells derived from the connective tissue of the ear skin and perichondrium that form a ring blastema under a thickened wound epidermis (Heber-Katz *et al* 2013, Gawriluk *et al* 2016). The p21 checkpoint protein is down regulated in 40% of the fibroblasts of the uninjured MRL/mpj ear, allowing them to proceed to G_2 arrest and be immediately available for mitosis after injury (Heber-Katz *et al* 2013). p21-null B6 mice acquire the ability to close ear holes completely (Bedelbaeva *et al* 2010). Consistent with G_2 arrest, the blastema cells express high levels of Evi5. Comparative transcriptomic analysis of regeneration-deficient *Mus* and regeneration-competent *Acomys* revealed a pro-fibrotic gene expression profile of collagen in *Mus*, as opposed to a profile dominated by fibronectin and tenascin in *Acomys* (Gawriluk *et al* 2016). These observations indicate that the ear hole ring blastema shares some similarities with the urodele limb regeneration blastema.

An interesting set of studies reported that strains and species such as the MRL/lpj mouse and *Acomys* may not be unique in having the ability to regenerate ear hole tissue, but rather may be a generic feature of mouse ear tissue. The regeneration-competent MRL/lpj and the regeneration-deficient BALB/c and B6 mice used in ear tissue regeneration studies were young (4–8 weeks old). BALB/c and B6 mice punched at 2 months of age were reported to average 40%–65% closure, whereas 'middle age' mice (8–9 months) averaged 90%–97%, many closing completely (Reines *et al* 2009). The average fell off to 70% when punches were made at 12 months. A correlation between age, body size and ability to close ear holes revealed that large strains such as MRL have the ability to close holes early in life, whereas smaller strains such as B6 and BALB/c acquire regenerative capacity later in life (Li *et al* 2001).

5.5 Regeneration of deer antlers

Antlers are twin, branched outgrowths of bone growing from the foreheads of male deer. Under the influence of rising testosterone levels, maturing males develop a first set of antlers that extend from the pedicles, bumps of trabecular bone protruding from the frontal bone of the skull. Each year thereafter, antlers follow a cycle where they are cast off over a period from December to March, regenerate during the summer, and are used for display and fighting with rivals for females during the fall mating season (Goss 1992, Price *et al* 1996).

Casting of the antler produces a circular wound that is healed by overgrowth of skin (figure 5.18). There is no histolysis and dedifferentiation of dermal or under-lying pedicle tissue, no blastema is formed, and there is no requirement for contact of underlying tissues with wound epidermis as in urodele limbs (Kierdorf and Kierdorf 2011, for a review). Antler regeneration is a process of modified endochondral bone regeneration in which MSCs in the lower layers of the periosteum covering the pedicle proliferate and differentiate (Li *et al* 2005, Kierdorf *et al* 2007, 2009). These cells express STRO-1, Oct-4, Nanog and CD9 that have been associated with other MSCs (Rolf *et al* 2008, Li *et al* 2009).

The tip of each branch of the growing antler has a growth zone of MSCs (Mount *et al* 2006, Berg and Goodell 2007, Li *et al* 2009), which can be induced *in vitro* to

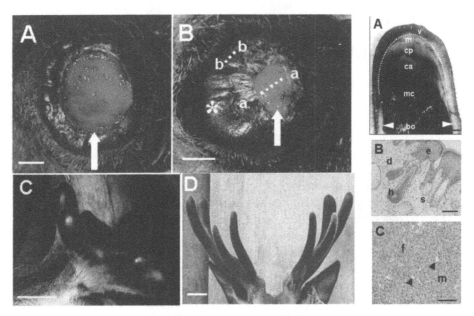

Figure 5.18. Regeneration of red deer antlers. Left: Anatomical views. (A) 24 h after casting the antler. (B) Seven days after casting. White arrows indicate the leading edge of the regenerating epidermis. aa and bb indicate where tissue sections were taken; aa sections are shown at right. (C, D) Four and eight weeks, respectively, of antler growth. Right: Longitudinal histological sections of growing blastema. (A) Blastema showing zones from least differentiated (top) to most differentiated (bottom). V = velvet epidermis; m = mesenchyme; cp = chondroprogenitor region; ca = non-mineralized cartilage; mc = mineralized cartilage; bo = bone. Arrowheads indicate intramembranous bone formation. (B) Velvet skin of blastema. E = epidermis; d = dermis; h = hair follicle; s = sebaceous gland. (C) Perichondrium surrounding cartilaginous region. F = fibrous zone; m = mesenchymal zone. Arrowheads indicate blood vessels. Reproduced with permission from (Faucheaux *et al* 2004). Copyright 2004 Wiley.

differentiate into osteogenic, chondrogenic and adipogenic lineages (Seo *et al* 2014). The molecular regulation of antler regeneration is similar to that of endochondral bone development, involving the PTHrP/hedgehog pathway and retinoic acid (Faucheux *et al* 2002, Allen *et al* 2002). Antler growth is terminated at the start of the mating season by an increase in testosterone level, and velvet epidermis and underlying periosteum of the antler are shed to expose fully ossified bone (Goss 1970, Bubenik 1983). A fall in testosterone level after the mating season leads to osteoclastic erosion of distal pedicle bone, allowing the antler to be cast (Goss 1992).

5.6 Enhancement of appendage regeneration

A goal of regenerative medicine is to regenerate an amputated human limb instead of replacing it with a prosthesis. We assume that mammalian appendages have a latent genetic circuitry for regeneration because mammalian digit tips and ear tissue regenerate, and because anuran limb buds can regenerate perfectly before they differentiate. The reasoning is that if we understand the differences between regeneration-competent and deficient appendages we can devise treatments to manipulate this circuitry therapeutically. Such interventions have been attempted for anuran limbs, mouse digits and neonatal mouse limbs.

5.6.1 Anuran limbs

Interventions to stimulate latent regenerative capacity of amputated anuran limbs have targeted the enhancement of histolysis/dedifferentiation, satellite cell contribution, fibroblastema cell proliferation, and patterning. To increase histolysis, the amputation surface has been subjected to repeated trauma by cutting and stabbing, exposure to hypertonic salt solutions, and retinoic acid solutions (Stocum 1995, for a review). Immersion of adult *R. catesbiana* and *R. pipiens* limbs amputated through the zeugopodium in retinoid solutions resulted in the regeneration of several digit-like structures (Cecil and Tassava 1986a, 1986b) (figures 5.19(A), (B)). Nerve augmentation and electrical stimulation resulted in several cases with segmented digits or muscle (Konieczna-Marczynska and Skowron-Cendrzak 1958, Borgens *et al* 1979). Implants of microbeads releasing Fgf-10 restored the regeneration of digits in stage 57 limbs amputated at the knee level, but without restoring more

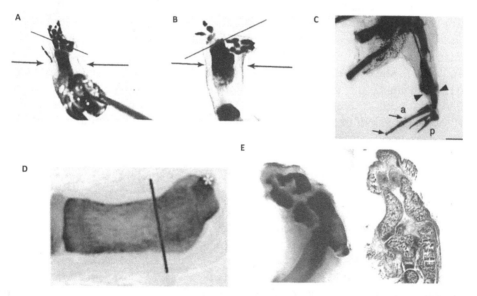

Figure 5.19. (A, B) Retinoic acid treatment of amputated adult frog limbs induced regeneration of cartilaginous structures resembling long bones (arrows) and digits. Lines indicate levels at which sections were taken (not shown). (C) Digits (arrows) and unidentified proximal cartilage regenerated from a stage 57 *Xenopus* hind limb after amputation at the knee (arrowhead) and implanting Fgf-10 beads at the amputation surface. A = anterior; p = posterior. (D) Regeneration of mouse 2nd phalange after amputation through the middle (line) and implanting BMP-2 beads on the amputation surface. The distal part of the phalange was regenerated. (E) Whole mount (left) and longitudinal section showing regeneration of zeugopodial cartilage elements after amputation through the mid-zeugopodial level of the neonatal mouse forelimb. (A, B) Reproduced with permission from (Cecil and Tassava 1986b). Copyright 1986 Wiley. (C) Reproduced with permission from (Yokoyama *et al* 2001). Copyright Elsevier 2001. (D) Reproduced from (Simkin *et al* 2015). Wiley, Open Access. (E) Reproduced with permission from (Ide 2012). Copyright 2012 Wiley.

proximal structure (Yokoyama *et al* 2001) (figure 5.19(C)), and induction of H^+ efflux/Na^+ influx by monensin at the wound surface of stage 57 limbs amputated at knee level likewise stimulated digit regeneration but not more proximal structure (Tseng and Levin 2013). BMP-2-soaked microbeads induced segmentation of the cartilage spike (Satoh *et al* 2005a) and implants of HGF-expressing cells into the fibroblastema attracted satellite cells into the regenerate where they formed patches of muscle (Satoh *et al* 2005b). It would be interesting to determine whether *Xenopus* limb regeneration (stage 57 and beyond) regeneration would be enhanced by injections of Shh and Fgf-8 gene constructs in baculovirus on the posterior and anterior side of the limb, respectively, or would evoke supernumerary limb regeneration as in the experiment performed by Nacu *et al* (2016).

5.6.2 Mouse digits and neonatal limbs

Regeneration does not take place at levels proximal to the distal two-thirds of the mouse terminal phalange. However, it has been stimulated after amputation through the second phalangeal segment (P2) by applying a microbead soaked in BMP-2 to the amputation surface (figure 5.19(D)). The regenerated bone is an extension of the remaining bone, but does not complete the digit distally (Yu *et al* 2010, 2012). Isolated P2 and P3 (terminal phalange) cells have position-specific characteristics that may be connected to their differential regenerative abilities. P2 cells will participate in blastema formation, however, when introduced into the amputated P3 digit tip, indicating that P2 cells can respond to regenerative signals in the P3 blastema (Wu *et al* 2013). Neonatal mouse limbs amputated through the mid-zeugopodium fail to regenerate, but if BMP7 beads are applied to the amputation surface the distal half of the radius and ulna regenerates along with more distal blobs of cartilage that might represent an attempt at regenerating autopodial structures (Masake and Ide 2007, Ide 2012) (figure 5.19(E)).

5.6.3 Mouse ear tissue

Two-month old Swiss Webster mice do not close ear hole punches compared to MRL/lpj mice of similar age. Zhang *et al* (2015) discovered that hypoxia-inducible factor 1α (HIF-1α) is required for closure of ear hole tissue in the MRL mouse. HIF-1α is normally degraded by prolyl hydrolases (PHDs), but is maintained at high expression levels in the MRL mouse compared to Swiss Webster; the MRL fails to close ear holes when *Hif1a* is inhibited by *Hif1a* siRNA. The PHD inhibitor 1,4-dihydrophenonthtolin-4-obne-3-carboxylic acid (1,4-DCPA) stabilizes HIF-1α. Injectable hydrogels containing 1,4-DCPA injected into Swiss Webster mice at multiple subcutaneous sites over a 10 day period promoted ear hole closure.

5.6.4 Are we missing something?

The idea of latent mammalian regeneration circuits that need only be properly stimulated to regenerate a limb has been challenged in an interesting way. Evolutionary surveys for Prod1, the blastema cell surface receptor for the AG protein that integrates proliferation and patterning in regenerating urodele limbs have revealed that Prod1 is unique to urodeles (Brockes and Gates 2014, Geng *et al* 2015), suggesting that local selective forces have left only urodeles with the genetic capacity for perfect limb regeneration throughout life. Regeneration genes may exist in anurans and higher taxa, but these genes enable regeneration only in undifferentiated anuran limb buds and are insufficient for the regeneration of differentiated limbs. Once differentiation of the anuran limb bud sets in, regeneration requires additional genes typical of the regenerating urodele limb, such as Prod1. Comparative molecular analyses of urodele versus anuran regenerating limbs may give insights into what additional genes might be needed (Rao *et al* 2014).

How soon might we see the regeneration of a human limb? We don't really know, but extreme optimists believe it might be within 15–20 years, while extreme pessimists believe the task is impossible. However, 'impossible' has been proven wrong many times before and fails to impress younger generations of scientists from different disciplines, who will continue the quest.

References

Adams D S, Masi A and Levin M 2007 H$^+$ pump-dependent changes in membrane voltage are an early mechanism necessary and sufficient to induce *Xenopus* tail regeneration *Development* **134** 1323–35

Allen S P, Maden M and Price J S 2002 A role for retinoic acid in regulating the regeneration of deer antlers *Dev. Biol.* **251** 409–23

Anastasia L *et al* 2006 Reversine-treated fibroblasts acquire myogenic competence *in vitro* and in regenerating skeletal muscle *Cell Death Differ.* **13** 2042–51

Bedelbaeva K *et al* 2010 Lack of p21 expression links cell cycle control and appendage regeneration in mice *Proc. Natl Acad. Sci. USA* **107** 5845–50

Berg J S and Goodell M A 2007 An argument against a role for Oct4 in somatic stem cells *Cell Stem Cell* **1** 359–60

Borgens R B, Vanable J W and Jaffe L F 1979 Role of subdermal current shunts in the failure of frogs to regenerate *J. Exp. Zool.* **209** 49–55

Borgens R B 1982 Mice regrow the tips of their foretoes *Science* **217** 747–50

Brockes J P 1984 Mitogenic growth factors and nerve dependence of limb regeneration *Science* **235** 1280–7

Brockes J P 2015 *Variation in salamanders: an essay on genomes, development, and evolution Salamanders in Regeneration Research: Methods and Protocols* ed A Kumar and A Simon (New York: Springer), pp 3–15

Brockes J P and Gates P B 2014 Mechanisms underlying vertebrate limb regeneration: lessons from the salamander *Biochem. Soc. Trans.* **42** 625–30

Brockes J P and Kintner C R 1986 Glial growth factor and nerve-dependent proliferation in the regeneration blastema of urodele amphibians *Cell* **45** 301–6

Bryant S V, French V and Bryant P J 1981 Distal regeneration and symmetry *Science* **212** 993–1002

Bryant D M *et al* 2017 A tissue-mapped axolotl *de novo* transcriptome enables identification of limb regeneration factors *Cell Rep.* **18** 762–76

Bubenik G A 1983 The endocrine control of the antler cycle *Antler Development in Cervidae* ed R D Brown (Kingsville: Caesar Kleberg Wildlife Institute), pp 73–107

Cameron J A and Fallon J F 1977 Evidence for polarizing zone in the limb buds of *Xenopus laevis* *Dev. Biol.* **55** 320–30

Carlson B M 1969 Inhibition of limb regeneration in the axolotl after treatment of the skin with Actinomycin D *Anat. Rec.* **163** 389–402

Carlson B M 1974 Morphogenetic interactions between rotated skin cuffs and underlying stump tissues in regenerating axolotl forelimbs *Dev. Biol.* **39** 263–85

Carlson B M 1975a The effects of rotation and positional change of stump tissues upon morphogenesis of the regenerating axolotl limb *Dev. Biol.* **47** 269–91

Carlson B M 1975b Multiple regeneration from axolotl limb stumps bearing cross-transplanted minced muscle regenerates *Dev. Biol.* **45** 203–8

Cecil M L and Tassava R A 1986a Vitamin A enhances forelimb regeneration in juvenile leopard frogs Rana pipiens *J. Exp. Zool.* **237** 57–61

Cecil M L and Tassava R A 1986b Forelimb regeneration in the postmetamorphic bullfrog: stimulation by dimethyl sulfoxide and retinoic acid *J. Exp. Zool.* **239** 57–61

Chen S *et al* 2004 Dedifferentiation of lineage-committed cells by a small molecule *J. Am. Chem. Soc.* **126** 410–1

Christen B, Robles V, Raya M, Paramonov I and Izpisua Belmonye J C 2010 Regeneration and reprogramming compared *BioMed. Central Biol.* **8** 5

Clark L D, Clark R K and Heber-Katz E 1998 A new murine model for mammalian wound repair and regeneration *Clin. Immunol. Immunopathol.* **88** 35–45

Cook A and Seifert A W 2016 Beryllium nitrate inhibits fibroblast migration to disrupt epimorphic regeneration *Development* **143** 3491–505

Crawford K and Stocum D L 1988a Retinoic acid coordinately proximalizes regenerate pattern and blastema differential affinity in axolotl limbs *Development* **102** 687–98

Crawford K and Stocum D L 1988b Retinoic acid proximalizes level-specific properties responsible for intercalary regeneration axolotl limbs *Development* **104** 703–12

Dahn R D and Fallon J F 2000 Interdigital regulation of digit identity and homeotic transformation by modulated BMP signaling *Science* **289** 438–41

De Both N J 1970 The developmental potencies of the regeneration blastema of the axolotl limb *Wilhelm Roux. Arch. Entwickl. Mech. Organ.* **165** 2452–276

Dent J N 1962 Limb regeneration in larvae and metamorphosing individuals of the South African clawed toad *J. Morphol.* **110** 61–77

Drossopoulou G *et al* 2000 A model for anteroposterior patterning of the vertebrate limb based on sequential long and short range Shh signaling and BMP signaling *Development* **127** 1377–1348

Duckmanton A, Kumar A, Chang Y-T and Brockes J 2005 A single cell analysis of myogenic dedifferentiation induced by small molecules *Chem. Biol.* **12** 1117–26

Dudley A T, Ros M A and Tabin C J 2002 A re-examination of proximodistal patterning vertebrate limb development *Nature* **418** 539–44

Echeverri K and Tanaka E M 2005 Proximodistal patterning during limb regeneration *Dev. Biol.* **279** 391–401

Egar M W 1993 Affinophoresis as a test of axolotl accessory limbs *Limb Development and Regeneration, Part B* ed J F Fallon, P F Goetinck, R O Kelley and D L Stocum (New York: Wiley-Liss), pp 203–11

Eldridge A G *et al* 2006 The EVI5 oncogene regulates cyclin accumulation by stabilizing the anaphase-promoting complex inhibitor EMI1 *Cell* **124** 367–80

Endo T, Yokoyama H, Tamura K and Ide H 1997 *Shh* expression in in developing and regenerating limb buds of *Xenopus laevis Dev. Dyn.* **209** 227–32

Fallon J F and Crosby G M 1975 The relationship of the zone of polarizing activity to supernumerary limb formation (twinning) in the chick limb bud *Dev. Biol.* **42** 24–34

Farkas J E, Freitas P D, Bryant D M, Whited J L and Monaghan J R 2016 Neuregulin-1 signaling is essential for nerve-dependent axolotl limb regeneration *Development* **143** 2724–31

Faucheaux C 2004 Recapitulation of the parathyroid hormone-related peptide-Indian hedgehog pathway in the regenerating deer antlerDevelop. Dynam. **231** 88–97

Faucheux C, Horton M A and Price J S 2002 Nuclear localization of type I parathyroid hormone/ parathyroid hormone-related protein receptors in deer antler osteoclasts: evidence for parathyroid hormone-related protein and receptor activator of NF-kB-dependent effects on osteoclast formation in regenerating mammalian bone *J. Bone Min Res.* **17** 455–64

Fekete D M and Brockes J P 1987 A monoclonal antibody detects a difference in the cellular composition of developing and regenerating limbs of newts *Development* **99** 589–602

Ferretti P and Brockes J P 1988 Culture of newt cells from different tissues and their expression of a regeneration-associated antigen *J. Exp. Zool.* **247** 77–91

Filoni S, Bernardini S and Cannata S M 1991 The influence of denervation on grafted hindlimb regeneration of larval *Xenopus laevis J. Exp. Zool. Part A* **260** 210–9

Filoni S, Velloso C P, Bernardini S and Cannata S M 1995 Acquisition of nerve dependence for the formation of a regeneration blastema in amputated hindlimbs of *Xenopus laevis*: the role of limb innervation and that of limb differentiation *J. Exp. Zool.* **273** 327–41

French V, Bryant P J and Bryant S V 1976 Pattern regulation in epimorphic fields *Science* **193** 969–81

Frobisch N B, Bickelmann C and Witzmann F 2014 Early evolution of limb regeneration in tetrapods: evidence from a 300-million-year-old amphibian *Proc. R. Soc. Lond.* B **281** 20141550

Gardiner D M and Bryant S V 1996 Molecular mechanisms in the control of limb regeneration: the role of homeobox genes *Int. J. Dev. Biol.* **40** 797–805

Gardiner D M, Muneoka K and Bryant S V 1986 The migration of dermal cells during blastema formation in axolotls *Dev. Biol.* **118** 488–93

Gawriluk T R *et al* 2016 Comparative analysis of ear-hole closure identifies epimorphic regeneration as a discrete trait in mammals *Nat. Commun.* **7** 11164

Geng J, Gates P B, Kumar A and Brockes J P 2015 Identification of the orphan gene Prod 1 in basal and other salamander familes *EvoDevo* **6** 9

Geraudie J and Ferretti P 1998 Gene expression during amphibian limb regeneration *Int. Rev. Cytol.* **180** 1–50

Ghosh S *et al* 2008 Analysis of the expression and function of Wnt-5a and Wnt-5b in developing and regenerating axolotl limbs *Dev. Growth Differ.* **50** 289–97

Globus M, Vethamany-Globus S and Lee Y C I 1980 Effect of apical epidermal cap on mitotic cycle and cartilage differentiation in regeneration blastemata in the newt, *Notophthalmus viridescens Dev. Biol.* **75** 358–72

Godwin J W, Pinto A R and Rosenthal N A 2013 Macrophages are required for adult salamander limb regeneration *Proc. Natl Acad. Sci. USA* **110** 99415–20

Godwin J W and Rosenthal N A 2014 Scar-free wound healing and regeneration in amphibians: immunological influences on regenerative success *Differentiation* **87** 66–75

Goss R J 1970 Problems of antlerogenesis *Clin. Orthoped.* **69** 227–38

Goss R J 1992 The mechanism of antler casting in the fallow deer *J. Exp. Zool.* **264** 429–36

Goss R J and Holt R 1992 Epimorphic vs tissue regeneration in *Xenopus* forelimbs *J. Exp. Zool.* **261** 451–7

Guimond J C *et al* 2010 BMP-2 functions independently of Shh signaling and triggers cell condensation and apoptosis in regenerating axolotl limbs *BioMed. Central Dev. Biol.* **10** 15

Han M, Yang X, Farrington J E and Muneoka K 2003 Digit regeneration is regulated by Msx1 and BMP4 in fetal mice *Development* **130** 5123–32

Hayashi S *et al* 2015 Epigenetic modification maintains intrinsic limb-cell identity in *Xenopus* limb bud regeneration *Dev. Biol.* **406** 271–82

Heber-Katz E *et al* 2013 Cell cycle regulation and regeneration *New Perspectives in Regeneration. Current Topics in Microbiology and Immunobiology* ed E Heberkatz and D L Stocum (Berlin: Springer), pp 253–76

Ide H 2012 Bone pattern formation in mouse limbs after amputation at the forearm level *Dev. Dyn.* **241** 435–41

Imokawa Y, Simon A and Brockes J P 2004 A critical role for thrombin in vertebrate lens regeneration *Philos. Trans. R. Soc. Lond.* B **369** 765–76

Imokawa Y and Yoshizato K 1997 Expression of *Sonic hedgehog* gene in regenerating newt limb blastemas recapitulates that in developing limb buds *Proc. Natl Acad. Sci. USA* **94** 9159–64

Iten L E and Bryant S V 1975 The interaction between the blastema and stump in the establishment of the anterior–posterior and proximal–distal organization of the limb regenerate *Dev. Biol.* **44** 119–47

Izpisua-Belmonte J C, Ede D A, Tickle C and Duboule D 1992 The mis-expression of posterior Hox-4 genes in talpid (ta3) mutant wings correlates with the absence of anteroposterior polarity *Development* **114** 959–63

Jenkins L S, Duerstock B S and Borgens R B 1996 Reduction of the current of injury leaving the amputation inhibits limb regeneration in the red spotted newt *Dev. Biol.* **178** 251–62

Jhamb D *et al* 2011 Network based transcription factor analysis of regenerating axolotl limbs *BioMed. Central Bioinformat.* **12** 80

Johnston A P *et al* 2016 Dedifferentiated Schwann cell precursors secreting paracrine factors are required for regeneration of the mammalian digit tip *Cell Stem Cell* **19** 433–48

Kelly D J and Tassava R A 1973 Cell division and ribonucleic acid synthesis during the initiation of limb regeneration in larval axolotls (*Ambystoma mexicanum*) *J. Exp. Zool.* **185** 45–54

Kierdorf U, Kierdorf H and Stewart T 2007 Deer antler regeneration: cells, concepts, and controversies *J. Morphol.* **268** 726–38

Kierdorf U, Li C and Price J 2009 Improbable appendages: deer antler renewal as a unique case of mammalian regeneration *Semin. Cell Dev. Biol.* **20** 535–42

Kierdorf U and Kierdorf H 2011 Deer antler—a model of mammalian appendage regeneration: an extensive review *Gerontology* **57** 53–65

King M W *et al* 2003 Identification of genes expressed during *Xenopus laevis* limb regeneration by using subtractive hybridization *Dev. Dyn.* **226** 398–409

King M W, Neff A W and Mescher A L 2009 Proteomics analysis of regenerating amphibian limbs: changes during the onset of regeneration *Int. J. Dev. Biol.* **53** 955–69

Knapp D *et al* 2013 Comparatie transcriptional profiling of the axolotl limb identifies a tripartite regeneration-specific gene program *PLoS One* **8** e61352

Konieczna-Marczynska B and Skowron-Cendrzak A 1958 The effect of the augmented nerve supply on the regeneration in postmetamorphic *Xenopus laevis Folia Biol.* **6** 37–56

Kragl M *et al* 2009 Cells keep a memory of their tissue origin during axolotl limb regeneration *Nature* **460** 60–5

Kumar A, Velloso C, Imokawa Y and Brockes J P 2004 The regenerative plasticity of isolated urodele myofibers and its dependence on MX1 *PLOS Biol.* **2** e218

Kumar A and Brockes J P 2007 Molecular basis for the nerve dependence of limb regeneration in an adult vertebrate *Science* **318** 772–7

Kumar A *et al* 2011 The aneurogenic limb identifies developmental cell interactions underlying vertebrate limb regeneration *Proc. Natl Acad. Sci. USA* **108** 13588–93

Lehcozky J A, Robert B and Tabin C J 2011 Mouse digit tip regeneration is mediated by fate-restricted progenitor cells *Proc. Natl Acad. Sci. USA* **108** 20609–14

Leitingtung Y *et al* 2002 *Shh* and Gli3 are dispensable for limb skeleton formation but regulate digit number and identity *Nature* **418** 979–83

Levesque M *et al* 2007 Transforming growth factor:β signaling is essential for limb regeneration in axolotls *PLoS One* **2** e1277

Lheureux E 1975 Regeneration des membres irradies de *Pleurodeles waltlii* Michah. (Urodele). Influence des qualites et orientations des greffons non irradies Roux's Archiv *Dev. Biol.* **176** 303–27

Lheureux E 1977 Importance des associations de tissus du member dans le developpement des membres surnumeraires induits par deviation de nerf chez le Triton *Pleurodeles waltlii* Michah *J. Embryol. Exp. Morphol.* **38** 151–73

Li X, Mohan S, Gu W and Baylink D J 2001 *Mamm. Genome* **12** 52–9

Li C, Suttie J M and Clark D E 2005 Histological examination of antler regeneration in red deer (*Cervus elaphus*) *Anat. Rec.* **282A** 163–74

Li C, Yang F and Sheppard A 2009 Adult stem cells and mammalian epimorphic regeneration-insights from studying annual renewal of deer antlers *Curr. Stem Cell Res. Ther.* **4** 237–51

Lin G and Slack J M 2008 Requirement for Wnt and FGF signaling in *Xenopus* tadpole tail regeneration *Dev. Biol.* **316** 323–35

Looso M *et al* 2013 A *de novo* assembly of the newt transcriptome combined with proteomic validation identifies new protein families expressed during tissue regeneration *Genome Biol.* **14** R16

Loyd R M and Tassava R A 1980 DNA synthesis and mitosis in adult newt limbs following amputation and insertion into the body cavity *J. Exp. Zool.* **214** 61–9

Lundkvist J and Lendhal U 2001 Notch and the birth of glial cells *Trends Neurosci.* **9** 492–4

MacCabe A B, Gasseling M T and Saunders J W Jr 1973 Spatiotemporal distribution of mechanisms that control outgrowth and and anterioposterior polarization of the limb bud in the chick embryo *Mech. Aging Dev.* **2** 1–12

Maden M 1977 The regeneration of positional information in the amphibian limb *J. Theor. Biol.* **69** 735–53

Maden M 1980a Structure of supernumerary limbs *Nature* **287** 803–5

Maden M 1980b Intercalary regeneration in the amphibian limb and the rule of distal trans-formation *J. Embryol. Exp. Morphol.* **56** 201–9

Maden M 1983 A test of the predictions of the boundary model regarding supernumerary limb structure *J. Embryol. Exp. Morphol.* **76** 147–55

Maden M 1998 Retinoids as endogenous components of the regenerating limb and tail *Wound Rep. Reg.* **6** 358–65

Maden M, Avila D, Roy M and Seifert A W 2015 Tissue-specific reactions to positional discontinuities in the regenerating axolotl limb *Regeneration* **2** 137–47

Makanae A, Mitogawa K and Satoh A 2014 Co-operative Bmp-and Fgf-signaling inputs convert skin wound healing to limb formation in urodele amphibians *Dev. Biol.* **396** 57–66

Masake H and Ide H 2007 Regeneration potency of mouse limbs *Dev. Growth Differ.* **49** 89–98

Matsuda H, Yokoyama H, Endo T, Tamura K and Ide H 2001 An epidermal signal regulates *Lmx*-1 expression and dorsal-ventral pattern during *Xenopus* limb regeneration *Dev. Biol.* **229** 351–62

McCusker C D, Diaz-Castillo C, Sosnik J and Gardiner D M 2016 Cartilage and bone cells do not participate in skeletal regeneration in *Ambystoma mexicanum* limbs *Dev. Biol.* **416** 26–33

McGann C J, Odelberg S J and Keating M T 2001 Mammalian myotube dedifferentiation induced by newt regeneration extract *Proc. Natl Acad. Sci. USA* **98** 13699–704

Meinhardt H 1983a A boundary model for pattern formation in vertebrate limbs *J. Embryol. Exp. Morphol.* **76** 115–37

Meinhardt H 1983b A bootstrap model for the proximodistal pattern formation in vertebrate limbs *J. Embryol. Exp. Morphol.* **76** 139–46

Mercader N, Tanaka E and Torres M 2005 Proximodistal identity during vertebrate limb regeneration is regulated by Meis homeodomain proteins *Development* **132** 4131–42

Mercer S E *et al* 2012 Multi-tissue microarray analysis identifies a molecular signature of regeneration *PLoS One* **7** e52375

Mescher A L 1976 Effects on adult newt limb regeneration of partial and complete skin flaps over the amputation surface *J. Exp. Zool.* **195** 117–28

Mescher A L 1996 The cellular basis of limn regeneration in urodeles *Int. J. Dev. Biol.* **40** 785–95

Mescher A L 2017 Macrophages and fibroblasts during inflammation and tissue repair in models of organ regeneration *Regeneration* **4** 39–53

Mescher A L and Tassava R A 1976 Denervation effects on DNA replication and mitosis during the initiation of limb regeneration in adult newts *Dev. Biol.* **44** 187–97

Mescher A L, White G W and Brokaw J J 2000 Apoptosis in regenerating and denervated nonregenerating urodele forelimbs *Wound Rep. Reg.* **8** 110–6

Mescher A L, Neff A W and King M W 2017 Inflammation and immunity in organ regeneration *Dev. Comput. Immunol.* **31** 383–93

Monaghan J R and Maden M 2012 Visualization of retinoic acid signaling in transgenic axolotls during limb development and regeneration *Dev. Biol.* **368** 63–75

Morais da Silva S M, Gates P B and Brockes J P 2002 The newt ortholog of CD59 is implicated in proximodistal identity during amphibian limb regeneration *Dev. Cell* **3** 547–55

Mosteiro L *et al* 2016 Tissue damage and senescence provide critical sihnals for cellular reprogramming *in vivo Science* **354** 6315

Mount J G *et al* 2006 Evidence that the canonical Wnt signaling pathway regulates deer antler regeneration *Dev. Dyn.* **235** 1390–9

Mullen L M, Bryant S V, Torok M A, Blumberg B and Gardiner D M 1996 Nerve dependency of regeneration: the role of *Distal-less* and FGF signaling in amphibian limb regeneration *Development* **122** 3487–97

Muneoka K, Fox W F and Bryant S V 1986 Cellular contribution from dermis and cartilage to the regenerating limb blastema in axolotls *Dev. Biol.* **116** 256–60

Nacu E *et al* 2013 Connective tissue cells, but not muscle cells, are involved in establishing the proximo-distal outcome of limb regeneration in the axolotl *Development* **140** 513–8

Nacu E *et al* 2016 FGF8 and SHH substitute for anterior–posterior tissue interactions to induce limb regeneration *Nature* **533** 407–10

Namenwirth M 1974 The inheritance of cell differentiation during limb regeneration in the axolotl *Dev. Biol.* **41** 42–56

Nardi J B and Stocum D L 1983 Surface properties of regenerating limb cells; evidence for gradation along the proximodistal axis *Differentiation* **25** 27–31

Niazi I A 1996 Background to work on retinoids and amphibian limb regeneration: Studies on anuran tadpoles—a retrospect *J. Biosci.* **21** 273–97

Nye H L D, Cameron J A, Chernoff E G and Stocum D L 2003 Regeneration of the urodele limb: a review *Dev. Dyn.* **226** 280–94

Odelberg S J, Kollhof A and Keating M 2001 Dedifferentiation of mammalian myotubes induced by msx-1 *Cell* **103** 1099–109

Ohgo S *et al* 2010 Analysis of hoxa11 and hoxa13 expression during patternless limb regeneration in *Xenopus Dev. Biol.* **338** 148–57

Pajcini K V *et al* 2010 Transient inactivation of *Rb* and *ARF* yields regenerative cells from postmitotic mammalian muscle *Cell Stem Cell* **7** 198–213

Pearl E J, Barker D, Day R C and Beck C W 2008 Identification of genes associated with regenerative success of *Xenopus laevis* hindlimbs *BMC Dev. Biol.* **8** 66

Pescitelli M J and Stocum D L 1980 The origin of skeletal structures during intercalary regeneration of larval *Ambystoma* limbs *Dev. Biol.* **79** 255–75

Pescitelli M J and Stocum D L 1981 Non-segmental organization of positional information in regenerating *Ambystoma* limbs *Dev. Biol.* **82** 69–85

Polezhaev L V 1937 Uber die Determination des Regenerats einer Extremitat bein Axolotl *C. R. Doklady Acad. Sci. USSR* **15** 387–90

Price J S *et al* 1996 Chondrogenesis in the regenerating antler tip in red deer: Expression of collagen types I, IIA, IIB, and X demonstrated by *in situ* nucleic acid hybridization and immunocytochemistry *Dev. Dyn.* **205** 332–47

Rao N *et al* 2009 Proteomic analysis of blastema formation in regenerating axolotl limbs *BMC Biol.* **7** 83

Rao N *et al* 2014 Proteomic analysis of fibroblastema formation in regenerating hind limbs of *Xenopus laevis* froglets and comparison to axolotl *BMC Dev. Biol.* **14** 32

Roy S, Gardiner D M and Bryant S V 2000 Vaccinia as a tool for functional analysis in regenerating limbs: ectopic expression of *Shh. Dev. Biol.* **218** 199–205

Reines B, Cheng L I and Matzinger P 2009 Unexpected regeneration in middle-aged mice *Rejuvenation Res.* **12** 45–51

Riddle R D, Johnson R L, Laufer E and Tabin C 1993 Sonic hedgehog mediates the polarizing activity of the ZPA *Cell* **75** 1401–16

Rinkevich Y, Lindau P, Ueno H, Lonngaker M T and Weissman I L 2013 Germ-layer and lineage-restricted stem/progenitors regenerate the mouse digit tip *Nature* **476** 409–13

Roensch K, Tazaki A, Chara O and Tanaka E M 2013 Progressive specification rather than intercalation of segments during limb regeneration *Science* **342** 1375–9

Rolf H J *et al* 2008 *PLoS One* **3** e2064

Rosania G R *et al* 2000 Myoseverin, a microtubule-binding molecule with novel cellular effects *Nat. Biotechnol.* **18** 304–8

Rosello-Diez A and Torres M 2014 Regulative patterning in limb bud transplants is induced by distalizing activity of apical ectodermal ridge signals on host limb cells *Dev. Dyn.* **240** 1203–11

Roy S and Gardiner D M 2000 Cyclopamine induces digit loss in regenerating axolotl limbs *J. Exp. Zool.* **293** 186–90

Saiz-Lopez P *et al* 2015 An intrinsic timer specifies distal structures of the vertebrate limb *Nat. Commun.* **6** 8108

Sandoval-Guzman T *et al* 2014 Fundamental differences in dedifferentiation and stem cell recruitment during skeletal muscle regeneration in two salamander species *Cell Stem Cell* **14** 174–87

Santosh N *et al* 2011 Matrix metalloproteinase expression during blastema formation in regeneration-competent versus regeneration-deficient amphibian limbs *Dev. Dyn.* **240** 1127–41

Satoh A, Ide H and Tamura K 2005a Muscle formation in regenerating *Xenopus* froglet limb *Dev. Dyn.* **233** 337–46

Satoh A, Suzuki M and Amano T *et al* 2005b Joint development in *Xenopus laevis* and induction of segmentation in regenerating froglet limb (spike) *Dev. Dyn.* **233** 1444–53

Satoh A, Cummings G M C, Bryant S V and Gardiner D M 2010 Neurotrophic regulation of fibroblast dedifferentiation during limb skeletal regeneration in the axolotl (*Ambystoma mexicanum*) *Dev. Biol.* **337** 444–57

Satoh A, Makanae A, Nishimoto Y and Mitogawa K 2016 FGF and BMP derived from dorsal root ganglia regulate blastema induction in *Ambystoma mexicanum Dev. Biol.* **417** 114–25

Schmidt A J 1968 *Cellular Biology of Vertebrate Regeneration and Repair* (Chicago: University of Chicago Press)

Seifert A W and Muneoka K 2017 The blastema and epimorphic regeneration in mammals *Dev. Biol.* **433** 190–9

Seo M-S *et al* 2014 Isolation and characterization of antler-derived multipotent stem cells *Cell Transpl.* **23** 831–43

Sessions S K and Bryant S V 1988 Evidence that regenerative ability is an intrinsic property of limb cells in *Xenopus J. Exp. Zool.* **247** 39–44

Shimizu-Nishikawa K S, Tsuji S and Yoshizato K 2001 Identification and characterization of newt RAD (ras associated with diabetes), a gene specifically expressed in regenerating limb muscle *Dev. Dyn.* **220** 74–86

Simkin J *et al* 2015 The mammalian blastema: regeneration at our fingertips *Regeneration* **2** 93–105

Simkin J, Gawriluk T R, Gensel J C and Seifert A W 2017 Macrophages are necessary for epimorphic regeneration in African spiny mice *eLife* **6** e24623

Simon H-G *et al* 1997 A novel family of T-box genes in urodele amphibian limb development and regeneration: candidate genes involved in vertebrate forelimb/hindlimb patterning *Development* **124** 1355–66

Singer M 1952 The influence of the nerve in regeneration of the amphibian extremity *Q. Rev. Biol.* **27** 169–200

Singer M 1978 On the nature of the neurotrophic phenomenon in urodele limb regeneration *Amer. Zoologist* **18** 829–41

Singer M and Craven L 1948 The growth and morphogenesis of the regenerating forelimb of adult *Triturus* following denervation at various stages of development *J. Exp. Zool.* **108** 279–308

Smith M J and Globus M 1989 Multiple interactions in juxtaposed monolayer of amphibian neuronal, epidermal, and mesodermal limb blastema cells *In Vitro Cell Dev. Biol.* **25** 849–56

Stark R J and Searls R L 1973 A description of chick wing development and a model of limb morphogenesis *Dev. Biol.* **33** 138–53

Stevenson T J, Vinarsky V, Atkinson D L, Keating M T and Odelberg S J 2006 Tissue inhibitor of metalloproteinase 1 regulates matrix metalloproteinase activity during newt limb regeneration *Dev. Dyn.* **235** 606–16

Stewart R *et al* 2013 Comparative RNA-seq analysis in the unsequenced axolotl: the oncogene burst highlights early gene expression in the blastema *PLOS Comput. Biol.* **9** e1002936

Stocum D L 1975 Regulation after proximal or distal transposition of limb regeneration blastemas and determination of the proximal boundary of the regenerate *Dev. Biol.* **45** 112–36

Stocum D L 1980 Autonomous development of reciprocally exchanged regeneration blastemas of normal forelimbs and symmetrical hindlimbs *J. Exp. Zool.* **212** 361–71

Stocum D L 1981 Distal transformation in regenerating double anterior axolotl limbs *J. Embryol. Exp. Morphol.* **65** 3–18

Stocum D L 1984 The urodele limb regeneration blastema: determination and organization of the morphogenetic field *Differentiation* **27** 13–28

Stocum D L 1995 *Wound Repair, Regeneration and Artificial Tissues* (Austin, TX: RG Landes)

Stocum D L 2011 The role of peripheral nerves in urodele limb regeneration *Eur. J. Neurosci.* **34** 908–16

Stocum D 2012 Regeneration of appendages Regenerative Biology and *Medicine* 2nd edn (New York: Academic) chapter 8

Stocum D L 2017 Mechanisms of urodele limb regeneration *Regeneration* **4** 159–200

Stocum D L and Dearlove G E 1972 Epidermal-mesodermal interaction during morphogenesis of the limb regeneration blastema in larval salamanders *J. Exp. Zool.* **181** 49–62

Stocum D L and Melton D A 1977 Self-organizational capacity of distally transplanted limb regeneration blastemas in larval salamanders *J. Exp. Zool.* **201** 451–62

Stocum D L and Cameron J A 2011 Looking proximally and distally: 100 years of limb regeneration and beyond *Dev. Dyn.* **240** 943–68

Stoick-Cooper C L *et al* 2007 Distinct Wnt signaling pathways have opposing roles in appendage regeneration *Development* **134** 479–89

Sugiura T *et al* 2016 MARCKS-like protein is an initiating molecule in axolotl appendage regeneration *Nature* **531** 237–40

Summerbell D, Lewis J H and Wolpert L 1973 Positional information in the chick limb bud *Nature* **244** 492–496

Takahashi K *et al* 2007 Induction of pluripotent stem cells from adult human fibroblasts by defined factors *Cell* **132** 861–72

Takeo M *et al* 2013 Wnt activation in nail epithelium couples nail growth to digit regeneration *Nature* **499** 228–32

Tamura K, Ohgo S and Yokoyama H 2009 Limb blastema cell: a stem cell for morphological regeneration *Dev. Growth Differ.* **52** 89–99

Tanaka E M, Gann F, Gates P B and Brocles J P 1997 Newt myotubes re-enter the cell cycle by phosphorylation of the retinoblastoma protein *J. Cell Biol.* **136** 155–65

Tanaka H V *et al* 2016 A developmentally regulated switch from stem cells to dedifferentiation for limb muscle regeneration in newts *Nat. Commun.* **7** 11069

Tarin D and Sturdee A P 1971 Early limb development of *Xenopus laevis J. Embryol. Exp. Morphol.* **26** 169–79

Tassava R A and Mescher A L 1975 The roles of injury, nerves and the wound epidermis during the initiation of amphibian limb regeneration *Differentiation* **4** 23–4

Thornton C S 1968 Amphibian limb regeneration *Adv. Morph.* **7** 205–49

Thornton C S and Thornton M T 1970 Recuperation of regeneration in denervated limbs of *Ambystoma* larvae *J. Exp. Zool.* **173** 293–301

Torok M A, Gardiner D M, Shubin N H and Bryant S V 1998 Expression of *HoxD* genes in developing and regenerating axolotl limbs *Dev. Biol.* **200** 225–33

Tschumi P A 1957 The growth of hindlimb bud of *Xenopus laevis* and its dependence upon the epidermis *J. Anat.* **91** 149–73

Tseng A S and Levin M 2013 Cracking the bioelectric code *Commun. Integr. Biol.* **6** e22595

Velloso C P, Simon A and Brockes J P 2001 Mammalian postmitotic nuclei reenter the cell cycle after serum stimulation in newt/mouse hybrid myotubes *Curr. Biol.* **11** 855–8

Vinarsky V *et al* 2005 Normal newt limb regeneration requires matrix metalloproteinase function *Dev. Biol.* **279** 86–98

Voss S R *et al* 2015 Gene expression during the first 28 days of axolotl limb regeneration: experimental design and global analysis of gene expression *Regeneration* **2** 120–36

Wada N 2011 Spatiotemporal changes in cell adhesiveness during vertebrate limb morphogenesis *Dev. Dyn.* **240** 969–78

Wada N *et al* 1998 Glycosylphosphatidylinositol-anchored cell surface proteins regulate position-specific cell affinity in the limb bud *Dev. Biol.* **202** 244–52

Wallace H 1980 Regeneration of reversed aneurogenic arms of the axolotl *J. Embryol. Exp. Morphol.* **56** 309–17

Wang L, Marchionni M A and Tassava R A 2000 Cloning and neuronal expression of a type III newt neuregulin and rescue of denervated nerve-dependent newt limb blastemas by rhGGF2 *J. Neurobiol.* **43** 150–8

Wang Y *et al* 2011 Gecko CD59 is implicated in proximodistal identity during tail regeneration *PLoS One* **6** e17878

Williams-Boyce P K and Daniel J C Jr 1980 Regeneration of rabbit ear tissue *J. Exp. Zool.* **212** 243–53

Wolfe A D, Nye H L D and Cameron J 2000 Extent of ossification at the amputation plane is correlated with the decline of blastema formation and regeneration in *Xenopus laevis* hindlmbs *Dev. Dyn.* **218** 681–97

Wu Y *et al* 2013 Connective tissue fibroblast properties are position-dependent during mouse digit tip regeneration *PLoS One* **8** e54764

Yajima H *et al* 1999 Role of N-cadherin in the sorting-out of mesenchymal cells and in the positional identity along the proximodistal axis of the chick limb bud *Dev. Dyn.* **216** 274–84

Yakushiji N *et al* 2007 Correlation between Shh expression and DNA methylation status of the limb-specific Shh enhancer region during limb regeneration in amphbians *Dev. Biol.* **312** 171–82

Yakushiji-Kaminatsui N, Yokoyama H and Tamura K 2009 Repatterning in amphibian limb regeneration: a model for study of genetic and epigenetic control of organ regeneration *Sem. Cell Dev. Biol.* **20** 565–74

Yntema C L 1959a Regeneration of sparsely innervated and aneurogenic forelimbs of *Ambystoma* larvae *J. Exp. Zool.* **140** 101–23

Yntema C L 1959b Blastema formation in sparsely innervated and aneurogenic forelimbs in *Amblystoma* larvae *J. Exp. Zool.* **142** 423–40

Yokoyama H *et al* 2000 Mesenchyme with *fgf*10 expression is responsible for regenerative capacity in *Xenopus* limb buds *Dev. Biol.* **219** 18–29

Yokoyama H, Ide H and Tamura K 2001 FGF-10 stimulates limb regeneration ability in *Xenopus laevis Dev. Biol.* **233** 72–9

Yokoyuchi Y, Sasaki H and Kuroiwa A 1991 Homeobox gene expression correlated with the bifurcation process of limb cartilage development *Nature* **353** 443–5

Yu J *et al* 2007 Induced pluripotent stem cells derived from human somatic cells *Science* **318** 1917–20

Yu L *et al* 2010 BMP signaling induces digit regeneration in neonatal mice *Development* **137** 551–9

Yu L *et al* 2012 BMP2 induces segment-specific skeletal regeneration from digit and limb amputations by establishing a new endochondral ossification center *Dev. Biol.* **372** 263–73

Yun M H, Gates P B and Brockes J P 2013 Regulation of p53 is critical for vertebrate limb regeneration *Proc. Natl Acad. Sci. USA* **110** 17392–7

Yun M H, Gates P B and Brockes J P 2014 Sustained ERK activation underlies reprogramming in regeneration-competent salamander cells and distinguishes them from their mammalian counterparts *Stem Cell Rep.* **3** 15–23

Yun M H, Davaapil H and Brockes J P 2015 Recurrent turnover of senescent cells during regeneration of a complex structure *eLife* **4** e05505

Zhang Y *et al* 2015 Drug-induced regeneration in adult mice *Sci. Transl. Med.* **7** 290–92

Zhao W and Neufeld D A 1995 Bone regrowth in young mice stimulated by nail organ *J. Exp. Zool.* **271** 155–9

Zielins E R, Ransom R C, Leavitt T E and Longaker M T 2016 The role of stem cells in limb regeneration *Organogenesis* **12** 16–27

Part II

Regenerative medicine

Chapter 6

Strategies of regenerative medicine

Summary

Regenerative medicine as a science emerged from ancient treatments of wounds. It has developed along two paths, one leading to prosthetic and biomimetic devices and organ transplantation, the other to the focus of this chapter, the regeneration of tissues by pharmaceutical induction, cell transplants and implants of biomimetic tissues. The pharmaceutical induction of regeneration involves the delivery of regeneration-promoting agents topically, by injection, or in a biodegradable scaffold, the use of biodegradable scaffolds alone to stimulate regeneration, as well pharmacogenetic approaches that transfect cells with regeneration-promoting gene constructs. The regenerative capacity of tissues declines with age due to accumulating genetic mutations affecting their stem cells, but also due to increasing levels in the circulation of factors inhibitory to regeneration. Identification of these factors could potentially lead to pharmaceutical therapies to neutralize them. Cells for transplant can be adult stem cells, or derivatives of ESCs or iPSCs. Adult stem cells can be difficult to harvest and expand, and thus derivatives of iPSCs may be a better alternative to adult stem cells for transplantation. The technology to produce iPSCs has been made possible by decades of research on embryonic stem cells, which has revealed the factors required to reprogram somatic cells to iPSCs. Increasing attention is being paid to biomimetic tissue implants, which are constructed either by seeding a scaffold with cells, or by scaffold-free methods. Major challenges to biomimetic tissue construction are the design of scaffolds that can mimic the normal cell niche to promote differentiation, and vascularization of large-volume constructs.

6.1 Historical notes

Regenerative medicine is the science of restoring the structure of a tissue or organ through interventions based on our understanding of regenerative biology. Regenerative medicine has been preceded and paralleled by prosthetic devices, and more recently by tissue and organ transplantation. It has ancient roots in the treatments of wounds suffered as a result of conflict and accidents. Cleansing and debridement of wounds was a common practice in all early civilizations, and the ancient Chinese and Egyptians used honey and wine as antiseptics (Fu *et al* 2001). Surgical interventions were central to the medicine of Sumerian, Egyptian, Chinese and Indian civilizations. Trephination to relieve intracranial pressure is a technique dating to the Neolithic and was brought to a high art by the Incas in Peru (Majno 1975, Brown 1992, Falabella 1998). The father of plastic surgery, Sushruta, who lived in India sometime between 1200 and 600 BCE developed methods of autogeneic skin transplantation to reconstruct severed noses and ears (Majno 1975). The first prostheses (for appendages) were developed by the Egyptians and have continued to evolve to the present day (Thurston 2007).

Greek and Roman physicians who made significant contributions to the treatment of wounds were Hippocrates (460–370 BCE), Celsus (25 BCE–50 AD), and particularly Galen (130–201 AD), who had extensive experience treating the wounds of gladiators (Brown 1992). Following collapse of the Western Roman Empire in the fifth century AD, Galen's teachings became the dominant guides for medical practice until the end of the medieval period. The Renaissance ushered in a new understanding of the anatomy and physiology of the human body, reflected in the book *Wounds and Fractures* published in 1363 by the surgeon Guy de Chruliac (1300–70), a text that detailed many kinds of wounds and how to treat them. At the intersection of the 16th and 17th centuries, Wilhelm Fabry (1560–1634) described nearly 70 topically applied formulations for the treatment of wounds, many of which have been re-examined and found to have true therapeutic value (Kirkpatrick *et al* 1996).

During the 17th and 18th centuries, surgery without anesthesia became a last-ditch intervention for otherwise untreatable conditions, although one needed a lot of luck to survive the operations. The introduction of aseptic technique and ether anesthesia in the 19th century increased the types of surgical interventions that could be made on the human body (Allan 1977, Brown 1992). The development of tissue and organ culture by Ross Harrison and Alexis Carrel in the first third of the 20th century, the discovery of stem cells, the crafting of biomimetic scaffolds, and understanding of the immune system, have enabled the translation of discoveries about mechanisms of natural regeneration to cellular and molecular regenerative therapies (Maienschein 2011, Calabrese 2013, Kaul and Ventikos 2015).

There are currently five major strategies for restoring tissue, organ and appendage structure and/or function. These are (1) the use of bionic/prosthetic devices; (2) organ and tissue transplants, (3) cell transplants, (4) implants of biomimetic tissues or organs, and (5) the pharmaceutical inhibition of scarring and/or induction of new tissue formation by cells at the site of injury. Only the last three are truly regenerative therapies (figure 6.1). Thus, following a brief review of bionic devices and organ transplants, this chapter will focus on cell transplants, biomimetic tissues and pharmaceutical induction of regeneration. The distinction between tissue regeneration by cell transplant and the pharmaceutical induction of regeneration *in situ* is not always clear-cut because transplanted cells can act as drug factories to secrete paracrine factors that improve host cell survival, modulate scarring and

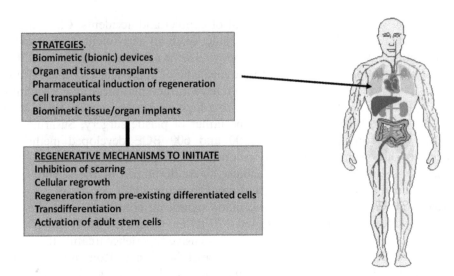

STRATEGIES.
Biomimetic (bionic) devices
Organ and tissue transplants
Pharmaceutical induction of regeneration
Cell transplants
Biomimetic tissue/organ implants

REGENERATIVE MECHANISMS TO INITIATE
Inhibition of scarring
Cellular regrowth
Regeneration from pre-existing differentiated cells
Transdifferentiation
Activation of adult stem cells

Figure 6.1. Strategies of regenerative medicine to initiate regenerative mechanisms. These strategies are augmented by organ and tissue transplants and biomimetic devices, which are ready-made replacements for failing organs.

chemically induce resident cells to regenerate new tissue. Not all of these strategies are applicable to every clinical situation. Which strategy is used depends on how well developed the strategy is, the tissue or organ type, and the nature and severity of tissue loss.

6.2 Biomimetic devices and organ transplants

Currently, the primary clinical solutions to tissue and organ loss are bionic/prosthetic devices and tissue and organ transplants. Examples of the former in clinical use are hearing aids, cochlear implants, insulin pumps, joint replacements, appendage prostheses, cardiac valves and left ventricular assist pumps. The DEKA or 'Luke' prosthetic arm (after Luke Skywalker of 'Star Wars' fame) has electrical circuits that potentially can be interfaced with brachial nerve endings to effect direct control of the arm from the motor cortex of the brain (see www.dekaresearch.com/deka_arm.shtml). Other devices still in various stages of development are total artificial hearts (TAHs) and retinas. The primary limitation of the TAH, of which a number have been designed, is the size, mobility and duration of the power source required. The most difficult organs to mimic are those that perform complex biochemical functions, such as the liver and pancreas. Given the development of new materials and miniaturized power sources, bionic devices will continue evolving to more closely mimic the original in size, function and durability.

Organ, tissue and cell transplants are a product of the second half of the 20th century. There are four types of transplants: autogeneic (donor and recipient are the same individual); isogeneic (donor and recipient are identical twins); allogeneic (donor and recipient are different individuals of the same species); and xenogeneic (donor and recipient are different species). Blood transfusion was made possible by

Karl Landsteiner's (1868–1943) discovery of the ABO blood antigens in 1901. Success in transplanting solid organs was not achieved, however, until Joseph Murray (1919–2012) made the first successful isogenic kidney transplant between identical twin brothers, Richard (recipient) and Ronald (donor) Herrick in 1954 at the Brigham and Women's Hospital in Boston. The development, in the 1980s, of effective anti-rejection drugs, such as cyclosporin A, made possible the transplant of allogeneic solid organs. Today, most organs can be transplanted with consistently high success as measured by an average survival rate for all grafts of around 50% at five years, and 40% at 10 years (see Unadjusted Graft and Patient Survival, 2009 Health Resources and Services Administration OPTN/SRTR Annual Report). The most spectacular transplants have been partial and total face transplants. Because the face is the primary feature by which human beings send and receive social signals, the success of face transplants has vastly improved both the physical and psychological integrity of their recipients (Smeets *et al* 2014).

Donor shortage and immunorejection remain major obstacles to organ transplantation. The incidence of immunological tolerance to transplanted organs is low (Orlando *et al* 2013). Trials to find ways of creating tolerance to allografts without immunosuppression are ongoing under the auspices of the Immune Tolerance Network (http://www.immunetolerance/org/). Promising findings are that recipients with lymphocyte chimerism (carrying lymphocytes derived from the donor) can render patients tolerant to the transplant (Alexander *et al* 2008, Scandling *et al* 2008), and particularly that regulatory T cells (Tregs) can promote bone marrow chimerism and prevent acute and chronic allograft rejection in mice (Joffre *et al* 2008).

One way to alleviate the donor shortage problem might be xenotransplantation (Esker *et al* 2015). Pig organs are anatomically and functionally similar to human organs, and pigs are therefore considered potential donors. Pig organs, however, are hyper-rejected by new world primates and humans because of preformed antibodies to $\alpha 1,3$-galactose residing on the surface of pig blood vessel endothelial cells, and require heavy immunosuppression to survive for any appreciable length of time. A kidney from an $\alpha 1,3$-galactose knockout pig grafted to a baboon, combined with an immunosuppressive regimen and anti-inflammatory agents, survived for 136 days, the longest survival reported to date (Iwase and Kobayashi 2015). Much more needs to be done, however, before we understand how to attain long-term survival of xenotransplants. Another solution to donor shortage may be the construction of biomimetic organs or organoids for transplant using autogeneic cells (Orlando *et al* 2013, see below). Such transplants would require neither immunosuppression nor induction of recipient immune tolerance.

6.3 Pharmaceutical induction of regeneration *in situ*

The 'holy grail' of regenerative medicine is to use molecular cocktails, gene therapy, or natural or artificial ECM scaffolds, singly or in combination, to inhibit scar formation and stimulate regeneration by healthy cells local to an injury (figure 6.2). These agents would be designed to initiate regeneration by one or more of the

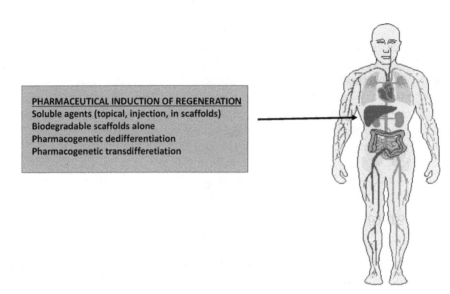

PHARMACEUTICAL INDUCTION OF REGENERATION
Soluble agents (topical, injection, in scaffolds)
Biodegradable scaffolds alone
Pharmacogenetic dedifferentiation
Pharmacogenetic transdifferetiation

Figure 6.2. Tactics for the pharmaceutical induction of regeneration.

regenerative mechanisms discussed in chapter 1: cellular re-growth, reproduction of pre-existing cells (by compensatory hyperplasia, EMT–MET, or dedifferentiation/redifferentiation), activation of adult stem cells, or transdifferentiation. The advantage of this *in situ* approach is that it eliminates problems associated with the logistics of expanding cell populations in culture for transplant, immunorejection and bioethical issues in one stroke and would be relatively low-cost compared to other strategies.

Soluble agents can be delivered topically, by injection, or in biodegradable scaffolds that act as delivery vehicles. For example, Akinc *et al* (2008) developed a combinatorial library of lipid-like materials to deliver therapeutic RNAi to cells. Nanoscale 'magic bullets' are being engineered to deliver drugs to cellular targets, such as cancer cells, that express specific sets of biomarkers on their surface (Zhu *et al* 2014). Biodegradable scaffolds by themselves can be used to induce regeneration or suppress scar formation, a good example being Oasis, a processed version of pig small intestine submucosa (SIS) that absorbs bioactive molecules and is used to promote the healing of chronic wounds (Nihsen *et al* 2007).

Pharmaceutical and pharmacogenetic approaches have also been used to dedifferentiate or transdifferentiate cells *in vitro* or *in vivo* to effect regeneration (figure 6.3). Cultured mouse C2C12 myofibers transfected with the *msx*-1 gene or exposed to protein extract from regenerating newt limbs were induced to dedifferentiate into cells with developmental potential similar to MSCs (McGann *et al* 2001, Odelberg 2002). Small molecules such as reversine induce dedifferentiation of fibroblasts *in vitro* (Ding and Schultz 2004). Hakelien (2002) reported the transdifferentiation *in vitro* of fibroblasts to T-lymphocytes by exposing them to nuclear and cytoplasmic extracts of human T-lymphocytes. Most spectacularly, transdifferentiation to cell types that need to be regenerated has been induced *in vivo* by gene transfection

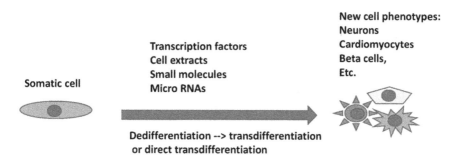

Figure 6.3. Use of transcription factors, cell extracts, small molecules or microRNAs to produce a range of cell phenotypes by indirect or direct transdifferentiation.

(Zhou and Melton 2008, Vierbuchen *et al* 2010, Ieda *et al* 2010, Sekiya and Suzuki 2011, Abad *et al* 2013). Such results suggest that it may someday be feasible to use genetic constructs, cell extracts, molecules isolated from cell extracts, or small synthetic molecules to activate regeneration *in situ*. A major problem in need of solution is how to precisely deliver these agents in sufficient volume to tissues *in vivo*.

6.4 Cell transplants

Cells can be transplanted either as suspensions or aggregates (figure 6.4). This strategy depends on two things: (1) having cells in numbers sufficient to replace damaged tissue and (2) an injury niche that provides appropriate biochemical and biophysical signals for the transplanted cells to differentiate and integrate with the surrounding tissue. Cells for transplant can be transplanted cells can be differentiated cells, adult stem cells, fetal cells embryonic stem cells (ESCs), or induced pluripotent stem cells (iPSCs). They can be autogeneic or allogeneic, genetically modified or not, fresh or expanded in culture (Stocum and Zupanc 2008, for a review). The ideal cell for transplant or bioartificial tissue construction would be (1) easy to harvest; (2) easy to expand to large numbers; (3) pluripotent, i.e. be able to differentiate into any of the more than 200 cell types of the human body; and (4) rejection-proof. Only derivatives of embryonic stem cells (ESCs) and induced pluripotent stem cells (iPSCs) meet the first three criteria, and only iPSCs meet all four, because they are donor somatic cells reprogrammed to an ESC state.

6.4.1 Adult stem cells

Currently, blood transfusion, bone marrow stem cell transplants for hematopoietic diseases, articular chondrocyte transplants, and transplants of epidermal stem cells and keratinocytes to resurface excisional wounds and burns are all in current clinical use. Success with other types of cell transplants has been more elusive, primarily due to difficulties in harvesting and growing the cells and delivering them to enough tissue volume to be effective, lack of niche factors to support survival and/or differentiation, and bioethical concerns (embryonic and fetal stem cells).

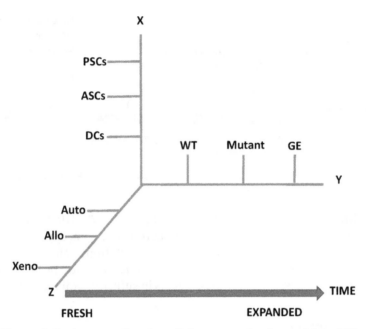

Figure 6.4. Diagram indicating ways of putting cell features together for transplant. PSC = pluripotent stem cells, ASCs = adult stem cells, DCs = differentiated cells, WT = wild type, GE = genetically engineered. Auto = autogeneic; Allo = allogeneic; Xeno = xenogeneic. Cells can be freshly isolated or expanded over time in culture.

6.4.1.1 Plasticity of adult stem cells

There have been many reports that adult stem cells, when transplanted into foreign injury niches, could transdifferentiate into out-of-lineage cell types. Several assays are used to test adult stem cell plasticity. Labeled cells can be exposed to foreign niche signals by injecting them into *scid* or lethally irradiated mice (bone marrow reconstitution assay), into mechanically injured tissues or organs, or into early embryos (chimeric embryo assay). Labeled cells are also cultured with our-of-lineage cell types *in vitro*, exposing them to medium conditioned by other cell types, or exposing them to sets of defined factors that specify non-homologous differentiation pathways. The adoption of out-of-lineage phenotypes by labeled test cells would be taken as evidence that the cells had undergone transdifferentiation.

Bone marrow cells in particular were reported able to transdifferentiate into a wide variety of out-of-lineage phenotypes, raising hopes that they could serve as a universal donor for regeneration. HSCs and MSCs each normally give rise to several cell types constituting the blood lineages and connective tissue and skeletal lineages, respectively (Fuchs and Segre 2000, Weissman 2000), but many studies have reported transdifferentiation of these cells to neural, myocardial, lung, pancreatic and other cells (Stocum and Zupanc 2008, for a review). Other studies, however, have failed to confirm these results (Laflamme and Murry 2005, Kanazawa and Verma 2003, Wang *et al* 2003, Menthena *et al* 2004, Lechner *et al* 2004). Other adult stem cells such as satellite cells, adipose-derived stem cells (ADSCs), neural stem cells, and rat liver stem cells were also reported to become cardiomyocytes, β-cells of the pancreas and other cell types. Thymic epithelial stem cells (TECs) were reported to transdifferentiate into epidermal structures (Bonfanti *et al* 2010). TECs express a wide variety of epidermal-type genes, suggesting that they might be primed to transdifferentiate when exposed to skin niche factors.

Most reports of transdifferentiation can be explained as artifacts (Rizzino 2007), including contamination of injected test cells with other cells (McKinney-Freeman *et al* 2002), entry of host immune cells into a transplant where they are mistaken for transdifferentiated cells (Laflamme and Murry 2005), fusion of donor cells with host cells to create a heterokaryon (Terada *et al* 2002, Ying *et al* 2002, Alvarez-Dolado *et al* 2003) and incomplete transdifferentiation. Incomplete transdifferentiation is an artifact in which donor cells may initially be incorporated into foreign tissues and induced temporarily to express molecular markers of host cells, but then fail to completely reprogram to the cell types of their host locations. The most conservative conclusion is that, within the limits of current experimentation, it is unlikely that ASCs for a given lineage engrafted into foreign niches outside that lineage can transdifferentiate, or are able to do so at such low frequencies as to be clinically irrelevant to regenerative medicine. This conclusion is strengthened by experiments using genetically marked cells, in which the marker was found only in cells of the labeled lineage (Laflamme and Murry 2011, for a review).

6.4.1.2 Effects of age on adult stem cells
The ability of tissues to regenerate declines with age (Nemoto and Finkel 2004, Abbot 2004). Oxidative stress takes place in the cells of aging tissues. There is an increasing imbalance between the intracellular levels of free radicals and cellular antioxidant systems due to both intrinsic changes and environmental insults; this imbalance is reflected in changes in cellular, tissue and organ system structure and function. An obvious question is whether the decline in regenerative capacity is due to a decline in the number of regeneration-competent cells, decline in the intrinsic ability of cells to respond to regenerative needs, a deteriorating niche environment, or some combination of these. The answer to this question is crucial for all regenerative medicine strategies based on the use of adult regeneration-competent cells. If the decline in regenerative capacity is due to declining cell numbers and intrinsic changes in the proliferative capacity of the cells, their potential therapeutic effectiveness will be progressively compromised with age. If the niche composition is at fault, it may be possible to reverse age-related regenerative decline by providing youthful niche factors to aging cells.

The evidence suggests that the cause of decline in regenerative potential with age is multifactorial. The accumulation of somatic mutations in the DNA of regeneration-competent cells is one factor. Somatic mutations accumulate with age to create 'clonal collapse', a shift from a diverse, polyclonal set of stem cells to dominance of a few or single clones by selection. In blood cells, the top somatic mutations are in epigenetic regulators that control DNA methylation status (Goodell and Rando 2015). HSCs exhibit reduced capacities for expansion into myeloid and lymphoid progenitors with age. The capacity for self-renewal of LT-HSCs does not appear to decline over the life of an animal, but the ability of the transit amplifying populations to proliferate in response to stress is diminished, primarily by accumulating deficiencies in DNA repair mechanisms (Rossi *et al* 2007, Nijnik *et al* 2007). In ageing mice, the blood vessels of the HSC niche show a substantial reduction in number and in endothelial Notch signaling that can be reversed by activating Notch signaling in genetic gain-of-function experiments (Ksumbe *et al* 2016).

There is evidence that reduced regenerative capacity with age involves either a decline in regeneration-promoting factors, the build-up of deleterious factors, or a combination of the two. These factors appear to circulate in the blood. They have been investigated in mice for liver, muscle and neural regeneration by the technique of heterochronic parabiosis, in which young animals are joined to old animals to create a common systemic circulation (figure 6.5). Isochronic parabionts of young to young and old to old serve as controls for comparison.

Liver regeneration

In rats, the liver transcription factor C/EBPα inhibits cyclin-dependent kinases and forms a complex with pRb, E2F transcription factors and the chromatin remodeling protein Brm to maintain the growth arrest of hepatocytes. Levels of C/EBPα are reduced after partial hepatectomy in young rats, allowing this complex to dissociate, initiating DNA synthesis. In aged rats, the level of C/EBPα is not sufficiently reduced after partial hepatectomy, and the complex continues to inhibit DNA synthesis, slowing the rate of regeneration (Wang *et al* 2001, 2002, Iakova *et al* 2003).

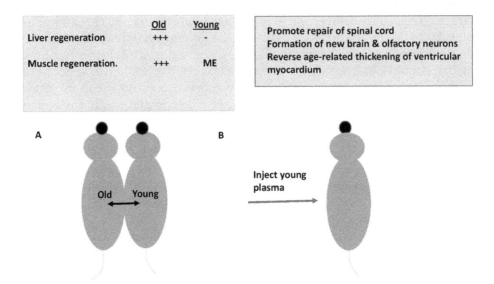

Figure 6.5. Fountain of youth experiments. (A) Parabiosis of old and young mice. Liver and muscle regeneration was enhanced in the old mice. In the young mice, liver regeneration was depressed somewhat, and there was minimal effect (ME) on muscle regeneration. (B) Injection of young plasma into old mice promoted repair of injured spinal cord, stimulated the formation of new brain and olfactory neurons, and reversed age-related thickening of ventricular myocardium.

Conboy *et al* (2005) reported that maintenance hepatocyte proliferation in aged mice after heterochronic parabiosis to young mice approached the level of proliferation in isochronic control parabionts of young mice. Conversely, hepatocyte proliferation in the young member of a heterochronic pair was reduced by nearly a third. This enhancement and reduction was not due to cross-circulating cells, as shown by parabionts in which one member of the pair was transgenic for GFP. GFP-expressing cells were virtually non-existent in the livers of the non-transgenic member. Furthermore, expression of the C/EBPα cell cycle inhibitory protein complex was diminished in the livers of the old mice after partial hepatectomy compared to isochronic controls and increased in the livers of the young mice of the pairs compared to their isochronic controls.

Skeletal muscle regeneration

Muscle regeneration is diminished in aged mice and rats. The regenerative capacity of aged muscle was increased when transplanted to young hosts, and that of young muscle was diminished when transplanted to old hosts (Carlson *et al* 2001). The number of satellite cells is not diminished with age, but their ability to proliferate declines significantly due to a deficiency in up regulating the Notch ligand Delta. This results in the activation of only 25% of the number of SCs activated in young muscle (Conboy *et al* 2003). In parabiosis experiments, the number of activated SCs and bulk muscle regeneration was restored in the old mice of heterochronic parabionts to nearly the level of young mice, while the number of activated SCs and bulk muscle regeneration in young mice of the pairs was only slightly reduced (Conboy *et al* 2005). Exposure of satellite cells from the muscles of old mice to the blood serum of young mice *in vitro* results in elevated expression of Delta, increased Notch activation, and enhanced proliferation (Conboy *et al* 2005).

The results of these experiments on hepatocytes and skeletal muscle were interpreted to mean that aging results in a loss of circulating factors that promote regeneration, setting off a search to identify the factors. A screen for proteins abundant in young mouse and human plasma identified growth and differentiation factor 11 (GDF11) as a particularly promising age-reversal molecule. The concentration of GDF11 appeared to decline with age, and infusing it into old mice was reported to activate SCs to rejuvenate injured skeletal muscles (Sinha *et al* 2014). This report, however, was contradicted by other findings that GDF11 level increases with age in the blood of both mice and humans, and infusing it into aged mouse muscle actually inhibits its regeneration (Egerman *et al* 2015).

There are caveats to heterochronic parabiosis experiments. First, it is difficult to precisely control or assess the amount of young versus old blood circulating in either member of a parabiont pair. Second, it takes weeks or months for the effects of the parabiotic exchange to take place and the precise timing is not known. Third, the parabionts share more than just blood. They are also sharing organ function (heart, kidneys, liver, lungs) that could be responsible for rejuvenation (old) or depression (young) of regenerative capacity. In light of these concerns, Rebo *et al* (2016) used a system that exchanges blood between young and old mice without parabiosis, and allows precise control of exchange volume and measurement of effects. In each experiment, blood was exchanged between young and old mice until half the blood of each was replaced by blood from the other, and then measurements of liver regeneration and muscle regeneration were made. The result was minimal improvement in the regenerative capacity of old mice receiving young blood, but large declines in regenerative capacity in young mice receiving old blood. These effects were seen within 24 h after blood exchange. The minimal benefits of young blood to old mice would be due to dilution of the concentration of regenerative inhibitors in old animals, and the declines observed in young animals receiving old blood would be due to an increase in the level of these inhibitors. These results suggest that an increase in regeneration-inhibitory factors is responsible for declining regenerative capacity with age.

Neurogenesis

The interpretation that decline in regenerative capacity is due to increase in the level of inhibitory factors in the blood is supported by data on neurogenesis. Microarray analysis of functional cells in the aging human frontal cortex suggests that after age 40, sets of genes that play central roles in synaptic plasticity, vesicular transport, mitochondrial function, stress response, anti-oxidation and DNA repair are down regulated in association with DNA damage in the promoters of these genes (Lu *et al* 2004). In the mouse and rat brain, there is a decline in the production of new neurons by NSCs in the two regions that normally maintain new neuron production, the subventricular zone of the lateral ventricles and the dentate gyrus of the hippocampus (Limke and Rao 2003). In the dentate gyrus, aging is associated with increased neuroinflammation, a continuous decline in the number of new neurons, and deficits in synaptic plasticity (Villeda *et al* 2011). The decline in new neuron production is the result of the differentiation, after a few rounds of division that produce new neurons, of NSCs into postmitotic astrocytes, gradually depleting the NSC population (Encinas *et al* 2011). This 'disposable NSC' mechanism might also be applicable to other ASCs that undergo reduction in number with age.

Parabiosis experiments revealed that hippocampal neurogenesis was enhanced in the older member of a heterochronic pair and decreased in the younger member (Villeda *et al* 2011). Injection of old plasma into *unpaired* young animals decreased neurogenesis and hippocampal-dependent learning and memory. The level of CCL11, a chemokine involved in allergic responses and not previously linked to ageing, neurogenesis, or cognition is increased in aging mice and humans, and in the younger member of a heterochronic parabiont. Intraperitoneal injection of CCL11 or stereotactic injection into the dentate gyrus of young mice decreased neurogenesis and exposure of young NSCs to old serum reduced their ability to form neurospheres. Injection of anti-CCL11 antibodies into the dentate gyrus of young animals, however, rescued neurogenesis. These data are consistent with an age-related increase in factors inhibitory to neurogenesis, and that restoring regenerative capacity may require neutralization of such factors. Rebo *et al* (2016) also found that the effects of old blood on unpaired young mice were particularly profound on hippocampal neurogenesis. Future research will focus on identifying other inhibitory factors.

Other systems

Injection of old mice with blood or plasma from young mice was reported to repair damaged spinal cord (Ruckh *et al* 2012), enhance the formation of new neurons in the brain and olfactory system (Villeda *et al* 2011, Katsimpardi *et al* 2014), improve performance on tests of learning and memory (Villeda *et al* 2014), and reverse age-related thickening of the ventricular myocardium (Loffredo *et al* 2013). These effects are presumably due to dilution of inhibitory factors circulating in old blood.

The seductive idea of a literal 'youth serum' has led to the initiation of clinical trials to test the potential rejuvenating effects of young human blood on older individuals (Kaiser 2016). These trials have been questioned on scientific grounds, and on bioethical grounds, since they charge a hefty fee to enroll elderly participants (see chapter 10).

6.4.2 Pluripotent stem cells

Pluripotent stem cells are considered ideal sources for the derivation of differentiated cells to be used for transplant or bioartificial tissue construction. Such cells can be cultured in large numbers and their differentiation theoretically can be directed into any of the body's approximately 200 cell types, which can then be used for transplant or incorporated into an implantable biomimetic tissue or organ (figure 6.6). There are two main types of pluripotent stem cells, embryonic stem cells (ESCs) and induced pluripotent stem cells (iPSCs), although some investigators have reported isolating pluripotent stem cells from tissues throughout the adult body (Young and Black 2004). The reality of these cells has been questioned, however, because other attempts to verify their existence have failed (Miyanishi *et al* 2013).

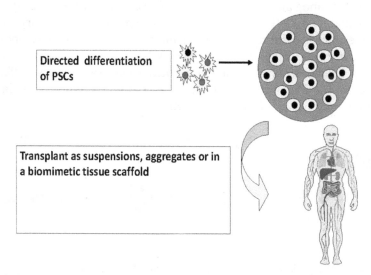

Directed differentiation of PSCs

Transplant as suspensions, aggregates or in a biomimetic tissue scaffold

Figure 6.6. Cells can be transplanted as suspensions, aggregates, or as part of a biomimetic tissue. The transplants require large numbers of cells that can be derived by the directed differentiation of pluripotent stem cells.

6.4.2.1 Embryonic stem cells (ESCs)

ESC cultures have been established from the early embryos of fish, birds and a variety of mammals (Smith 2001, Rippon and Bishop 2004). The mammalian blastocyst is an early-stage embryo that has not yet implanted into the uterine wall, which in humans is reached at about five days after conception. The blastocyst consists of a trophoblast that will contribute to the placenta, and an inner cell mass (ICM) of 13-25 pluripotent cells that will give rise to all the differentiated cells of the animal body (Jaenisch and Young 2008).

Human ESC cultures were first established from unused frozen blastocysts produced by *in vitro* fertilization in assisted reproduction facilities (Thomson *et al* 1998) (figure 6.7). Mouse and human ESC colonies differ in morphology and requirements for maintenance of pluripotency. Mouse ESCs form domed colonies and require leukemia enhancing factor (LIF) acting through the STAT3 pathway to maintain the pluripotent state. BMP4 works in concert with LIF to stabilize mouse ESCs by inducing inhibitors of differentiation (Id) genes. Human ESCs form flat colonies and require FGF-2, Activin, Lefty2 and IGF signaling for their maintenance (Silva and Smith 2008, Jaenisch and Young 2008). The pluripotency of mouse ESCs has been demonstrated by grafting them into blastocysts where they contribute to the development of all tissues, and by their differentiation into teratomas containing representatives of all three germ layers when implanted into the gastrocnemius muscle of immunodeficient mice. Human ESCs behave the same way in the teratoma test, but for obvious reasons a chimeric embryo test has not been applied to human ESCs.

Epigenetic basis of pluripotency

The core transcription factors maintaining pluripotency in mouse and human ESCs are Oct4, Sox2, Nanog and Tcf3 (Boyer *et al* 2005), along with Klf4, cMyc, ESRRB, Sall4, Tbx3, and STAT3 (Hanna *et al* 2010). The epigenetic regulatory circuitry for ICM cells is established during fertilization and pre-implantation development of the zygote. As the zygote develops to the blastocyst stage, the chromatin undergoes epigenetic modifications that activate pluripotency genes and repress differentiation genes (Stocum and Zupanc 2008, for a review). These modifications involve the addition and subtraction of acetyl or methyl groups to specific amino acids of

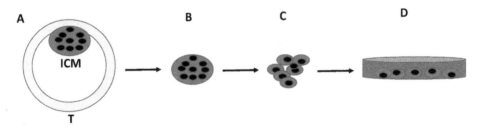

Figure 6.7. Technique of isolating and culturing mammalian embryonic stem cells from a blastocyst (A). ICM = inner cell mass; T = trophoblast. (B, C) enzymatic isolation of the ICM and dissociation of ICM cells. (D) Expansion of dissociated cells on feeder fibroblasts or adhesive substrate in culture medium containing the appropriate growth factors.

histones and addition or subtraction of methyl groups at CpG (cytidine-phosphate-guanine) islands of the DNA by histone and DNA methyltransferases, demethylases and histone acetyltransferases and deacetylases (Jaenisch and Young 2008, Hawkins *et al* 2010). Combinations of these 'epigenetic marks' determine the degree of compaction of the chromatin and thus transcriptional activity of genes (Morgan *et al* 2005, Barrero *et al* 2010). Pluripotency is in general associated with hypomethylation and acetylation, in contrast to differentiation, which is associated with hypermethylation and deacetylation. In addition, microRNAs (miRNAs) are associated with pluripotency and must be degraded for differentiation to occur (Stadler and Ruohola-Baker 2008). How pluripotency genes are epigenetically silenced and differentiation genes activated during the transition from pluripotency to differentiation is not well understood.

'Designer' ESCs

Derivatives of ESCs will be immunorejected unless transplanted into a non-immunoprivileged site. Different hESC lines have distinct HLA profiles (Carpenter *et al* 2004), thus it might be possible to establish an extensive bank of hESC lines to increase the probability of a match for a given individual within a population. A more direct solution to the immunorejection problem, however, is the derivation of patient-specific (autogeneic) ESCs by somatic cell nuclear transfer (SCNT), a technique pioneered by Gurdon in amphibians (Gurdon *et al* 1975) and by Wilmut and Campbell in mammals (Wilmut *et al* 2002). In this technique, a somatic cell nucleus is introduced into an enucleated egg (figure 6.8). The donor nucleus comes with a full set of epigenetic marks and organization that define its phenotype-related transcriptional activity. The egg cytoplasm reprograms the epigenetic status of the donor nucleus to that of the zygote and the egg undergoes cleavage. At the blastocyst stage, the inner cell mass is removed to create an autogeneic ESC line. Mammalian autogeneic ESC cultures were first created for mice and shown capable of differentiating into neurons and muscle *in vitro* (Munsie

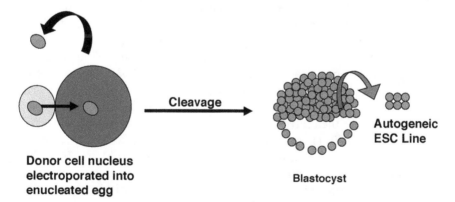

Figure 6.8. Technique for generating a 'designer' embryonic stem cell line. After removal of an egg nucleus, nuclear transfer is made from a fibroblast of the donor who will be the recipient of the cells. The egg is donated by a second party. SCNT mimics *in vitro* fertilization. Cleavage produces a blastocyst within a few days. The ICM of the blastocyst is removed and the cells are cultured to produce a designer autogeneic cell line.

et al 2000, Wakayama *et al* 2001). Autogeneic ESCs that differentiate into cell types representative of all three germ layers were then derived from SCNT blastocysts of monkeys (Byrne *et al* 2007). The standard SCNT protocol did not give viable human blastocysts, but these were successfully produced by adding a diploid adult somatic nucleus to the haploid nucleus of the egg without enucleation (Noggle *et al* 2011) or treating the egg with caffeine to prevent completion of meiosis, followed by removal of the meiotic spindle and introduction of a diploid fetal or neonatal somatic cell nucleus (Tachibana *et al* 2013). ESC cultures were derived from the ICM in both cases. The latter technique was then used to make ESCs from SCNT blastocysts created with somatic nuclei from 35 and 75 year old males and a 32 year old diabetic woman, showing that neither the age of the nucleus nor the donor's medical condition was a barrier to reprogramming (Chung *et al* 2014, Yamada *et al* 2014).

Three drawbacks to the creation of human ESCs by SCNT are (1) difficulties in procuring human eggs, (2) a low efficiency of reprogramming and (3) bioethical concerns about the morality of human ESC research (report from the President's Council on Bioethics 2002). The bioethical arguments are based on the belief that the embryo has a moral right to exist and attain personhood. The counter argument is that derivatives of human ESCs have the potential to alleviate human suffering from various diseases and injuries. These issues have been largely bypassed by technologies that convert adult somatic cells to pluripotent cells, called induced pluripotent cells (iPSCs). This does not mean that research on SCNT-derived ESCs should cease, however, because the creation of human iPSCs would not have been possible were it not for research on human ESCs, and there are still many questions to be answered by this research.

6.4.2.2 Induced pluripotent stem cells (iPSCs)

In 2006, Takahashi and Yamanaka reprogrammed adult mouse dermal fibroblasts to iPSCs by retrovirally transfecting them with a combination of the ESC pluripotency transcription factor genes *Oct4*, *Sox2*, *Klf4* and *cMyc* (OSKM, called 'Yamanaka factors'). A year later, Takahashi *et al* (2007) and Yu *et al* (2007) created iPSCs from human dermal fibroblasts. Takahashi *et al* used their original combination of genes, whereas Yu *et al* used a combination of *Oct4*, *Sox2*, *Nanog*, and *Lin* 28 (OSNL). The technique is illustrated in figure 6.9. Similar results were quickly published from other laboratories (Yamanaka 2012, for a review). The efficiency of reprogramming fibroblasts with OSKM and OSNL retroviral constructs was initially low (<0.1%), and slow, taking 2–4 weeks (Mikkelsen *et al* 2008). The OSKM proteins themselves (Zhou *et al* 2009, Kim *et al* 2009), microRNAs (Anokye-Danso *et al* 2011), and synthetic mRNAs for OSKML (Warren *et al* 2010) have all been shown to be capable of reprogramming fibroblasts. In addition to sources of cells for tissue replacement, cultured iPSCs are potentially useful for analysis *in vitro* of the genetic basis of disease ('disease in a dish'), birth defects, and for drug screening.

Comparative analysis of hiPSCs and hESCs under carefully standardized conditions of iPSC production indicate that they are virtually identical in morphology, growth characteristics, global patterns of gene activity and DNA methylation status

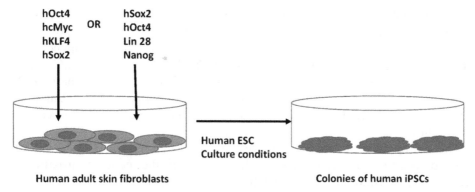

Figure 6.9. Technique for generating human induced pluripotent stem cells. In the original work by Takahashi *et al* (2007) and Yu *et al* (2007), two overlapping suites of transcription factors were employed to convert adult skin fibroblasts to iPSCs.

of pluripotency gene promoters (Guenther *et al* 2010, Boulting *et al* 2011). Successfully reprogrammed mouse iPS cells up regulated telomerase, and contributed to tissues derived from all three germ layers in the chimeric embryo assay and after intramuscular injection into SCID mice (Maherali *et al* 2007, Okita *et al* 2007, Wernig *et al* 2007). Mouse iPSCs could generate a complete mouse when injected into the blastocyst of a tetraploid embryo (Boland *et al* 2009, Kang *et al* 2009). In this technique, called tetraploid blastocyst complementation, tetraploid embryos are created by fusing two blastomeres of an early-stage embryo to create a tetraploid complement of DNA. These embryos develop a trophoblast, but no inner cell mass; the inner cell mass is provided by injection of the diploid iPSCs.

Many small molecules have been identified that enhance reprogramming efficiency when introduced with Yamanaka factors (Feng *et al* 2009, Lin and Wu 2015, Ma *et al* 2017, for reviews). Small molecules that enhance reprogramming fall into several categories: methylation or deacetylation inhibitors, epigenetic modifiers, Wnt signal modulators, modulators of cell senescence, modulators of metabolism, and inhibitors of TGF-β signaling to enhance MET, many of which can replace individual Yamanaka factors. The histone deacetylase2 inhibitor valproic acid (VPA) improves efficiency over 100-fold in mouse embryonic fibroblasts transfected by OSK (Huangfu *et al* 2008a). VPA also promotes efficient reprogramming of human fibroblasts using only OS (Huangfu *et al* 2008b). Vitamin C enhances the reprogramming efficiency of mouse and human fibroblasts transfected with OSK or OSKM (Esteban *et al* 2010). *Oct*4 transfection in combination with the small molecules PS48 (activator of 3-phosphoinositide-dependent protein kinase, PDK1), A-83-01 (an inhibitor of certain TGFβ type 1 receptors), tranylcypromine (parnate, an anti-depressant) and CHIR99021 (an inhibitor of GSK3) reprogrammed keratinocytes, human umbilical vein endothelial cells, and adipose-derived stem cells to iPSCs (Zhu *et al* 2010).

A cocktail of six small molecules (VC6TFZ) has been developed that can reprogram embryonic mouse fibroblasts without introduction of any Yamanaka factors (Hou *et al* 2008). These molecules are VPA (V), CHIR99021 (C), 616452

(6, induces polyploidization), tranylcypromine (T), forskolin (F) and 3-deazaneplanocin (Z, an inhibitor of the EZH2 histone methyltransferase). Pluripotency was induced in three successive incubation steps: (1) in mouse ESC medium containing VC6FT; (2) in ESC medium containing VC6FTZ; (3) and in ESC medium containing LIF, PD0325901 (inhibits mitogen-activated protein kinase, MAK) and C (called 2i medium). Some progress has been reported on generating human iPSCs with this molecular cocktail, but success has not yet been achieved (Valamehr *et al* 2014).

The reprogramming process involves rewriting the epigenetic code of differentiated cells to duplicate the ESC code, reflected in the methylation patterns of promoters, patterns of histone methylation and acetylation, and degree of chromatin compaction (Hanna *et al* 2010, Singhal *et al* 2010, Bhutani *et al* 2010, Koche *et al* 2011). Because these changes take place prior to any cell division or changes in transcription, it is likely that they are a critical step toward acquisition of pluripotency. Suppression of the p53 pathway and inactivation of the pRb protein also appear to be important for reprogramming, suggesting a relationship between reprogramming and entry into the cell cycle (Li *et al* 2010, Loewer *et al* 2010). Mouse and human iPSCs have been reported to have an epigenetic bias toward differentiation of their parent cells, compared to ESCs and isogenic non-β iPSCs (Kim *et al* 2010, Polo *et al* 2010, Bar-Nur *et al* 2011). If the goal of making a hiPSC culture is to differentiate copies of the cell type from which the culture is made, carryover of epigenetic memory could be an advantage.

Protocols have been developed to specifically direct the differentiation of both mouse and human hESCs and hiPSCs to specific cell phenotypes by signaling molecules and substrates, based on knowledge of the developmental pathways that lead to those phenotypes (Murry and Keller 2008). Directed differentiation of ESCs has been achieved for virtually all classes of tissues in the human body.

6.4.2.3 Impediments to clinical use of pluripotent stem cells
A number of problems must be overcome before PSCs are ready for clinical use, such as reprogramming and culture-induced chromosomal abnormalities (Spitz *et al* 2008, Narva *et al* 2010, Ma *et al* 2014), and the presence of undifferentiated cells that differentiate incorrectly with respect to their location, produce excessive numbers of progeny, or form tumors (Kiuru *et al* 2009). Methods will need to be developed to limit chromosomal abnormalities, to rigorously test differentiation capacity and long-term functional vigor of transplanted PSC derivatives, and to eliminate post-transplant cells that are not precisely localized and appropriately differentiated. Another question has been whether derivatives of PSCs evoke any immune reaction when transplanted. Careful comparisons indicate that derivatives of SCNT-ESCs or iPSCs have little or no immunogenicity when transplanted autogeneically (Araki *et al* 2013). The cost of producing PSCs and their derivatives on demand will be a factor in determining their clinical usefulness. In addition, we will need to insure safe ways to deliver and home PSC derivatives to damaged tissues (Laird *et al* 2008).

6.4.3 Induced multipotent stem cells

Transient treatment with the nucleoside analog and demethylating compound 5-Azacytidine in combination with the growth factor PDGF-AB was reported to convert mouse osteocytes and adipocytes to multipotent stem cells, called induced multipotent stem cells (iMSCs) (Chandrakanthan *et al* 2016). The cells were reported to possess a distinct transcriptome, self-renew, have immunosuppressive capacity, and multi germ layer differentiation potential. When grafted in collagen sponges next to decorticated lumbar vertebrae, the cells participated in the formation of bone, cartilage and skeletal muscle. These cells may represent a viable alternative to iPSCs for the derivation of transplantable cells.

6.5 Biomimetic tissues

Biomimetic tissues are constructs in which cells are incorporated into scaffolds that attempt to mimic the regenerative niche(s) of the cells that the construct is to replace (Rustad *et al* 2010, Orlando *et al* 2013). Biomimetic tissues and organs are being developed to replace natural ones lost to trauma or disease, and to act as platforms for drug testing and disease modeling (Fisher and Mauck 2013, Harrison *et al* 2014, Womba and Vunjak-Novakovic 2016). Ideally, the scaffolds of bioartificial tissues would mimic the geometry, biophysical and biochemical properties of the ECM *in vivo*, but also sequester and release biological signals essential for cell proliferation and differentiation (Langer and Tirell 2004). In reality, we are far from achieving this goal. Type I collagen alone or in combination with other ECM molecules, or decellularized connective tissue matrix are the most widely used natural scaffold materials, while ceramics and polymers such as polydioxanone, poly (epsilon-caprolactone), poly (glycolic acid), and poly (lactic acid) are widely used synthetic materials (Langer and Tirell 2004).

6.5.1 Types of biomimetic tissues

Biomimetic tissues can be open (vascularized by the host) or closed (cells encapsulated in a biomaterial and dependent on diffusion for survival). Allogeneic or xenogeneic cells of an open construct will be subject to immunorejection; those of a closed construct will be immunoprotected. In closed biomimetic tissues, cells are placed into porous microspheres or microcapsules <0.5 mm diameter, porous macrocapsules shaped as rods, sheaths or disks 0.5–1 mm in diameter, or vascular devices in which cells are placed in an extracellular sheath surrounding a tubular membrane ~1 mm in diameter that can be attached to blood vessels. The matrix supporting a closed construct must be resistant to degradation. Open systems are preferable to closed, and are constructed of cells attached to biodegradable polymer (natural or artificial) matrices. The matrix supports the proliferation, vascularization and differentiation of the cells into tissue that becomes integrated into the natural tissues as the matrix degrades. Scaffolds of open constructs must be biodegradable so that they are replaced over weeks or months with the natural matrix made by the cells.

6.5.2 Challenges to biomimetic tissue construction

The vision of tissue engineers is that open biomimetic tissues can be designed and constructed to replace large volumes of tissue or even replace organs. Technical challenges specific to biomimetic tissue construction that must be overcome are three-fold. The first is how to provide a three-dimensional niche that promotes regeneration. The second is how to provide biomimetic tissues with enough vascularization to ensure cell survival. The third challenge is to develop standardized procedures for producing and testing tissue-engineered products and components.

6.5.2.1 Mimicking a niche that can direct cell differentiation

The cells of all tissues are embedded in a tissue-specific ECM engineered by evolution and manufactured by the cells. The structures of ECMs fall into two general categories, 'soft' (non-mineralized) structures and 'hard' (mineralized) structures. Soft materials are composed of fibrous constituents such as collagen, keratin, elastin, chitin, lignin and other biopolymers, of which collagen is the most important biopolymer (Bauer *et al* 2014), and hydrogels (e.g. alginate) whereas 'hard' (mineralized) structures consist of mineral composites such as hydroxyapatite, calcium carbonate plus organic fibrous components (primarily collagen and chitin). A major goal of materials science with regard to tissue and organ regeneration is to mimic the structure, mechanical properties and function of natural ECMs (Bauer *et al* 2014, for a review). This biomimicry is driven by manipulating the biochemistry and physical chemistry of natural materials and creating various synthetic polymers that can replicate the structural and functional properties of natural materials, including their micro and nano-architecture, porosity, adhesion, and ability to sequester and release signaling molecules. Depending on the tissue to be regenerated, scaffold biomaterials can be made as two-dimensional sheets that are folded or rolled into tubes, or as three-dimensional constructs from the outset to mimic organ structure. Fibrous or mineralized biomaterials are used to encourage cell adhesion and proliferation for building sheets (e.g. skin, heart muscle), tubes (e.g. blood vessels, peripheral nerves), and mineralized biomaterials in the form of rods (e.g. long bones, skeletal muscle), while materials in the form of hydrogels are more often used as matrices to incorporate and release bioactive molecules (e.g. skin, spinal cord). Cells sense features of the ECM from nanoscale to microscale levels of organization (Stevens and George 2005).

The natural matrix is 'smart'; i.e. it releases the appropriate biological signaling information at the right times, places and concentrations to promote and maintain cell adhesion, differentiation and tissue organization. Thus, processed natural biomaterials such as cadaver dermis and pig small intestine submucosa have been logical choices for use as regeneration templates and scaffolds for bioartificial tissues. The processing removes cells, eliminating the immunorejection response, but it also changes the properties of the matrix, and may thus remove biochemical and biophysical information essential for cell differentiation and organization as well. The use of synthetic biomaterials is advantageous because they can be manufactured in virtually unlimited quantities to specified standards, with additional shape-shifting features built in, such as liquidity and small volume at room temperature, changing within the tissue space to gelation, expansion and space-filling at body temperature.

The newest generation of biomaterials involves the tailoring of biomimetic degradable polymers to elicit specific cellular responses by immobilizing cell-specific adhesive and signaling molecules onto the polymers that can biomechanically organize cells in three dimensions and promote their differentiation into organized tissues (Lutolf *et al* 2009). The focus is on micro- and nanofibrillar biomaterial gels, including self-assembling peptide and non-biological amphiphiles, and non-fibrillar synthetic hydrophilic polymer hydrogels that have the physical and chemical properties of natural ECM (Cai and Heilshorn 2014). A number of biologically important signaling and enzyme-sensitive entities can be incorporated into these hydrogels, including recognition sequences for cell adhesion proteins, soluble growth factors, and protease-sensitive oligopeptide or protein elements. For example, Gerecht *et al* (2007) have described a hyaluronic acid hydrogel that maintains encapsulated hESCs in their undifferentiated state. Differentiation to endothelial cells within the hydrogel was demonstrated by culturing the gel in medium containing VEGF. Derivatized amino reactive polyethylene glycols (PEG) containing both peptide substrates for proteases and binding peptides for soluble factors or cell adhesion molecules appear to have promise for creating mimics of ECM-cell interactive processes. Equally important are advances in configuring biomaterials into macro and micro lightweight architectures that mimic the configurations seen in natural CM (Cheung and Gershenfeld 2013).

6.5.2.2 Vascularization

A major technical obstacle to replacing more than small amounts of damaged tissue by cell transplants or biomimetic tissue constructs is the massive amount of cell death that occurs due to hypoxia and lack of nutrients, particularly toward the center of a cell mass. Thus, another active area of research is to find ways to quickly provide cell transplants and biomimetic constructs with vascularization (Rehman 2014). Vascularization of implanted biomimetic tissues has been enhanced by the addition of endothelial cells and/or VEGF to the construct. VEGF has been covalently bound to one arm of a hydrogel made from four-armed PEG, with the other arms linked to protease-sensitive peptides or cell adhesive peptides (Zisch *et al* 2003) or an engineered variant of VEGF has been covalently linked to fibrin for cell-induced release (Ehrbar *et al* 2004). Human umbilical vein endothelial cells seeded in the gels showed active movement, and VEGF-bound hydrogel or fibrin grafted to the chick chorioallantoic membrane or subcutaneously in rats elicited the invasion of host capillaries. The local concentration of VEGF is very important in determining the stability of the ingrowing vessels. Low concentrations favor stable vessels, whereas high concentrations lead to aberrantly formed and leaky vessels (Ozawa *et al* 2004).

6.5.2.3 Standardization of products and process

Scaffolds, cells and biomolecules need to meet rigid standards for characterization, processing and functionality in order to validate the performance of the end product

(Picciolo and Stocum 2001). Furthermore, developing standards for interactions among the components of a bioartificial tissue and with the host tissue will also be crucial to product validation. Test methods also need to be developed to assess the impact of product performance over the long term, something that has only rarely been done in animal experiments or clinical trials so far, and results need to be reported and thoroughly disseminated.

References

Abad M *et al* 2013 Reprogramming *in vivo* produces teratomas and iPS cells wth totipotency features *Nature* **502** 340–5

Abbot A 2004 Growing old gracefully *Nature* **428** 116–8

Akinc A *et al* 2008 A combinatorial library of lipid-like materials for delivery of RNA1 therapeutics *Nat. Biotechnol.* **26** 561–9

Alexander S I *et al* 2008 Chimerism and tolerance in a recipient of a deceased-donor liver transplant *New Eng. J. Med.* **358** 369–74

Allan G 1977 *Life Science in the Twentieth Century* (Cambridge: Cambridge University Press)

Alvarez-Dolado M *et al* 2003 Fusion of bone-marrow-derived cells with Purkinje neurons, cardiomyocytes and hepatocytes *Nature* **425** 968–73

Anokye-Danso F *et al* 2011 Highly efficient miRNA-mediated reprogramming of mouse and human somatic cells to pluripotency *Cell Stem Cell* **8** 376–88

Araki R *et al* 2013 Negligible immunogenicity of terminally differentiated cells derived from induced pluripotent or embryonic stem cells *Nature* **494** 100–4

Bar-Nur O, Russ H A, Efrat S and Benvenisty N 2011 Epigenetic memory and preferential lineage-specific differentiation in induced pluripotent stem cells derived from human pancreatic islet beta cells *Cell Stem Cell* **9** 17–23

Barrero M J, Boue S and Izpisua Belmonte J C 2010 Epigenetic mechanisms that regulate cell identity *Cell Stem Cell* **7** 565–70

Bauer A J P *et al* 2014 Current development of collagen-based biomaterials for tissue repair and regeneration *Soft Mater.* **12** 359–70

Bhutani N *et al* 2010 Reprogramming towards pluripotency requires AID-dependent DNA demethylation *Nature* **463** 1042–7

Boland M J *et al* 2009 Adult mice generated from induced pluripotent stem cells *Nature* **461** 91–4

Bonfanti P *et al* 2010 Microenvironmental reprogramming of thymic epithelial cells to skin multipotent stem cells *Nature* **466** 978–82

Boulting G *et al* 2011 A functionally characterized test set of human induced pluripotent stem cells *Nat. Biotechnol.* **29** 279–86

Boyer L *et al* 2005 Core transcriptional regulatory circuitry in human embryonic stem cells *Cell* **122** 947–56

Brown H 1992 Wound healing research through the ages *Wound Healing: Biochemical and Clinical Aspects* ed I K Cohen, R F Diegelmann and W J Lindblad (Philadelphia: WB Saunders), pp 5–18

Byrne J A *et al* 2007 Producing primate embryonic stem cells by somatic cell nuclear transfer *Nature* **450** 497–502

Cai L and Heilshorn S C 2014 Designing ECM-mimetic materials using protein engineering *Acta Biomater.* **10** 1751–60

Calabrese E J 2013 Historical foundations of wound healing and its potential for acceleration: dose-response considerations *Wound Rep. Reg.* **21** 180–93

Carlson B M, Dedkov E I, Borisov A B and Faulkner J A 2001 Skeletal muscle regeneration in very old rats *J. Gerontol.* **A56** B224–33

Carpenter M K *et al* 2004 Properties of four human embryonic stem cell lines maintained in a feeder-free culture system *Dev. Dyn.* **229** 243–58

Chandrakanthan V *et al* 2016 PDGF-AB and 5-azacytidine induce conversion of somatic cells into tissue-regenerative multipotent stem cells *Proc. Natl Acad. Sci. USA* **113** E2306–15

Cheung K C and Gershenfeld N 2013 Reversibly assembled cellular composite materials *Science* **341** 1219–21

Chung Y G *et al* 2014 Human somatic cell nuclear transfer using adult cells *Cell Stem Cell* **14** 1–4

Conboy I M, Conboy M J, Smythe G M and Rando T A 2003 Notch-mediated restoration of regenerative potential to aged muscle *Science* **302** 1575–7

Conboy I M *et al* 2005 Rejuvenation of aged progenitor cells by exposure to a young systemic environment *Nature* **433** 760–4

Ding S and Schultz P G 2004 A role for chemistry in stem cell biology *Nat. Biotechnol.* **22** 833–40

Egerman M A *et al* 2015 GDF11 increases with age and inhibits skeletal muscle regeneration *Cell Metabol.* **22** 164–74

Ehrbar M *et al* 2004 Cell-demanded liberation of $VEGF_{121}$ from fibrin implants induces local and controlled blood vessel growth *Circ. Res.* **94** 1124–32

Encinas J M *et al* 2011 Division-coupled astrocytic differentiation and age-related depletion of neural stem cells in the adult hippocampus *Cell Stem Cell* **8** 566–79

Esker B, Cooper D K C and Tector A J 2015 The need for xenotransplantation as a source of organs and cells for clinical transplantation *Int. J. Surg.* **23** 199–204

Esteban M A *et al* 2010 Vitamin C enhances the generation of mouse and human induced pluripotent stem cells *Cell Stem Cell* **6** 71–9

Falabella A 1998 Debridement of wounds *Wounds* **10** 1C–9C

Feng B, Ng J-H, Heng J-C D and Ng H-H 2009 Molecules that promote or enhance reprogramming of somatic cells to induced pluripotent stem cells *Cell Stem Cell* **4** 301–12

Fisher M B and Mauck R L 2013 Tissue engineering and regenerative medicine: recent innovations and the transition to translation *Tissue Eng.* **B 19** 1–13

Fu X, Wang Z and Sheng Z 2001 Advances in wound healing research in China: from antiquity to the present *Wound Rep. Reg.* **9** 2–10

Fuchs E and Segre J A 2000 Stem cells: a new lease on life *Cell* **100** 143–56

Gerecht S *et al* 2007 Hyaluronic acid hydrogel for controlled self-renewal and differentiation of human embryonic stem cells *Proc. Natl Acad. Sci. USA* **194** 11298–303

Goodell M A and Rando T A 2015 Stem cells and healthy aging *Science* **350** 1199–204

Guenther M G, Frampton G M and Soldner F 2010 Chromatin structure and gene expression programs of human embryonic and induced pluripotent stem cells *Cell Stem Cell* **7** 249–57

Gurdon J, Laskey R and Reeves O 1975 The developmental capacity of nuclei transplanted from keratinized skin cells of adult frogs *J. Embryol. Exp. Morphol.* **34** 93–112

Hakelien A-M 2002 Reprogramming fibroblasts to express T-cell functions using cell extracts *Nat. Biotechnol.* **20** 460–6

Hanna J H, Saha K and Jaenisch R 2010 Pluripotency and cellular reprogramming: facts, hypotheses, unresolved issues *Cell* **143** 508–24

Harrison R H, St-Pierre J-P and Stevens M M 2014 Tissue engineering and regenerative medicine: a year in review *Tissue Eng.* B **20** 1–16

Hawkins R D *et al* 2010 Distinct epigenomic landscapes of pluripotent and lineage-committed human cells *Cell Stem Cell* **6** 479–91

Hou P *et al* 2008 Pluripotent stem cells induced from mouse somatic cells by small-molecule compounds *Science* **341** 651–4

Huangfu D *et al* 2008a Induction of pluripotent stem cells by defined factors is greatly improved by small-molecule compounds *Nat. Biotechnol.* **26** 795–7

Huangfu D *et al* 2008b Induction of pluripotent stem cells from primary human fibroblasts with only *Oct*4 and *Sox*2 *Nat. Biotechnol.* **26** 1269–74

Iakova P, Awad S S and Timchenko N A 2003 Aging reduces proliferative capacities of liver by switching pathways of C/EBPα growth arrest *Cell* **113** 495–506

Ieda M *et al* 2010 Direct reprogramming of fibroblasts into functional cardiomyocytes by defined factors *Cell* **142** 375–86

Iwase H and Kobayashi T 2015 Current status of pig kidney xenotransplantation *Int. J. Surg.* **23** 229–33

Jaenisch R and Young R 2008 Stem cells, the molecular circuitry of pluripotency and nuclear reprogramming *Cell* **132** 567–82

Joffre O *et al* 2008 Prevention of acute and chronic allograft rejection with CD4+CD25+Foxp3+ regulatory T lymphocytes *Nat. Med.* **14** 88–92

Kaiser J 2016 Antiaging trial using young blood stirs concerns *Science* **353** 527–8

Kanazawa Y and Verma I M 2003 Little evidence of bone marrow-derived hepatocytes in the replacement of injured liver *Proc. Natl Acad. Sci. USA* **100** 11850–3

Kang L *et al* 2009 iPS cells can support full-term development of tetraploid blastocyst-complemented embryos *Cell Stem Cell* **5** 135–8

Katsimpardi L *et al* 2014 Vascular and neurogenic rejuvenation of the aging mouse brain by young systemic factors *Science* **344** 630–4

Kaul H and Ventikos Y 2015 On the genealogy of tissue engineering and regenerative medicine *Tissue Eng.* B **21** 203–17

Kim D *et al* 2009 Generation of human induced pluripotent stem cells by direct delivery of reprogramming proteins *Cell Stem Cell* **4** 472–6

Kim K *et al* 2010 Epigenetic memory in induced pluripotent stem cells *Nature* **467** 285–90

Kirkpatrick J J R, Curtis B and Naylor I L 1996 Back to the future for wound care? The influences of Padua on wound management in Renaissance Europe *Wound Rep. Reg.* **4** 326–34

Kiuru M, Boyer J L, O'Connor T P and Crystal R G 2009 Genetic control of wayward pluripotent stem cells and their progeny after transplantation *Cell Stem Cell* **4** 289–98

Koche R P *et al* 2011 Reprogramming factor expression initiates widespread targeted chromatin remodeling *Cell Stem Cell* **8** 96–105

Ksumbe A P *et al* 2016 Age-dependent modulation of vascular niches for haematopoietic stem cells *Nature* **532** 380–4

Laflamme M A and Murry C E 2005 Regenerating the heart *Nat. Biotechnol.* **23** 845–56

Laflamme M A and Murry C E 2011 Heart regeneration *Nature* **473** 326–35

Laird D J, von Andrian U H and Wagers A 2008 Stem cell trafficking in tissue development, growth and disease *Cell* **132** 612–30

Langer R and Tirell D A 2004 Designing materials for biology and medicine *Nature* **428** 487–92

Lechner A *et al* 2004 No evidence for significant transdifferentiation of bone marrow into pancreatic *β*-cells *in vivo Diabetes* **53** 616–23

Li R *et al* 2010 A mesenchymal-to-epithelial transition initiates and is required for the nuclear reprogramming of mouse fibroblasts *Cell Stem Cell* **7** 51–63

Limke T L and Rao M S 2003 Neural stem cell therapy in the aging brain:pitfalls and possibilities *J. Hematother. Stem Cell Res.* **12** 615–23

Lin T and Wu S 2015 Reprogramming with small molecules instead of exogenous transcription factors *Stem Cells Int.* **2015** 794632

Loewer S *et al* 2010 Large intergenic non-coding RNA-RoR modulates reprogramming of human induced pluripotent stem cells *Nat. Genet.* **42** 1113–9

Loffredo F S *et al* 2013 Growth differentiation factor 11 is a circulating factor that reverses aging related cardiac hypertrophy *Cell* **153** 828–39

Lu T, Pan Y, Kao S-Y, Li C, Kohane I, Chan J and Yankner B A 2004 Gene regulation and DNA damage in the ageing human brain *Nature* **429** 883–91

Lutolf M P, Gilbert P M and Blau H M 2009 Designing materials to direct stem cell fate *Nature* **462** 433–41

Ma H *et al* 2014 Abnormalities in human pluripotent cells due to reprogramming mechanisms *Nature* **511** 177–83

Ma X, Kong L and Zhu S 2017 Reprogramming cell fates by small molecules *Protein Cell* **8** 328–48

Maherali N *et al* 2007 Directly reprogrammed fibroblasts show global epigenetic remodeling and widespread tissue contribution *Cell Stem Cell* **1** 55–70

Maienschein J 2011 Regenerative medicine's historical roots in regeneration, transplantation, and translation *Dev. Biol.* **358** 278–84

Majno 1975 *The Healing Hand: Man and Wound in the Ancient World* (Cambridge, MA: Harvard University Press), pp 600

McGann C J, Odelberg S J and Keating M T 2001 Mammalian myotube dedifferentiation induced by newt regeneration extract *Proc. Natl Acad. Sci. USA* **98** 13699–704

McKinney-Freeman S L *et al* 2002 Muscle-derived hematopoietic stem cells are hematopoietic in origin *Proc. Natl Acad. Sci. USA* **99** 1341–6

Menthena A *et al* 2004 Bone marrow progenitors are not the source of expanding oval cells in injured liver *Stem Cells* **22** 1049–61

Mikkelsen T S *et al* 2008 Dissecting direct reprogramming through integrative genetic analysis *Nature* **454** 49–55

Miyanishi M *et al* 2013 Do pluripotent stem cells exist in adult mice as very small embryonic stem cells? *Stem Cell Rep.* **1** 198–208

Morgan H D *et al* 2005 Epigenetic reprogramming in mammals *Hum. Mol. Genet.* **14** R47–58

Munsie M *et al* 2000 Isolation of pluripotent embryonic stem cells from reprogrammed adult mouse somatic cell nuclei *Curr. Biol.* **10** 989–92

Murry C E and Keller G 2008 Differentiation of embryonic stem cells to clinically relevant populations: Lessons from embryonic development *Cell* **132** 661–80

Narva E *et al* 2010 High-resolution DNA analysis of human embryonic stem cell lines reveals culture-induced copy number changes and loss of heterozygosity *Nat. Biotechnol.* **28** 371–7

Nemoto S and Finkel T 2004 Ageing and the mystery at Arles *Nature* **429** 149–52

Nihsen E S *et al* 2007 Absorption of bioactive molecules into OASIS wound matrix *Dev. Skin Wound Care* **20** 541–8

Nijnik A *et al* 2007 DNA repair is limiting for haematopoietic stem cells during ageing *Nature* **447** 686–90

Noggle S *et al* 2011 Human oocytes reprogram somatic cells to a pluripotent state *Nature* **478** 70–5

Odelberg S J 2002 Inducing cellular dedifferentiation: a potential method for enhancing endofenous regeneration in mammals *Sem. Cell Dev. Biol.* **13** 335–43

Okita K, Ichisaka T and Yamanaka S 2007 Generation of germline-competent induced pluripotent stem cells *Nature* **448** 313–7

Orlando G, Soker S and Stratta R J 2013 Organ bioengineering and regeneration as the new Holy Grail for organ transplantation *Ann. Surg.* **258** 221–32

Ozawa C R *et al* 2004 Microenvironmental VEGF concentration, not total dose, determines a threshold between normal and aberrant angiogenesis *J. Clin. Invest.* **113** 516–27

Picciolo G L and Stocum D L 2001 ASTM lights the way for tissue engineered medical products standards *ASTM Stand. News* **29** 30–5

Polo J M *et al* 2010 Cell type of origin influences the molecular and functional properties of mouse induced pluripotent stem cells *Nat. Biotechnol.* **28** 848–55

President's Council on Bioethics 2002 The ethics of cloning-for-biomedical research *Human Cloning and Human Dignity: An Ethical Inquiry* ch 6 www.bioethics.gov/reports/cloning-report/research.html

Rebo J *et al* 2016 A single heterochronic blood exchange reveals rapid inhibition of multiple tissues by old blood *Nat. Commun.* **7** 13363

Rehman J 2014 Building flesh and blood *Scientist* **28** 48–53

Rippon H J and Bishop A E 2004 Embryonic stem cells *Cell Proliferat.* **37** 23–34

Rizzino A 2007 A challenge for regenerative medicine: proper genetic reprogramming, not cellular mimicry *Dev. Dyn.* **236** 3199–207

Rossi D J *et al* 2007 Deficiencies in DNA damage repair limit the function of haematopoietic stem cells with age *Nature* **447** 725–9

Ruckh J M *et al* 2012 Rejuvenation of regeneration in the aging central nervous system *Cell Stem Cell* **10** 96–103

Rustad K C *et al* 2010 Strategies for organ level tissue engineering *Organogenesis* **6** 151–7

Scandling J D *et al* 2008 Tolerance and chimerism after renal and hematopoietic-cell transplantation *New Eng. J. Med.* **358** 362–8

Sekiya S and Suzuki A 2011 Direct conversion of mouse fibroblasts to hepatocyte-like cells by defined factors *Nature* **475** 390–3

Silva J and Smith A 2008 Capturing pluripotency *Cell* **132** 532–6

Smith A 2001 Embryo-derived stem cells: of mice and men *Ann. Rev. Cell Dev. Biol.* **17** 435–62

Singhal N *et al* 2010 Chromatin-remodeling components of the BAF complex facilitate reprogramming *Cell* **141** 943–55

Sinha M *et al* 2014 Restoring systemic GDF levels reverses age-related dysfunction in mouse skeletal muscle *Science* **344** 649–52

Smeets R *et al* 2014 Face transplantation: on the verge of becoming clinical routine? *Biomed. Res. Int.* **2014** 907272

Spitz C *et al* 2008 Recurrent chromosomal abnormalities in human embryonic stem cells *Nat. Biotechnol.* **26** 1361–6

Stadler B M and Ruohola-Baker H 2008 Small RNAs: keeping stem cells in line *Cell* **132** 563–6

Stevens M M and George J H 2005 Exploring and engineering the cell surface interface *Science* **310** 1135–43

Stocum D L and Zupanc G K H 2008 Stretching the limits: stem cells in regeneration science *Dev. Dyn.* **237** 3648–71

Tachibana M *et al* 2013 Human embryonic stem cells derived by somatic cell nuclear transfer *Cell* **153** 1228–36

Takahashi K *et al* 2007 Induction of pluripotent stem cells from adult human fibroblasts by defined factors *Cell* **131** 861–72

Terada N *et al* 2002 Bone marrow cells adopt the phenotype of other cells by spontaneous cell fusion *Nature* **416** 542–5

Thomson J A *et al* 1998 Embryonic stem cell lines derived from human blastocysts *Science* **282** 1145–7

Thurston A J 2007 Pare' and prosthetics: the early history of artificial limbs *ANZ J. Surg.* **77** 1114–9

Valamehr B *et al* 2014 Platform for induction and maintenance of transgene-free hiPSCs resembling ground state pluripotent stem cells *Stem Cell Rep.* **2** 366–81

Vierbuchen T *et al* 2010 Direct conversion of fibroblasts to functional neurons by defined factors *Nature* **463** 1035–40

Villeda S A *et al* 2011 The ageing systemic milieu negatively regulates neurogenesis and cognitive function *Nature* **477** 90–4

Villeda S A *et al* 2014 Young blood reverses age-related impairments in cognitive function and synaptic plasticity in mice *Nat. Med.* **20** 659–63

Wakayama T *et al* 2001 Differentiation of embryonic stem cell lines generated from adult somatic cells by nuclear transfer *Science* **292** 740–2

Wang H *et al* 2001 C/EBPα arrests cell proliferation through direct inhibition of Cdk2 and Cdk4 *Mol. Cell* **8** 817–28

Wang H *et al* 2002 C/EBPα triggers proteasome-dependent degradation of Cdk4 during growth arrest *EMBO J.* **21** 930–41

Wang X *et al* 2003 The origin and liver repopulating capacity of murine oval cells *Proc. Natl Acad. Sci.* **100** 11881–8

Warren L *et al* 2010 Highly efficient reprogramming to pluripotency and directed differentiation of human cells with synthetic modified RNA *Cell Stem Cell* **7** 618–30

Weissman I L 2000 Stem cells: units of development, units of regeneration, and units in evolution *Cell* **100** 157–68

Wernig M *et al* 2007 *In vitro* reprogramming of fibroblasts into a pluripotent ES-cell-like state *Nature* **448** 318–24

Wilmut I *et al* 2002 Somatic cell nuclear transfer *Nature* **419** 583–6

Womba H and Vujak-Novakoic G 2016 Tissue engineering and regenerative medicine 1015: a year in review *Tissue Eng. B* **22** 101–13

Yamada M *et al* 2014 Human oocytes reprogram adult somatic nuclei of a type 1 diabetic to diloid pluripotent cells *Nature* **510** 533–6

Yamanaka S 2012 Induced pluripotent stem cells: past, present, and future *Cell Stem Cell* **10** 678–84

Ying Q-L, Nichols J, Evans E P and Smith A G 2002 Changing potency by spontaneous fusion *Nature* **416** 545–8

Young H E and Black A C Jr 2004 Adult stem cells *Anat. Rec. Part A* **276A** 75–102

Yu J *et al* 2007 Induced pluripotent stem cell lines derived from human somatic cells *Science* **318** 1917–24

Zhou Q and Melton D M 2008 Extreme makeover: converting one cell into another *Cell Stem Cell* **3** 382–8

Zhou H *et al* 2009 Generation of induced pluripotent stem cells using recombinant proteins *Cell Stem Cell* **4** 381–378

Zhu S *et al* 2010 Reprogramming of human primary somatic cells by Oct4 and chemical compounds *Cell Stem Cell* **7** 651–5

Zhu G, Mei L and Tan W 2014 From bioimaging to drug delivery and therapeutics, nano-technology is poised to change the way doctors practice medicine *Scientist* **28** 8

Zisch A H *et al* 2003 Cell-demanded release of VEGF from synthetic, biointeractive cell ingrowth matrices for vascularized tissue growth *FASEB J.* **17** 2260–2

IOP Publishing

Foundations of Regenerative Biology and Medicine

David L Stocum

Chapter 7

Pharmaceutical therapies for wound repair and regeneration

Summary

Pharmaceutical enhancement of regeneration *in situ* from remaining healthy cells employs a variety of approaches, including the application of soluble topical agents, the use of acellular scaffolds, gene therapy, and directed transdifferentiation *in situ*. The growth factors FGF-2, KGF, and PDGF are FDA-approved therapies for maximizing the repair of acute and chronic skin wounds, and many other topical agents have been shown to have positive effects on skin wound repair in animal studies. Several acellular dermal templates and hydrogels are FDA-approved for treatment of excisional wounds, burns and chronic wounds. Biomaterial conduits are commercially available to promote peripheral nerve regeneration. Biological molecules that neuroprotect severed or damaged spinal cord axons, promote axon sprouting, or reduce glial scarring, have been assessed for their efficacy in promoting spinal cord regeneration, but with only modest success. As with peripheral nerves, the most success in bridging spinal cord lesions has been obtained using biomaterial conduits. Pharmacogenetic approaches to correct the dystrophin gene have successfully treated some forms of Duchenne muscular dystrophy. Biomimetic scaffolds, with or without incorporated growth factors, have shown promise for regenerating bone across critical size defects, and for repair of lesions in articular cartilage. Thymosin B4 has been shown in animal studies to be cardioprotective after myocardial infarct by increasing the survival of cardiomyocytes, leading to diminished scarring of the heart muscle. Pharmacogenetic transdifferentiation of heart fibroblasts to cardiomyocytes *in situ* has also been shown to improve myocardial function.

7.1 Skin

We cannot yet induce the perfect regeneration of skin by pharmaceutical means, but we have been able to accelerate and/or enhance fibrosis of acute wounds, burns and chronic wounds, or reduce scarring by treatment with a variety of topical agents, and natural and synthetic dermal templates.

7.1.1 Topical agents

7.1.1.1 Acute wounds

A wide variety of topical agents have been tested for their efficacy in accelerating wound repair or reducing fibrosis in acute wounds (figure 7.1). These agents limit inflammation and enhance structural repair. The general measure of efficacy is more rapid wound closure and enhanced collagen I deposition with respect to untreated controls.

The first step to limiting inflammation in acute excisional wounds and deep burns of patients is debridement. Debridement is the removal of damaged and/or necrotic tissue by the application of papain/urea (Wang *et al* 2008) or medical maggots, the secretions of which liquefy necrotic tissue and have antibiotic and antifungal activity (including versus methicillin resistant *Staphylococcus aureus*, MRSA) (Cazander *et al* 2010, Evans *et al* 2015). Extracts of medical maggots enhance the repair of wounds through the TGF-β pathway (Li *et al* 2015). Inflammation and scarring due to excessive TGF-$\beta 1$ activity is reduced by Celicoxib (Wilgus *et al* 2003), chitosan (Diegelmann *et al* 1996), FGF-2, FGF-2 in combination with HGF (Akita *et al* 2008, Xie *et al* 2008), anti-TGF-$\beta 1$ antibodies (Ferguson and O'Kane 2004), Il-10 (Kieran *et al* 2013) and topical insulin (Chen *et al* 2012). All of these agents have been used as clinical therapies. TGF-$\beta 3$ has also been reported to reduce scarring (Occleston *et al* 2008), but efforts to transform this result into a clinical therapy have failed. Experimental wound dressings based on generating oxygen from sodium percarbonate and calcium peroxide in a polycaprolactone and polyvinyl alcohol polymer matrix (Chandra *et al* 2015) or a membrane composed of silicone-coated polyethylene terephthalate film and dried collagen vitrigel (Aoki *et al* 2015) decreased inflammation and enhanced repair of skin wounds in pigs and mice, respectively.

A variety of growth factors have been shown to enhance structural repair (Fu *et al* 2005). Re-epithelialization is enhanced by TGF-$\beta 1$ (Roberts 1995). Angiogenesis

Acute wounds	Chronic Wounds
Celicoxib	Growth factors
Growth factors	Plant resins/extracts
Oxygen	Autogeneic skin mince
Plant extracts	Hyaluronic acid
Dermal templates	Connexin 43, C-terminus
Hydrogels	Thymosin b4
Medical maggots	Infrared light
Alprostadil	Oxygen
TM	Platelet rich plasma
	Fibrin patches

Figure 7.1. Topical agents used to accelerate the healing of acute and chronic wounds. Some agents can be used for either.

is stimulated by VEGF and MMP-2 in excisional skin wounds of rabbits (Mirastschijski *et al* 2004). Fibroblast proliferation and collagen synthesis are enhanced by FGF-2 (Ichioka *et al* 2005), TGF-β (Franzen 1995, Roberts 1995) and human growth hormone (Gilpin *et al* 1994), resulting in a higher density of granulation tissue. FGF-2 is FDA-approved for clinical use in the US as Trafermin™ and in Japan as Fiblast™. Human amniotic membrane and MSCs seeded on scaffolds as wound dressings enhance wound repair by paracrine secretion of growth factors, MMPs and TIMPs, and hyaluronan (Litwiniuk and Grzela 2014, Formigli *et al* 2015).

Examples of plant extracts that enhance the formation of granulation tissue in acute wounds of animals are extract of *Celosia argentia* leaf (Priya *et al* 2004), fruit extracts of the tropical evergreen *Emblica officinalis* (Sumitra *et al* 2009), and a poly-herbal formulation (PHF) prepared by combining leaf extracts of *Hippophae rhamnoides* and *Aloe vera* (Gupta *et al* 2008). Extracts of these plants have been shown to promote angiogenesis during the formation of granulation tissue (Majewska and Gendaszewska-Darmach 2011, for a review). One substance that has been widely used and studied is curcumin, a yellow pigment derived from the curry spice turmeric, obtained by grinding the roots of the plant *Curcuma longa* (Singh 2007, for a review). Intravenous treatment of rabbit ear wounds with curcumin significantly enhanced the formation of granulation tissue that was associated with decreases in the pro-inflammatory cytokines IL-1, IL-6 and the chemokine IL-8 (Jia *et al* 2014).

Libraries of pharmacological compounds used topically for human purposes have been screened for their efficacy in enhancing the proliferation and self-renewal of dermal SKPs *in vitro* (Naska *et al* 2016). These dermal stem cells (Toma *et al* 2005) cells can reconstitute the dermis and induce morphogenesis of hair follicles (Biernaskie *et al* 2009). Five such compounds were identified and two of them, alprostadil (used to treat erectile dysfunction) and trimebutane maleate (a spasmo-lytic), were found to enhance the repair of full-thickness punch wounds in middle-aged mice by acting through the MEK-ERK pathway. These drugs may also be useful for the treatment of chronic wounds.

7.1.1.2 Chronic wounds

Treatment of chronic wounds is a greater challenge in terms of biology and cost due to the metabolic and circulatory problems underlying repair failure (Sen *et al* 2009). Many topical agents have been tested for their efficacy in healing chronic wounds (figure 7.1) Medical maggots and salve made from Norway spruce resin provide antibiotic activity (Mudge *et al* 2013, Jokinen and Sipponen 2013). Propolis, an anti-inflammatory resin collected from plants by worker bees to protect hives from bacterial and fungal infections accelerated repair of skin wounds in diabetic rats (McLennan *et al* 2008, Martinotti and Ranzato 2015, for reviews). Application of a bacteriophage cocktail effectively decreased bacterial colony counts and improved wound healing in diabetic rats and pigs (Mendes *et al* 2013). Several growth factors and cytokines that are deficient in chronic wounds have shown limited efficacy to correct faulty structural repair. The data suggest that GM-CSF, PDGF, FGF and

VEGF have the greatest potential for clinical use (Barrientos *et al* 2008). Most studies, however, have been small, with different endpoints and modes of administration of the factors; larger randomized trials and standardized study protocols are needed to support efficacy, long-term outcomes and side effect profiles.

Platelet-rich plasma and fibrin patches are reported to have positive effects on the healing of chronic wounds in patients (Roy *et al* 2011, Lundquist *et al* 2013, Picard *et al* 2015). PDGF is the effective agent of both. Re-epithelialization is also enhanced by KGF. KGF-2 and PDGF-BB are FDA-approved as (Repifermin™) and (Regranex™). An amino acid domain of placental growth factor (PlGF-$2_{123-144}$), when fused to VEGF-A and PDGF-BB, was shown to confer super-affinity to ECM on these growth factors, enabling low combination doses of the two to significantly enhance closure of skin wounds in diabetic mice over that seen with much higher doses of the unmodified growth factors (Martino *et al* 2014). Spreading a mince of healthy autogeneic skin over venous ulcers of patients enhanced their repair (Boggio *et al* 2008). Medium conditioned by culture of skin mince revealed higher levels of TNF-α, IL-1α, PDGF and FGF-2 than medium conditioned by pieces of intact skin (Pertusi *et al* 2012). In a clinical study of diabetic foot ulcers treated with a suspension of processed fat cells that release paracrine factors, 100% of patients achieved complete healing versus 62% of untreated control patients (Han *et al* 2010).

Other agents reported to enhance the repair of chronic wounds in mice are the angiogenic factor thymosin $\beta4$ (Philp *et al* 2003), copper oxide (Borkow *et al* 2010), and a material named SBD.4a extracted from the plant *Angelica sinensis* (Zhao *et al* 2006). Treatment of patients with hyperbaric oxygen increased the number and recruitment of circulating vascular stem/progenitor cells to the wound (Londahl *et al* 2010, Thom *et al* 2011). The chaperone protein of the endoplasmic reticulum, calreticulin, enhanced the migration and proliferation of fibroblasts in diabetic mouse wounds (Greives *et al* 2012). Erythropoietin (EPO) enhanced all phases of chronic wound repair in both animal and clinical studies (Hamed *et al* 2014). A meta-analysis of eight randomized controlled clinical trials indicated that hyaluronic acid promoted complete healing or a significant reduction in the size of chronic wounds (Voight and Driver 2012), and a multicenter, randomized clinical trial showed that administration of a peptide mimic of the C-terminus of connexin-43, alpha connexin carboxy-terminal (ACT1), accelerated the healing of chronic diabetic foot ulcers (Grek *et al* 2015). Venous ulcers have been effectively treated by injection of veins with a sclerosing microfoam (Lloret *et al* 2015).

Infrared (700 nm–1200 nm wavelength) and near infrared (600–700 nm) light delivered through lasers or LEDs (Ross and Domankevitz 2005) have been reported to enhance the repair of chronic wounds in animal and clinical studies. The effect of the light may be to stimulate cytochrome *c* oxidase in the mitochondria, resulting in increased oxygen consumption and production of ATP (Karu 1999). Studies on diabetic rats indicated significant increases in the amount of collagen and in tensile strength of light-treated wounds over controls (Reddy *et al* 2001). A disposable organic light emitting diode (OLED) has been devised from glass and organic components that emits light at peak wavelength of 623 nm and can be molded to fit wounds of various shapes and sizes. This diode increased FGF-2 expression and enhanced macrophage activation in a diabetic rat model (Wu *et al* 2015).

7.1.2 Dermal templates and hydrogels

Dermal templates are sheets of natural or synthetic ECM used to cover a wound and encourage the ingrowth of fibroblasts and capillaries. The templates can be overlaid with a keratinocyte suspension, or a perforated membrane such as Silastin to allow release of exudates. Van der Veen *et al* (2011) and Sharokhi *et al* (2014) have reviewed the characteristics of various dermal templates and their ability to repair excisional wounds, burns, and chronic wounds. Alloderm®, Oasis®, Permacol™ and PriMatrix™ are examples of FDA-approved templates of decellularized human, pig and bovine skin ECM that enhance the healing of excisional wounds (Wainwright *et al* 1996, Sheridan and Choucair 1997). Integra® is an FDA-approved construct of bovine collagen I plus chondroitin 6-sulfate for burns that has achieved good results, but vascularizes slowly (Druecke *et al* 2004). Integra® decreased time to closure of diabetic foot wounds, increased the rate of closure, and increased quality of life with fewer adverse effects (Driver *et al* 2015). Matriderm^R is a collagen-elastin construct containing bovine collagens type I, III and V, and elastin that promotes more rapid vascularization (Haslik *et al* 2010). Selig *et al* (2013) reported that Suprathel, a polylactide-based copolymer, performed as well as full-thickness autogeneic skin grafts in terms of scar formation and skin quality. On the other hand, Greenalgh (2014) has pointed out that most dermal templates are simply variations on a theme that has made no significant advances over the last three decades.

Hydrogels are commercially available wound dressings to deliver soluble molecules to wounds. They consist of a cross-linked solid material (continuous phase) gelled in physiological saline (discontinuous phase). The continuous phase is commonly made from chitosan (a high molecular weight polysaccharide extracted from crab shells), gelatin, alginate, collagen, GAGs or combinations thereof. Hydrogels make good wound dressings because their high capacity to retain water allows them to maintain a moist wound environment, and they are easy to remove from the wound bed (Lu *et al* 2010). Growth factors and other molecules can be easily incorporated into hydrogels for delivery to tissues. Multipurpose chitosan hydrogels have been developed that are hemostatic, bactericidal and contain lidocaine for pain relief in excisional wounds (Du *et al* 2012, Liu *et al* 2016) and moderate the inflammatory response and accelerate wound closure in burn wounds (Ribeiro *et al* 2009). In rats with mitomycin C-impaired wounds, the formation of granulation tissue was significantly enhanced by hydrogel sheets composed of alginate, chitin/chitosan and fucoidin (Murakami *et al* 2010).

Fibroblasts grown on a chitosan/gelatin scaffold loaded with FGF-2 microspheres increased their proliferation and synthesis of GAGs and laminin transcripts (Liu *et al* 2007). An alginate hydrogel impregnated with SDF-1 enhanced angiogenesis and accelerated wound closure in back wounds of mice (Henderson *et al* 2011), and a collagen–glycosaminoglycan hydrogel incorporating SDF-1 accelerated re-epithelialization of back wounds in mice (Sarkar *et al* 2010). Recombinant human GM-CSF delivered in a hydrogel to deep second-degree burns reduced healing time by a third (Zhang *et al* 2009). Sun and Mal (2011) reported that application of dextran-based hydrogels to mouse third degree burn wounds stimulated new vascularization within a week, and a mature epithelial structure with hair follicles and sebaceous glands in three weeks. At five weeks, dermal differentiation was advanced and the thickness of the skin was equivalent to that of normal skin.

7.2 Neural tissues

7.2.1 Peripheral nerve axons

Injuries to peripheral nerves that sufficiently disrupt the internal architecture of the nerve (severe crush, transection) often are unable to regenerate and result in long-term disability. In such cases, regeneration requires that the injury be bridged with a regeneration-promoting material (Dodla *et al* 2011, for a review). The gold standard is an autograft of the sural nerve or saphenous branch of the femoral nerve, which allows axon regeneration through endoneurial tubes vacated by axon degeneration. However, only small segments of these nerves can be harvested as autografts; these are sufficient for gaps of slightly less than 1 cm. Allografts of cadaver nerve provide greater lengths of nerve and avoid harvesting a patient's own nerve, but must be procured from a biobank. Research is thus focused on the development of biomaterial conduits, with or without added soluble factors.

A non-biodegradable silicon tube is the reference standard for efficacy of bio-material nerve conduits. Silicon tubes promote regeneration across a 1 cm gap in a standard rat sciatic nerve transection model, virtually the same as a sural nerve autograft. The tube first fills with endoneurial-derived hypertonic wound fluid containing PDGF, FGF-1 and NGF that gels into a fibrin matrix containing trapped red blood cells by 7 days (Yannas 2001, for a review). Schwann cells, fibroblasts and capillaries grow from the proximal and distal cut ends of the nerve to meet in the middle of the tube, and non-myelinated and myelinated axons grow through the tube from the proximal nerve stump, using dedifferentiated Schwann cells as adhesive substrates. Schwann cells remyelinate the axons, though the myelin is thinner than normal. The regenerated nerve does not synthesize an epineurium and never reaches the diameter of the normal nerve, resulting in a lower conduction velocity.

Many different kinds of biomaterial conduits and fills have been tested for their efficacy in promoting peripheral nerve regeneration (figure 7.2). These include ethylene-vinyl acetate (EVAc), poly (lactic-co-glycolic acid) (PLGA), polyhydrox-ybutyrate (PHB) and type I collagen, with or without supplementation with adhesion, growth and neurotrophic factors (Constans 2004, DeRuiter *et al* 2009), a keratinocyte hydrogel made from human hair proteins (Sierpinski *et al* 2008), a tyrosine-derived polycarbonate terpolymer (Ezra *et al* 2013), BDNF secreted by adipose-derived stem cells (Lopatina *et al* 2011), and silk fibroin (Madduri *et al* 2010, Teuschi *et al* 2015). Several biomaterial nerve conduits have been commer-cialized (Kehoe *et al* 2010). Most of these perform as well as a nerve autograft, but none have proved better than autografts.

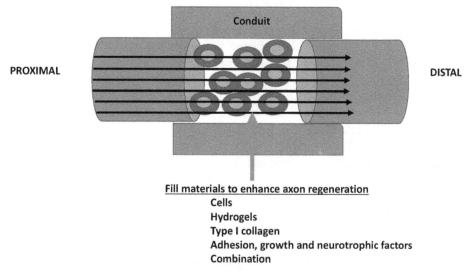

Figure 7.2. Diagram illustrating the use of a biomaterial conduit to stimulate the regeneration of peripheral nerve axons across a crush or transected region of the nerve. Arrows indicate direction of axon regeneration; cells are illustrated in the transection space.

Functional assessments of conduit-promoted rat sciatic nerve regeneration indicate that normal walking patterns are not restored (Ijkema-Paassen *et al* 2004). There are long-term abnormalities in walking and electromyographic patterns, as well as abnormalities in neuromuscular contacts and shifts in the histochemical properties of target muscles. These deficits are likely due to lack of specificity of the conduits for guiding regenerating axons into their previous endoneurial tubes distally, resulting in random innervation of target muscles, and failure to synthesize endoneurium and epineurium. We do not yet have sufficient knowledge of the requirements for peripheral regeneration to design conduits with the necessary features that will promote robust axon regeneration to target tissues.

7.2.2 Mammalian CNS axons

7.2.2.1 Response of mammalian spinal cord to injury
The mammalian CNS has very limited ability to regenerate neurons or axons lost to crush injury or transection. The limitation is not due to an intrinsic inability of the axons to regenerate, but to a difference in the niche provided by oligodendrocytes, the counterparts of Schwann cells that myelinate CNS axons. This was shown by experiments in which the axons of spinal nerves failed to regenerate when myelinated by brain oligodendrocytes and some CNS axons regenerated when myelinated by Schwann cells (David and Aguayo 1981, 1985).

Clinical manifestations of spinal cord injury are sensory deprivation and paralysis below the level of injury, muscle atrophy and spasticity, and bone loss (Eser *et al* 2004). Functional deprivation is proportional to how high in the cord the injury occurs. Cervical injuries cause the greatest degree of impairment, including paralysis and loss of autonomic control over breathing, blood pressure, heart rate and temperature. Most SCIs are crush injuries; complete transections are rare. The annual incidence of acute traumatic spinal cord injuries (SCI) worldwide is estimated to be 15–40 per million; in the United States the injury rate has been estimated at 54 per million. As of 2016, about 282 000 people were living with SCI in the United States. Fifty-eight percent of the injuries involved some degree of tetraplegia (inability to move any limbs). The lifetime direct medical care expense in the US of all forms of SCI in a 25 year old patient (in 2015 dollars) was estimated to be $12.1 million (Statistics are from the National Spinal Cord Injury Statistical Center at uab.edu/NSCISC). Animal studies, conducted primarily on rats and mice, have suggested that substantial neurological function can be preserved if as few as 5% of the original number of axons survive (Kakulas 2004).

The human spinal cord responds to damage in two phases, primary and secondary (figure 7.3) (Rowland *et al* 2008, for a review). The primary phase is characterized by neuronal and glial cell death by compression as excess plasma leaking from damaged vessels causes swelling of the cord tissue, interrupting blood flow and depriving the injured area of oxygen and glucose. Undamaged neurons are killed by glutamate excitotoxicity, in which overexcited neurons release excess amounts of the neurotransmitter glutamate and free radicals generated by lipid peroxidation. Excess glutamate opens calcium channels via degradation of cyclic AMP by the enzyme phosphodiesterase IV (PPD-IV), allowing the influx of toxic amounts of calcium. Axons crossing the lesion in the subpial zone are frequently spared, but are demyelinated.

In the secondary phase, myelin breakdown proteins derived from oligodendrocytes of surviving axons spread the damage to neurons outside the area of primary damage and causing growth cone collapse, a process that can go on for months. Inflammation plays a major role in secondary damage. Microglia are activated during the primary phase and up regulate the pro-inflammatory factors TNFα and IL-1β. PDGF and TGF-β from platelets attract neutrophils and macrophages into

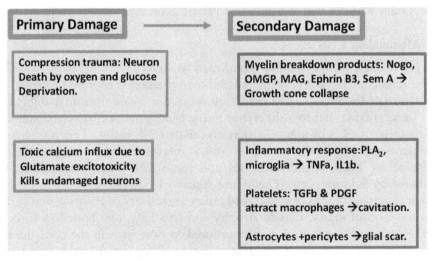

Figure 7.3. Scheme illustrating the events following a mammalian spinal cord injury.

the lesion to ingest bacteria and debris. Phospholipase A_2 (PLA_2) is a molecule that mediates free radical induced cell death, glutamate excitotoxicity, and inflammatory processes in the secondary phase of spinal cord injury (Liu *et al* 2006). The PLA_2 family of enzymes hydrolyzes glycerophospholipids to precursors of pro-inflammatory molecules. After SCI in rats, PLA_2 expression and activity rises significantly in neurons and oligodendrocytes. Both PLA_2 and its activator melittin induced neuronal death associated with inflammation, local demyelination, and necrosis when injected into uninjured cord, leading to functional impairment. Demyelination could be significantly reversed by injection of mepacrine, a drug that inhibits PLA_2 activity. A prominent feature of the second phase is cavitation and glial scar tissue formation by resident reactive astrocytes, ependymal-derived astrocytes and perhaps capillary pericytes (Busch and Silver 2007, Barnabe-Heider *et al* 2010, Goritz *et al* 2011). Cavitation and glial scar tissue is substantial in rodents, but is not as prominent in human cord (Rowland *et al* 2008, for a review).

Oligodendrocyte breakdown proteins involved in growth cone collapse include 'Nogo', oligodendrocyte myelin glycoprotein (Omgp), myelin-associated glycoprotein (MAG) and Ephrin 3B (Ramer *et al* 2005, Yiu and He 2006, Rowland *et al* 2008). These and other inhibitory molecules act on neurons through a variety of receptors (figure 7.4). Nogo, MAG and Omgp act through the Nogo-66 receptor (Ng-66R) and its co-receptors, p75/TROY and LINGO-1a. Ephrin-B3 causes growth cone collapse with an activity equivalent to that of Nogo, MAG and Ompg combined and signals via the Eph4A receptor. Another inhibitory protein is the neural guidance factor, semaphorin 3A (Sem3A) (Wu *et al* 2005), which signals

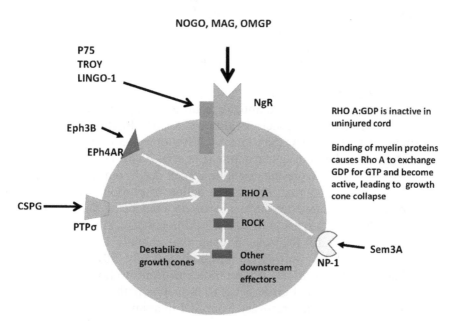

Figure 7.4. Diagram illustrating myelin breakdown molecules and their receptors that cause growth cone collapse after SCI.

through the neuropilin (NP-1) receptor (He and Tessier-Lavigne 1997). Chondroitin sulfate proteoglycans (CSPGs) of the ECM produced by cells of the glial scar also cause growth cone collapse (Niederost *et al* 1999, Ramer *et al* 2005). Neurocan, versican and NG2 CSPGs, the heparan sulfate PG syndecan-1, and the keratan sulfate PG, lumican, have been identified as potential inhibitors of axon elongation (Rowland *et al* 2008). CSPGs exert their effect by binding with high affinity to the transmembrane protein tyrosine phosphatase receptor, PTPσ (Shen *et al* 2009). The intracellular signals produced by Nogo, MAG, Ompg and CSPGs all activate the Rho A GTPase-ROCK pathway, leading to depolymerization of the growth cone actin cytoskeleton (Ramer *et al* 2005, Rowland *et al* 2008).

7.2.2.2 Therapies for mammalian spinal cord regeneration

Most experiments aimed at enhancing spinal cord axon regeneration have been done on rats and mice. The results are usually presented in terms of partial functional recovery from paralysis, defined as regaining the ability to bear weight on the hind limbs and restoration of partly coordinated stepping motions. This kind of recovery, however, is often observed even without intervention after complete spinal cord transection, due to stereotypical central locomotor patterns generated within the cord itself and not under voluntary control (Steward *et al* 1999). Rats do not regenerate the corticospinal tract, which is the primary conduit for signals controlling *voluntary* movements (Wakabayashi *et al* 2001), which too often are indistinguishable from recovery of involuntary movement (Steward *et al* 1999).

Pharmaceutical therapies for spinal cord regeneration aim to neutralize the niche injury factors that cause neuron death, demyelination, growth cone collapse and scarring, and/or provide factors to the niche that drive regeneration (Ramer *et al* 2005, Rowland *et al* 2008, Plemel and Craig 2012, for reviews). These therapies include soluble agents that protect surviving neurons, prevent growth cone collapse and stimulate axon extension, and limit, modify or bypass glial scar.

Neuroprotective agents

Neuroprotective agents reported to mitigate neuron damage and enhance functional recovery after SCI in animal models are dibutryl cAMP (Bhatt *et al* 2004), inositol polyphosphatase 4A (INPP4A) (Sasaki *et al* 2010), omega-3 fatty acids (King *et al* 2006) and the anti-malarial drug mepacrine (Liu *et al* 2006). Dibutryl cAMP closes calcium channels; INPP4A prevents excitotoxic cell death, omega-3 fatty acids act as antioxidants, and mepacrine inhibits the action of phospholipase A_2. Clinical trials of these agents, however, have shown no convincing functional benefits (Rowland *et al* 2008, Hawryluk *et al* 2008). Additional clinical trials are underway to evaluate the effects of Riluzole, a drug that inhibits toxic calcium influx (Samantaray *et al* 2010), anti-Nogo antibodies (Cully 2013) and SUN1387 (Asubio Pharmaceuticals). SUN1387 promotes neuroprotection and axonal outgrowth by modulation of the FGFR signal transduction pathway and is reported to promote significant functional recovery from SCI in animal models.

Agents that promote axon sprouting and elongation

Growth factors such as BDNF, GDNF and NGF block the inhibitory effect of MAG (Cai *et al* 1999). Rolipram, a drug that neutralizes PPD-IV enhanced the sprouting of central axons following lesions to the dorsal columns, leading to some functional recovery (Cai *et al* 2001, 2002, Nikulina *et al* 2004). Inhibition of myelin proteins and CSPGs has been explored as a regenerative therapy. Targeting the Rho-ROCK pathway with bacterial C3 transferase (Cethrin®) or with specific inhibitors of Rho kinase in SCI rats (Dergham *et al* 2002, Fournier *et al* 2003), preventing the binding of Nogo to the NgR with competitive synthetic amino-terminal peptide fragments of Nogo-66 (GrandPre *et al* 2002), and administration of anti-Nogo antibodies combined with injection of NT3 promoted axon regeneration in the lesioned corticospinal tracts of rats (Schnell *et al* 1994). While these treatments had some positive effects on axon regeneration and partial functional recovery, coordinated locomotion was not achieved. None have been effective in clinical trials, but new trials are being conducted with anti-Nogo antibodies and other drugs (Cully 2013).

Newer agents may prove more promising. Modulating the PTPσ receptor with a peptide mimetic that binds the receptor prevents CSPG-mediated inhibition of axon extension and improves functional recovery of locomotor and urinary systems in SCI rats (Lang *et al* 2015). Epothilone (epoB) is a newer small molecule drug that can cross the CNS blood–brain barrier after intraperitoneal injection. A single injection of epoB improved motor function in SCI rats by inhibiting fibroblast migration and promoting axon extension by stabilizing microtubules (Ruschel *et al* 2015).

Agents that reduce CSPGs of glial scar

The glial scar has been thought to constitute a physiological and mechanical barrier to axon elongation via the production of CSPGs. Chondroitinase ABC has been used to digest chondroitin sulfate GAGs from the core protein of CSPGs in the glial scar of SCI rats, increasing their adhesiveness for elongating axons and improving functional recovery (Bradbury *et al* 2002). Lee *et al* (2010) reported that delivery of thermostabilized ChABC in hydrogel microtubes to SCI rats maintained CSPG at low levels for up to six weeks post-injury and allowed significant axon elongation and functional recovery. Microtubule stabilization with taxol in the SCI of rats prevented the accumulation of CSPGs and led to functional improvement (Hellal *et al* 2011). Another approach was to inhibit the PTPσ receptor for CSPGs in SCI lesions by systemic administration of a compound called intracellular sigma peptide (ISP) coupled to a shuttle molecule. ISP rendered the glial scar susceptible to penetration by axons regenerating from spared tracts (Lang *et al* 2015). No clinical trials have yet emerged from these studies.

Many other studies suggest that the glial scar *per se* is not an impediment, but has positive effects on axon regeneration after SCI in mice and rats (reviewed by Sofroniew 2018). Anderson *et al* (2016) created mice transgenic for the diphtheria toxin receptor in spinal cord astrocytes. After injuring the cord to evoke glial scar formation, they ablated the astrocytes with very low doses of the toxin. Surprisingly, this resulted in inhibition of all axon regrowth and there was pronounced tissue degeneration. Furthermore, ablating the astrocytes did not decrease CSPG levels, suggesting non-astrocyte sources of CSPGs. Injecting a gel containing the neurotrophins NT3 and BDNF into SCI lesions stimulated axon regrowth through the glial scar, but these neurotrophins did not promote axon regrowth if the astrocytes were ablated. These findings strongly suggest that at least some reactive astrocytes of the glial scar promote axon regrowth. It is possible that different subsets of reactive astrocytes (and different kinds of CSPGs) promote and inhibit axon regrowth and that ChABC improves axon regeneration through a SCI lesion because it targets the CSPGs produced by inhibitory astrocytes. Identifying these astrocyte subsets and establishing the functions of other cells such as pericytes and fibroblasts in the glial scar may allow a deeper understanding of how to manipulate the environment of the SCI lesion to permit better axon regeneration.

Implants of regeneration conduits
Many different biodegradable natural and synthetic materials have been tested alone or in combination with other agents for their ability to promote axon regeneration, re-myelination and functional recovery in animal studies of SCI (Gumera *et al* 2011, Tsintou *et al* 2015, Fuhrmann *et al* 2017). Natural biodegradable materials tested include collagen, chitosan, fibronectin, fibrin, hyaluronate, silk fibroin, xyloglucan (a thermally gelling polysaccharide derived from tamarind seeds) and Matrigel. Synthetic biodegradable materials include the poly(α-hydroxy acids) PGA, PLA and copolymers (PLGA), polycaprolactone (PCL), PHB, and peptide amphiphiles. The latter are molecules that contain hydrophobic and hydrophilic domains enabling them to self-assemble into cylindrical nanofibers upon change in the ionic strength of their environment. Regeneration-promoting agents added to these conduits include neurotrophins (NT3, BDNF, NGF, GDNF), antibodies to the Nogo-66 receptor, chABC, and laminin adhesive domains (IKVAV, YIGSR, RGD, RNIAEIIKDI).

Most animal studies suggest that although these conduits promote axon regeneration across lesions, there is little or no regeneration to targets beyond the lesion. Functional recovery is modest, with little improvement beyond spontaneous recovery. The best results have been obtained with implants into rat SCIs of self-assembled peptide amphiphile nanofibers incorporating the laminin epitope IKVAV (Tysseling-Mattiace *et al* 2008) and implants of chitosan tubes filled with semi-fluid collagen (Li *et al* 2008). The latter promoted functional recovery that was statistically significant compared to controls (figure 7.5). Histological analysis indicated that axons had regenerated across the lesions and beyond, suggesting acquisition of distal targets.

SCI above the level of C4 in rats disrupts axons connecting brainstem neurons that control breathing to the phrenic motor nuclei (PMN) of the cervical spinal cord (C3–C6). After disruption of these axons on one side of the cord to paralyze the diaphragm on that side, Alilain *et al* (2011) grafted a segment of tibial nerve to serve as a conduit between C2 and the PMN at C4. In addition, they injected chABC in the PMN area and at both ends of the graft to prevent CSPG accumulation (figure 7.6). The rats recovered activity of the paralyzed diaphragm as assessed by

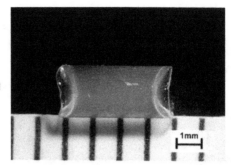

CHITOSAN TUBE FILLED WITH
SEMIFLUID TYPE I COLLAGEN

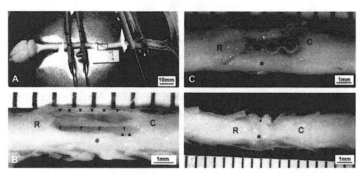

Figure 7.5. Stimulation of spinal cord regeneration with a chitosan tube conduit (top) filled with type 1 collagen. The cord was transected across two-thirds of its diameter. (A) View of brain and spinal cord 12 months after injury. The boxed area indicates the lesion site. (B) Magnification of the lesion. Arrows indicate regenerated tissue visible through the transparent wall of the conduit. Arrowheads outline the outer edge of the conduit wall. R = rostral; C = caudal. (C) Implant of chitosan tube alone. A small amount of tissue regenerated at the rostral end of the tube (R). (D) Control, no implant. The lesion was repaired with dense connective tissue. Asterisks indicate intact cord tissue (the one-third of the diameter not transected). Reproduced with permission from (Li *et al* 2008). Copyright 2008 Elsevier.

electromyelogram. Anterograde tracing of dextran amine Texas Red or biotinylated dextran amine (BDA) injected into the medulla showed that axons had regenerated into the graft and to the gray matter of the cord.

A few biomaterial conduits have been tried on human SCI. Segments of intercostal nerve embedded in a fibrin matrix loaded with FGF-2 allowed regenerating axons to bypass glial scar and promote some functional recovery in rat SCIs (Cheng *et al* 1996), but a sural nerve autograft and FGF-1 failed to improve sensory or motor function over a five-year recovery period in patients (Steeves *et al* 2004). However, Oppenheim *et al* (2009) reported a case of T-11 SCI in which the patient recovered partial motor and sensory function after rerouting intercostal nerves originating above the level of injury to the spinal canal below the level of injury. A case of partial recovery from paralysis caused by an 8 mm SCI has also been reported in which autogeneic olfactory epithelial cells were injected on either end of the lesion to induce axons to regenerate through nerve segments implanted in the gap. Two years after treatment the patient was reported able to walk using a walking frame (Tabakow *et al* 2014).

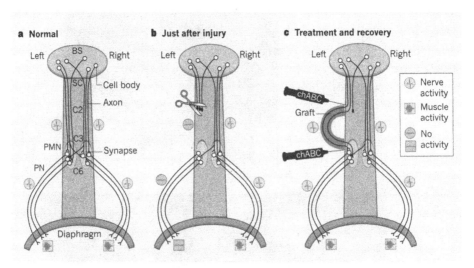

Figure 7.6. Stimulation of the transected phrenic nerve in rats. Left: Normal wiring diagram. Green circles and squares indicate normal function (impulse transmission and contraction of diaphragm). Middle: Just after paralyzing the left side of the diaphgram by transecting nerves descending from the brainstem (BS) that synapse with neurons of the phrenic nucleus in the cervical spinal cord. Impulse transmission over the left phrenic nerve (PN) is lost and the diaphgram on that side cannot contract (brown squares). Right: A segment of tibial nerve was grafted into the gap to serve as a regeneration conduit between the cut ends of the descending axons and the PMN. Chondroitinase ABC was injected into both ends of the graft and into the PMN to prevent accumulation of CSPGs. This restored the connection between the brainstem and the PMN, with return of normal function. SC = spinal cord; C = cervical vertebrae. Reproduced with permission from (Zukor and He 2011), after (Alilain 2011). Copyright 2011 Nature.

Overall, despite thousands of experiments, few of the pharmaceutical interventions tried for SCI in experimental animals or humans have resulted in significant recovery from paralysis. What is still missing is the growth of substantial numbers of axons into endoneurial sheaths on either side of the lesion, and their elongation to targets that results in the restoration of a functional circuitry. We do not yet understand the complex cellular and molecular interactions required to regenerate spinal cord axons. Highlighting our ignorance, Lu *et al* (2012) have carried out experiments on SCI rats using conduits containing multiple growth factors, but with the disappointing result that motor outcomes were worse than single molecule controls. A more comprehensive approach that involves the deployment of neuroprotective agents, neurotrophic factors, chemoattractants and adhesive factors in the right combination in a biomaterial conduit might give better results. Finding that combination is the challenge.

Useful approaches may also emerge from comparative studies of regeneration-competent versus regeneration-deficient species or mutants. Like skin, the fetal spinal cord of birds and mammals can regenerate (Nicholls and Saunders 1996, Iwashita *et al* 1994). Delaying myelination of spinal cord axons in the chick embryo extends the period of development over which regeneration is possible, suggesting that regenerative inhibition of the spinal cord coincides with oligodendrocyte differentiation (Kierstead *et al* 1995). Consistent with this notion, Nogo only becomes inhibitory when the locus of its expression shifts from neurons to oligodendrocytes later in development (O'Neill *et al* 2004). Furthermore, injured spinal cords of newborn rats and the South American opossum, *Monodelphus domestica* regenerate with complete functional recovery (Treherne *et al* 1992, Saunders *et al* 1995, Wakabayashi *et al* 2001). In the newborn opossum, microarray analysis showed changes in the expression patterns of 129 genes after SCI (Farlow *et al* 2000). More comparative studies of regeneration-competent versus regeneration-deficient stages of spinal cord development made give insights into potential interventions for SCI.

7.3 Musculoskeletal tissues

7.3.1 Volumetric muscle loss

A major cause of falls in the elderly is sarcopenia, the loss of muscle mass and loss of muscle strength. Sarcopenia begins at about age 30 and results in a 3%–5% loss of muscle mass per decade thereafter. The muscles of old rats and humans are deficient in the production of MGH (Owino *et al* 2001, Hameed *et al* 2004). Resistance training retards the onset of sarcopenia in aging humans by inducing myofiber hypertrophy and significantly increases MGH production, particularly in combination with growth hormone administration (Goldspink 2012). Thus, a deficiency of MGH may exert a significant negative effect on the ability of aging muscle to maintain mass or regenerate (Goldspink 2012).

Ruas *et al* (2012) identified an isoform, PGC-1α4, of the transcriptional co-activator PGC-1α that is highly and preferentially expressed in muscle exercised by resistance training. PGC-1α4 represses myostatin, an inhibitor of muscle growth, and induces MGF expression, leading to robust myofiber hypertrophy. These results suggest that administration of PGC-1α4, combined with strength training, might provide a therapy that would retard sarcopenia due not only to age, but also to muscle wasting associated with diseases such as cancer. Exercise physiology and physical therapy are rapidly advancing clinical sciences that will, along with studies of aging muscle, play key therapeutic roles in combating sarcopenia.

Another approach to restoring volumetric muscle loss in wounded muscle is the implantation of hydrogels into the muscle to promote ingression of satellite cells and muscle regeneration. Keratin (from human hair) hydrogels loaded with FGF-2 plus IGF-1 enabled recovery of muscle (latissmus dorsi) contractile force in a mouse model of volumetric muscle loss that was greater than that observed with hydrogels loaded with either growth factor alone, or with hydrogels loaded with both growth factors plus muscle progenitor cells (Baker *et al* 2017).

7.3.2 Duchenne muscular dystrophy

Duchenne muscular dystrophy (DMD) is a genetic early-onset muscle degenerative disease of males characterized by progressive muscle weakness, atrophy and replacement of myofibers by fat and scar tissue (Burton and Davies 2002). There are many mutations in the dystrophin (*Dmd*) gene that result in a lack of functional dystrophin, disrupting the dystrophin–glycoprotein complex (DGC) linking the cytoskeleton of the myofiber to the basement membrane (chapter 3). The disruption causes sarcolemmal instability and structural damage leading to increased calcium influx and myofiber necrosis (Bushby 2000, Cohn and Campbell 2000). The pathology of DMD is mimicked in the dystrophin-deficient *mdx* mouse. Muscle mass in this mouse, as in human DMD, is maintained by regeneration from satellite cells during the early stage of the disease, but since the SCs also carry the genetic defect, the regenerated myofibers are dystrophic as well. Eventually the muscles exhaust their pool of SCs and become fatty and fibrotic (Cossu and Mavilio 2000, Heslop *et al* 2000). Drugs such as losartan and BGP-15 can reduce fibrosis and enhance regeneration in *mdx* muscles (Gehrig *et al* 2012) but they cannot halt the eventual exhaustion of SCs.

Pharmacogenetic approaches may be more promising. Most mutations in the *Dmd* gene create a frameshift in the dystrophin mRNA. For example, one mutation in the mouse is a change from C to T in exon 23 of the gene that creates a stop codon. Rarely, this exon is skipped during transcription, giving rise to revertant myofibers expressing truncated, but functional, dystrophin protein (Lu *et al* 2000). Exon skipping has been achieved experimentally by excising the mutated mRNA with antisense oligomers (morpholinos) during pre-mRNA splicing. The skipped mRNA was placed in adenoviral vectors that were injected directly into the tibialis anterior muscle. The result was the production of a truncated but functional dystrophin that restored the histology and function of *mdx* myofibers (Goyenvalle *et al* 2004). Benchaouir *et al* (2007) reported a significant recovery of muscle morphology, function and dystrophin expression in *scid/mdx* mouse muscles transplanted with human CD133$^+$ dystrophic stem cells corrected by exon 23 skipping using lentiviral antisense oligonucleotides.

Three new morpholinos have been designed to skip other mutated exons in the dystrophin gene (Sazani *et al* 2015). Eteplirsen (AVI-4658) skips exon 51, which is mutated in 13% of the DMD population. SRP-4045 and SRP-4053 skip exons 45 and 53, respectively. Mutations in these exons account for an additional 8% each of DMD cases. Eteplirsen, given by intravenous infusion, passed a phase II clinical trial and is now the subject of a phase III clinical trial, with remarkable results. Ten of 12 boys retained the ability to walk compared to boys the same age whose walking ability has declined over the same time period. Since there is a large range of mutations over 79 exons in the longest splice form of dystrophin mRNA, it is likely that a wide variety of morpholinos will be needed to address the remaining 70% of DMD cases.

7.3.3 Critical size defects in bone

Mammals are unable to regenerate bone across a critical size gap (CSD), which in mammals is about 20% of the total bone length, but even smaller gaps remain a therapeutic challenge. CSDs result from the necessity of surgically removing a segment from a bone severely damaged by trauma or disease. Data from the Joint Theater Trauma Registry on wounding patterns among warfighters in Iraq and

Afghanistan indicate that the most prevalent combat wounds (54%) are to extremities (Owens *et al* 2007, 2008). These wounds account for approximately 65% of total patient resource use, and have projected disability benefit costs approaching \$2 billion (Masini *et al* 2009).

The current standard of treatment for CSDs is autologous or allogeneic bone grafting. Autografts are not ideal because of the limited source of bone and potential chronic pain to the patient. Processed bone allografts can circumvent this limitation, but these have limitations such as infection, non-union and stress fracture (Stevenson 1999). Therefore, pharmaceutical strategies have been sought to improve regeneration across CSDs. The principal strategy is the implantation of biodegradable osteoconductive scaffolds, with or without osteogenic inducing proteins or gene expression constructs, to promote the migration and differentiation of local cells with osteogenic capacity into the scaffold (Cameron *et al* 2013, Fisher and Mauck 2013, Harris *et al* 2013, for reviews). The factors most commonly delivered to promote bone regeneration are BMPs, which commit immigrating MSCs to osteoblastic differentiation.

7.3.3.1 Scaffolds alone

The least complex and most cost-effective method to induce regeneration across a CSD is to implant a scaffold that by itself has the necessary osteoconductive and osteoinductive properties to attract host cells capable of osteogenesis. Scaffold materials fall into four general categories (Seeherman *et al* 2002): (1) inorganics such as calcium-phosphate ceramics, calcium-phosphate-based cements and bioactive glass, (2) natural polymers, of which collagen is the most widely used, (3) synthetic polymers, of which the poly(α-hydroxy acids) are most commonly used, and (4) combinations of these.

Several of these scaffold types have shown efficacy in promoting osteogenesis. Sintered porous hydroxyapatites induced bone regeneration in calvarial defects of adult baboons by adsorbing BMPs from the host environment (Ripamonte 2010). Wojtowicz *et al* (2011) coated polycaprolactone (PCL) scaffolds with GFOGER, a synthetic triple helical collagen-mimetic peptide that binds to the integrin $\alpha 2\beta 1$ receptor. GFOGER-coated PCL scaffolds induced a greater volume of bone regeneration across a rat femoral CSD than was induced by PCL alone. However, no differences were detected in torsional strength of the regenerated bone, indicating no difference in bone quality. Suckow *et al* (1999) found that small intestine submucosa (SIS) promoted regeneration across large segment defects in the radius of rats. The SIS was unprocessed and therefore may have contained growth factors.

7.3.3.2 Scaffolds with incorporated growth factors

Scaffolds incorporating growth factors have frequently been used to induce bone regeneration across CSDs in experimental animals. Lutolf *et al* (2003) devised a polyethylene glycol (PEG) scaffold that slow-released bioactive factors via the degradative activity of the immigrating cells. RGD adhesion peptides were coupled to PEG chains, which were then cross-linked with a MMP cleavage site peptide to form a hydrogel in the presence of BMP-2. Incoming cells adhered to the RGD sites and released BMP-2 by their production of MMPs that degraded the gel by digesting the cross-linking cleavage peptide. This scaffold/BMP-2 combination promoted efficient and highly localized membrane bone regeneration across a CSD in rat calvarium.

A combination of MSCs and endothelial stem cells transgenic for the BMP-2 gene significantly increased bone regeneration and vascularization in a rat calvarial CSD (He *et al* 2013). Shah *et al* (2014) reported that a PLGA scaffold coated with several layers of polyelectrolyte carrying BMP-2 and PDGF-BB in physiological amounts (200 ng each) promoted robust bone formation in a CSD of rat calvarium. Other strategies have used a polymethylmethacrylate (PMMA) spacer to induce formation of a vascularized membrane (Viateau *et al* 2007, Klaue *et al* 2009) or poly tetrafluoroethylene (PTFE) membranes (El-Fayomi *et al* 2003) that compartmentalize the bone-inducing molecules within the CSD.

Bioactive factors can also be delivered to host osteogenic cells by gene-activated matrices (GAMS) (Pelled *et al* 2010). Viral or plasmid vectors containing genes for bioactive factors are incorporated into the scaffold and are taken up by host osteogenic cells, where they direct synthesis of the factors. Plasmids containing the hrPTH (aa 1-34) and BMP-4 genes in a collagen sponge elicited new bone formation after implanting the sponge into a segment defect in the rat femur (Fang *et al* 1996, Goldstein and Bonadio 1999). The hrPTH 1-34 plasmid also elicited bone regeneration when administered in a collagen carrier to cylindrical defects drilled into the femur or tibia of dogs (Goldstein and Bonadio 1999).

Although hundreds of similar experiments with scaffolds, with or without growth factors, have been conducted over the years on mice and rats, there has been no breakthrough that has justified long-term clinical trials with any of the methods described. We are still waiting for the optimum construct of osteoinductive scaffold and soluble factors that will initiate the cascade of events essential for the bridging of CSDs. A new model to investigate how to induce regeneration across CSDs is the amphibian long bone. Feng *et al* (2011) used 1,6 hexanedioldiacrylate (HDDA) scaffolds loaded with BMP-4 and VEGF to induce bridging of a *Xenopus* tarsal CSD by a cartilage template, which was followed by the beginning of osteogenesis in the mid-region of the cartilage. BMP-2 delivered by microbeads induced bridging of CSDs in the axolotl radius (Hutchison *et al* 2007, Satoh *et al* 2010), and a combination of BMP-4 and HGF stimulated regeneration across large gaps in the axolotl fibula (figure 7.7) (Chen *et al* 2015).

7.3.4 Articular cartilage defects

Chondral defects are refractive to treatment by growth factors, biomimetic scaffolds or combinations thereof unless first subjected to microfracture to create an osteochondral defect. BST-CarGel® (Smith and Nephew) is a liquid chitosan repair solution approved for use in Europe, Australia and Canada to treat chondral defects after microfracture. An international randomized controlled trial of BST-CarGel® conducted on 80 patients age 18–55 years with chondral lesions on the femoral

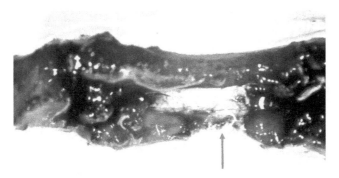

Figure 7.7. Cartilage (blue, arrow) stimulated by limb tissue extract to regenerate across a critical size gap in the fibula of an axolotl hind limb.

condyles was reported to show better quantity and quality of repair tissue and fewer structural failures than microfracture alone at five years post-treatment (Shive *et al* 2015). PDGF-AA and TGF-β, released from a heparin-conjugated fibrin scaffold implanted into a microfractured chondral defect, and covered by a chitosan bioadhesive, promoted the recruitment and retention of MSCs at the defect site and enhanced their differentiation into fibrocartilage (Lee *et al* 2015). Transfection of human, rat and mouse MSCs and chondrocytes with growth factors and chondrogenic transcription factor genes such as *Sox*9 have been reported to enhance differentiation of fibrocartilage from MSC implants in osteochondral defects (Im 2016).

7.4 Cardiac muscle

A soluble factor that has shown success in regenerating mouse cardiac muscle is thymosin β4 (TB4), a pro-angiogenic factor enriched in endothelial cells and cells involved in the formation of cardiac valves, ventricular trabeculae and outflow tract (Bock-Marquette *et al* 2004). TB4 is up regulated in the infarcted myocardium and enhances cardiomyocyte survival. Administration of TB4 intraperitoneally or directly to the mouse heart immediately after coronary artery ligation doubled the ejection fraction of the infarcted hearts (from 28% to 58%) over controls, though this was still short of the 75% observed in sham-operated mice. This recovery was associated with a significant reduction in scarring and a marked decrease in cardiomyocyte apoptosis.

TB4 works by activating the IGF-1 regulated Akt enzyme (protein kinase B), which interferes with apoptotic pathways and renders cardiomyocytes resistant to hypoxia (Fujio *et al* 2000). Akt activation is dependent on an interaction of TB4 with integrin-linked kinase (ILK). Exposure of hypoxic rat ventricular cardiomyocytes *in vitro* to conditioned medium derived from hypoxic cardiomyocytes transfected with the *Akt* gene increased the number of surviving cardiomyocytes by 40% compared to controls (Gnecchi *et al* 2005) (figure 7.8). Moreover, *in vivo* injection of concentrated conditioned medium from *Akt*-transfected hypoxic cardiomyocytes into five different sites of the infarct border zone 30 min after left coronary artery occlusion reduced the size of the infarct by 40% and apoptosis of cardiomyocytes by

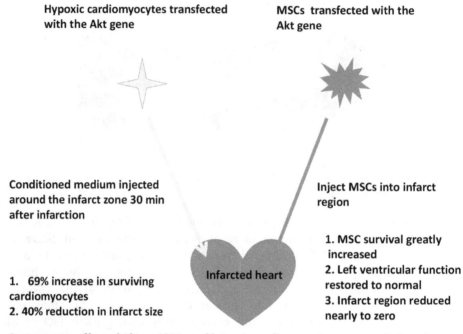

Hypoxic cardiomyocytes transfected
with the Akt gene

MSCs transfected with the
Akt gene

Conditioned medium injected
around the infarct zone 30 min
after infarction

Inject MSCs into infarct
region

1. 69% increase in surviving
cardiomyocytes
2. 40% reduction in infarct size

Infarcted heart

1. MSC survival greatly
   increased
2. Left ventricular function
   restored to normal
3. Infarct region reduced
   nearly to zero

Downstream effect of Akt on MSCs and hypoxic cardiomyocytes = up regulation of
paracrine factors that have cardioprotective effects and reduce scarring

Figure 7.8. Paracrine effects of hypoxic cardiomyocytes and mesenchymal stem cells transfected with the Akt gene greatly improve cardiac function in the infarcted hearts of mice.

69%. Other cell types transfected with the *Akt* gene also promote cardiomyocyte survival. GFP-labeled rat Akt-MSCs from male donors injected into the infarcted myocardium of female hosts greatly increased survival of the transplanted cells (Mangi *et al* 2003). Left ventricular function was restored and the infarct region was reduced to nearly zero. These results are consistent with the idea that Akt expression allows cells to up regulate paracrine factors that diminish scarring and perhaps promote regeneration of cardiac muscle.

Full-length TB4 stimulates the migration and differentiation of epicardial cells into fibroblasts and smooth muscle cells. TB4 has endopeptidase activity that produces the peptide cleavage product *N*-acetyl-seryl-aspartyl-lysyl-proline (AcSDKP), which induces the differentiation of epicardial cells into endothelial cells. The amount of AcSDKP, and thus endothelial cell differentiation is significantly reduced by knock down of TB4 (Smart *et al* 2007). Collectively, these results suggest that TB4 and AcSDKP might be a promising cocktail of molecules for clinical trial to promote cardiomyocyte survival and reduce scarring after myocardial infarction.

Improvement of infarcted heart function in mice has also been achieved by the pharmacogenetic reprogramming of heart fibroblasts, which make up half of the cells of the heart. Genetically marked cardiac fibroblasts were transdifferentiated to cardiomyocytes *in vivo* by delivering retroviral constructs of cardiac transcription factors to the border zone of the infarct. Fibroblasts were transfected with the genes for Gata-4, Mef2, and Tbx5 (GMT) (Qian *et al* 2012), or Gata-4, Hand2, Mef2, and Tbx5 (GHMT) (Song *et al* 2012). One month later reprogrammmed fibroblasts represented up to 35% (Qian *et al* 2012) and 2.4%–6.5% (Song *et al* 2012) of the cardiomyocytes in the border zone. TB4 significantly improved transfection and thus the number of cardiomyocytes generated by fibroblast transdifferentiation, and likely had a positive effect on cardiomyocyte survival and angiogenesis as well. Mathison *et al* (2012) found that injection of a lentiviral GMT construct into rat hearts three weeks after infarction improved myocardial function through trans-differentiation of scar fibroblasts to cardiomyocytes. Injection of VEGF at the time the infarction was made, followed by delivery of GMT three weeks later improved ejection fraction by a factor of four over GMT alone, again demonstrating the importance of vascularization in restoring myocardial function.

In these studies, it was unclear whether or not the transdifferentiated cardiomyo-cytes proliferated. Pharmaceutical therapies for myocardial infarct would benefit from agents that could significantly induce cardiomyocyte proliferation. To this end, Eualio *et al* (2012) identified 40 human micro RNAs that more than doubled the rate of division of postnatal rat cardiomyocytes *in vitro*. They injected retroviral constructs for two of these miRNAs (hsa-miR-590 and hsa-miR-199a) into infarcted rat hearts and reported that they stimulated cardiac regeneration with almost complete recovery of cardiac function. A single intramyocardial injection of the ECM protein agrin significantly reduced scar formation and improved cardiac function in infarcted mouse hearts (Bassat *et al* 2017). No clinical applications have yet resulted from any of these studies.

References

Akita S, Akino K, Imaizumi T and Hirano A 2008 Basic fibroblast growth factor accelerates and improves second-degree burn wound healing *Wound Rep. Reg.* **16** 635–41

Alilain W J *et al* 2011 Functional regeneration of respiratory pathways after spinal cord injury *Nature* **475** 196–200

Anderson M A *et al* 2016 Astrocyte scar formation aids central nervous system axon regeneration *Nature* **532** 195–200

Aoki S *et al* 2015 A new cell-free bandage-type artificial skin for cutaneous wounds *Wound Rep. Reg.* **23** 819–29

Baker H B *et al* 2017 Cell and growth factor-loaded keratin hydrogels for treatment of volumetric muscle loss in a mouse model *Tissue Eng.* A **23** 572–84

Barnabe-Heider F *et al* 2010 Origin of new glial cells in intact and injured spinal cord *Cell Stem Cell* **7** 470–82

Barrientos S *et al* 2008 Growth factors and cytokines in wound healing *Wound Rep. Reg.* **16** 585–601

Bassat E *et al* 2017 The extracellular protein agrin promotes heart regeneration in mice *Nature* **547** 179–84

Benchaouir R *et al* 2007 Restoration of human dystrophin following transplantation of exon-skipping-engineered DMD patient stem cells into dystrophic mice *Cell Stem Cell* **1** 646–57

Bhatt D H, Otto S J, Depoister B and Fetcho J R 2004 Cyclic AMP-induced repair of zebrafish spinal circuits *Science* **305** 254–8

Biernaskie J *et al* 2009 SKPs derive from hair follicle precursors and exhibit properties of adult dermal stem cells *Cell Stem Cell* **5** 610–23

Bock-Marquette I *et al* 2004 Thymosin $\beta4$ activates integrin-linked kinase and promotes cardiac cell migration, survival and cardiac repair *Nature* **432** 466–72

Boggio P *et al* 2008 Is there an easier way to autograft skin in chronic leg ulcers? 'Minced micrografts', a new technique *J. Eur. Acad. Dermatol. Venereol.* **22** 1168–72

Borkow G, Okon-Levy N and Gabbay J 2010 Copper oxide impregnated wound dressing: biocidal and safety studies *Wounds* **22** 301–10

Bradbury E J *et al* 2002 Chondroitinase ABC promotes functional recovery after spinal cord injury *Nature* **416** 636–40

Burton E A and Davies K E 2002 Muscular dystrophy—reason for optimism? *Cell* **108** 5–8

Bushby K M 2000 Genetics and the muscular dystrophies *Dev. Med. Child Neurol.* **42** 780–4

Busch S A and Silver J 2007 The role of extracellular matrix in CNS regeneration *Curr. Opin. Neurobiol.* **17** 120–7

Cai D *et al* 1999 Prior exposure to neurotrophins blocks inhibition of axonal regeneration by MAG and myelin via a cAMP-dependent mechanism *Neuron* **22** 89–101

Cai D *et al* 2001 Neuronal cycle cAMP controls the developmental loss in ability of axons to regenerate *J. Neurosci.* **21** 4731–9

Cai D *et al* 2002 Arginase I and polyamines act downstream from cyclic AMP in overcoming inhibition of axonal growth by MAG and myelin *in vitro Neuron* **35** 711–19

Cameron J A *et al* 2013 Employing the biology of successful fracture repair to heal critical size bone defects *New Perspectives in Regeneration*Current Topics in Microbiology and Immunology vol 367 ed E Heber-Katz and D L Stocum (Berlin: Springer), 113–32 pp

Cazander G *et al* 2010 Synergism between maggot excretions and antibiotics *Wound Rep. Reg.* **18** 637–42

Chandra P K *et al* 2015 Peroxide-based oxygen generating topical wound dressing for enhancing healing of dermal wounds *Wound Rep. Reg.* **23** 830–41

Chen X, Liu Y and Zhang X 2012 Topical insulin application improves healing by regulating the wound inflammatory response *Wound Rep. Reg.* **20** 425–34

Chen X *et al* 2015 The axolotl fibula as a model for the induction of regeneration across large segment defects in long bones of the extremities *PLoS One* **10** e0130819

Cheng H, Cao Y and Olson L 1996 Spinal cord repair in adult paraplegic rats: partial restoration of hind limb function *Science* **273** 510–3

Cohn R D and Campbell K P 2000 The molecular basis of muscular dystrophy *Muscle Nerve* **23** 1456–71

Constans A 2004 Neural tissue engineering *Scientist* 40–2

Cossu G and Mavilio F 2000 Myogenic stem cells for the therapy of primary myopathies: wishful thinking or therapeutic perspective? *J. Clin. Invest.* **105** 1669–74

Cully M 2013 Drug development: Chemical brace *Nature* **503** S10–2

David S and Aguayo A J 1981 Axonal elongation into peropheral nervous system 'bridges' after central nervous system injury in adult rats *Science* **214** 931–3

David S and Aguayo A J 1985 Axonal regeneration after crush injury of rat central nervous system fibres innervating peripheral nerve grafts *J. Neurocytol.* **14** 1–12

Dergham P *et al* 2002 Rho signaling pathway targeted to promote spinal cord repair *J. Neurosci.* **22** 6570–7

De Ruiter G C W *et al* 2009 Designing ideal conduits for peripheral nerve repair *Neurosurg. Focus* **26** 1–9

Diegelmann R F, Dunn J D, Lindblad W J and Cohen I K 1996 Analysis of the effects of chitosan on inflammation, angiogenesis, fibroplasias, and collagen deposition in polyvinyl alcohol sponge implants in rat wounds *Wound Rep. Reg.* **4** 48–52

Dodla M C, Mukhatyar V J and Bellamkonda R V 2011 Peripheral nerve regeneration *Principles of Regenerative Medicine* 2nd ed ed A Atala, R Lanza, J A Thomposn and R Nerem (San Diego: Elsevier/Academic), 1047–62

Driver V R *et al* 2015 A clinical trial template for diabetic foot ulcer treatment *Wound Rep. Reg.* **23** 891–900

Druecke D *et al* 2004 Modulation of scar tissue formation using different dermal regeneration templates in the treatment of experimental full-thickness wounds *Wound Rep. Reg.* **12** 518–27

Du L, Tong L, Jin Y and Li X 2012 A multifunctional in situ-forming hydrogel for wound healing *Wound Rep. Reg.* **20** 904–10

El-Fayomi A, El-Shahat A, Omara M and Safe I 2003 Healing of bone defects by guided bone regeneration (GBR): an experimental study *Egypt. J. Plast. Reconstr. Surg.* **27** 159–66

Eser P, Schiessl H and Willnecker J 2004 Bone loss and steady state after spinal cord injury: a cross-sectional study using pQCT *J. Musculoskel. Neuron Interact.* **4** 197–8

Eualio A *et al* 2012 Functional screening identifies miRNAs inducing cardiac regeneration *Nature* **492** 376–81

Evans R, Dudley E and Nigam Y 2015 Detection and partial characterization of antifungal bioactivity from the secretions of the medicinal maggot, *Lucilia sericata Wound Rep. Reg.* **23** 361–8

Ezra M *et al* 2013 Enhanced femoral nerve regeneration after tubulization with a tyrosine-derived polycarbonate terpolymer: effects of protein adsorption and independence of conduit porosity *Tissue Eng.* A **20** 1–11

Fang J *et al* 1996 Stimulation of new bone formation by direct transfer of osteogenic plasmid genes *Proc. Natl Acad. Sci. USA* **93** 5753–8

Farlow D N *et al* 2000 Gene expression monitoring for gene discovery in models of peripheral and central nervous system differentiation, regeneration, and trauma *J. Cell. Biochem.* **80** 171–80

Feng L *et al* 2011 Long bone critical size defect repair by regeneration in adult *Xenopus laevis* hindlimbs *Tissue Eng. Part* A **17** 691–701

Ferguson M W J and O'Kane S 2004 Scar-free healing: from embryonic mechanisms to adult therapeutic intervention *Philos. Trans. R. Soc. Lond.* B **359** 839–50

Fisher M B and Mauck R L 2013 Tissue engineering and regenerative medicine: recent innovations and the transition to translation *Tissue Eng.* B **19** 1–13

Formigli L *et al* 2015 MSCs seeded on bioengineered scaffolds improve skin wound healing in rats *Wound Rep. Reg.* **23** 115–23

Fournier A E, Takizawa B T and Strittmatter S M 2003 Rho kinase inhibition enhances axonal regeneration in the injured CNS *J. Neurosci.* **23** 1416–23

Franzen L E, Ghassemifar N, Nordman J, Schultz G and Skogman R 1995 Mechanisms of TGF-b action in connective tissue repair of rat mesenteric wounds *Wound Rep. Reg.* **3** 322–9

Fu X, Li X, Chen W and Sheng Z 2005 Engineered growth factors and cutaneous wound healing: success and possible questions in the past 10 years *Wound Rep. Reg.* **13** 122–30

Fuhrmann T, Anandakumaran N and Shoichet M S 2017 Combinatorial therapies after spinal cord injury: how can biomaterials help? *Adv. Healthc. Mater* **6** 1601130

Fujio Y *et al* 2000 Akt promotes survival of cardiomyocytes in vitro and protects against ischemia-reperfusion injury in mouse heart *Circulation* **101** 660–7

Gehrig S M *et al* 2012 Hsp72 preserves muscle function and slows progression of severe muscular dystrophy *Nature* **484** 394–8

Gilpin D A *et al* 1994 Recombinant human growth hormone accelerates wound healing in children with large cutaneous burns *Ann. Surg.* **220** 19–24

Goldspink G 2012 Age-related loss of muscle mass and strength *J. Aging Res.* **2012** 158279

Goldstein S A and Bonadio J 1999 Potential role for direct gene transfer in the enhancement of fracture healing *Clin. Orthopaed. Rel. Res.* **355S** S154–62

Goritz C *et al* 2011 A pericyte origin of spinal cord scar tissue *Science* **333** 238–42

Goyenvalle A *et al* 2004 Rescue of dystrophic muscle through U7 snRNA-mediated exon skipping *Science* **306** 1796–9

GrandPre T, Li S and Strittmatter S M 2002 Nogo-66 receptor antagonist peptide promotes axonal regeneration *Nature* **417** 547–55

Gnecchi M *et al* 2005 Paracrine action accounts for marked protection of ischemic heart by Akt-modified mesenchymal stem cells *Nat. Med.* **11** 367–8

Grek C L *et al* 2015 Topical administration of a connexin43-based peptid augments healing of chronic neuropathic diabetic foot ulcers: a multicenter, randomized trial *Wound Rep. Reg.* **23** 203–12

Greenalgh D G 2014 Editorial comments: the use of dermal substitutes in burn surgery *Wound Rep. Reg.* **22** 1–2

Greives M R *et al* 2012 Exogenous calreticulin improves diabetic wound healing *Wound Rep. Reg.* **20** 715–30

Gumera C, Rauck B and Wang Y 2011 Materials for central nervous system regeneration: bioactive cues *J. Mater. Chem.* **21** 7033–51

Gupta A, Upadhyay N K, Sawhney R C and Kumar R 2008 A poly-herbal formulation accelerates normal and impaired diabetic wound healing *Wound Rep. Reg.* **16** 784–90

Hamed S *et al* 2014 Erythropoietin, a novel repurposed drug: an innovative treatment for wound healing in patients with diabetes mellitus *Wound Rep. Reg.* **22** 23–33

Hameed M *et al* 2004 The effect of recombinant human growth hormone and resistance training on IGF-I mRNA expression in the muscles of elderly men *J. Physiol.* **555** 231–40

Han S-K, Kim H-R and Kim W-K 2010 The treatment of diabetic foot ulcers with uncultured, processed lipoaspirate cells: a pilot study *Wound Rep. Reg.* **18** 342–8

Harris J S, Bemenderfer T B, Wessel A R and Kacena M A 2013 A review of mouse critical size defect models in weight bearing bones *Bone* **55** 241–7

Haslik W *et al* 2010 Management of full-thickness skin defects in the hand and wrist region:first long-term experiences with the dermal matrix Matriderm *J. Plast. Reconstr. Aesthet. Surg.* **63** 360–4

Hawryluk G W J, Rowland J, Kwon B K and Fehlings M G 2008 Protection and repair of the injured spinal cord: a review of completed, ongoing, and planned clinical trials for acute spinal cord injury *Neurosurg. Focus* **25** E14

He Z and Tessier-Lavigne M 1997 Neuropilin is a receptor for the axonal chemorepellant semaphoring III *Cell* **90** 739–51

He X *et al* 2013 BMP2 genetically engineered MSCs and EPCs promote vascularized bone regeneration in rat critical-sided calvarial bone defects *PLoS One* **8**(4) e60473

Hellal F *et al* 2011 Microtubule stabilization reduces scarring and enables axon regeneration after spinal cord injury *Science* **331** 928–31

Henderson P W *et al* 2011 Stroma-derived factor-1 delivered via hydrogel drug-delivery vehicle accelerates wound healing in vivo *Wound Rep. Reg.* **19** 420–5

Heslop L, Morgan J E and Partridge T A 2000 Evidence for a myogenic stem cell that is exhausted in dystrophic muscle *J. Cell Sci.* **113** 2299–308

Hutchison C, Mireille P and Roy S 2007 The axolotl limb: a model for bone development, regeneration and fracture healing *Bone* **40** 45–56

Ichioka S, Ohura N and Nakatsuka T 2005 The positive experience of using a growth-factor product on deep wounds with exposed bone *J. Wound Care* **14** 105–9

Ijkema-Paassen J, Jansen K, Gramsbergen A and Meek M F 2004 Transection of peripheral nerves, bridging strategies and effect evaluation *Biomaterials* **25** 1583–92

Im G-I 2016 Gene transfer strategies to promote chondrogenesis and cartilage regeneration *Tissue Eng.* B **22** 136–48

Im G-I 2016 Endogenous cartilage repair by recruitment of stem cells *Tissue Eng.* B **22** 160–71

Iwashita Y, Kawaguchi S and Murata M 1994 Restoration of function by replacement of spinal cord segments in the rat *Nature* **367** 167–70

Jia S *et al* 2014 Intravenous curcumin efficacy on healing and scar formation in rabbit ear wounds under nonischemic, ischemic, and ischemia-reperfusion conditions *Wound Rep. Reg.* **22** 730–9

Jokinen J J and Sipponen A 2013 Refined spruce resin to treat chronic wounds: rebirth of an old folkloristic therapy *Adv. Wound Care* **5** 198–207

Kakulas B A 2004 Neuropathology: the foundation for new treatments in spinal cord injury *Spinal Cord* **42** 549–63

Karu T 1999 Primary and secondary mechanisms of action of visible to near-IR radiation on cells *J. Photochem. Photobiol.* B **49** 1–17

Kehoe S, Zhang X F and Boyd D 2010 FDA approved guidance conduits and wraps for peripheral nerve injury: a review of materials and efficacy *J. Injury* **43** 553–72

Kieran I *et al* 2013 Interleukin-10 reduces scar formation in both animal and human cutaneous wounds: results of two preclinical and phase II randomized control studies *Wound Rep. Reg.* **21** 428–36

Kierstead H S *et al* 1995 Axonal regeneration and physiological activity following transection and immunological disruption of myelin within the hatchling chick spinal cord *J. Neurosci.* **15** 6963–74

King V R, Huang W L, Dyall S C, Curran O E, Pro estly J V and Michael-Titus A T 2006 Omega-3 fatty acids improve recovery, whereas omega-6 fatty acids worsen outcome, after spinal cord injury in the adult rat *J. Neurosci.* **26** 4672–80

Klaue K *et al* 2009 Bone regeneration in long-bone defects: tissue compartmentalisation? *In vivo* study on bone defects in sheep *Injury* **40S4** S95–102

Lang B T *et al* 2015 Modulation of the proteoglycan receptor PTPσ promotes recovery after spinal cord injury *Nature* **518** 404–8

Lee H, McKeon R J and Bellamkonda R V 2010 Sustained delivery of thermostabilized chABC enhances axonal sprouting and functional recovery after spinal cord injury *Proc. Natl Acad. Sci. USA* **107** 3340–5

Lee C H *et al* 2015 Protein-releasing polymeric scaffolds induce fibrochondrocytic differentiation of endogenous cells for knee meniscus regeneration in sheep *Sci. Transl. Med.* **6** 266ra171

Li X *et al* 2008 Repair of thoracic spinal cord injury by chitosan tube implantation *Biomaterials* **30** 1121–32

Li P-N *et al* 2015 Molecular events underlying maggot extract promoted rat *in vivo* and human *in vitro* skin wound healing *Wound Rep. Reg.* **23** 65–73

Litwiniuk M and Grzela T 2014 Amniotic membrane: new concepts for an old dressing *Wound Rep. Reg.* **22** 451–6

Liu N-K *et al* 2006 A novel role of phospholipase A_2 in mediating spinal cord secondary injury *Ann. Neurol.* **59** 606–19

Liu H *et al* 2007 Effects of the controlled-released basic fibroblast growth factor from chitosan-gelatin microspheres on human fibroblasts cultured on a chitosan-gelatin scaffold *Biomacromolecules* **8** 1446–55

Liu L, Gao Q, Lu X and Zhou H 2016 In situ forming hydrogels based on chitosan for drug delivery and tissue regeneration *Asian J. Pharm. Sci.* **11** 673–83

Lloret P, Redondo P, Cabrera and Sierra A 2015 Treatment of venous leg ulcers with ultrasound-guided foam sclerotherapy: Healing, long-term recurrence and quality of life evaluation *Wound Rep. Reg.* **23** 369–78

Londahl M, Katzman P, Nilsson A and Hammerlund C 2010 Hyperbaric oxygen therapy facilitates healing of chronic foot ulcers in patients with diabetes *Diabetes Care* **33** 998–1003

Lopatina T *et al* 2011 Adipose-derived stem cells stimulate regeneration of peripheral nerves: BDNF secreted by these cells promotes nerve healing and axon growth de novo *PLoS One* **6** e17899

Lu Q L *et al* 2000 Massive idiosyncratic exon skipping corrects the nonsense mutation in dystrophic mouse muscle and produces functional revertant fibers by clonal expansion *J. Cell Biol.* **148** 985–95

Lu G *et al* 2010 A novel in situ-formed hydrogel wound dressing by the photocross-linking of a chitosan derivative *Wound Rep. Reg.* **18** 70–9

Lu P *et al* 2012 Motor axonal regeneration after partial and complete spinal cord transection *J. Neurosci.* **32** 8208–18

Lundquist R *et al* 2013 Characteristics of an autologous leukocyte and platelet-rich fibrin patch intended for the treatment of recalcitrant wounds *Wound Rep. Reg.* **21** 66–75

Lutolf M P *et al* 2003 Repair of bone defects using synthetic mimetics of collagenous extracellular matrices *Nat. Biotechnol.* **21** 513–8

Madduri S, Papaloizos M Y and Gander B 2010 Trophically and topographically functionalized silk fibroin nerve conduits for guided peripheral nerve regeneration *Biomats* **31** 2323–34

Majewska I and Gendaszewska-Darmach E 2011 Proangiogenic activity of plant extracts in accelerating wound healing—a new face of old phytomedicines *Acta Biochim. Pol.* **58** 449–60

Mangi A A *et al* 2003 Mesenchymal stem cells modified with Akt prevent remodeling and restore performance of infarcted hearts *Nat. Med.* **9** 1195–201

Martino M M *et al* 2014 Growth factors engineered for super-affinity to the extracellular matrix enhance tissue healing *Science* **343** 885–8

Martinotti S and Ranzato E 2015 Propolis: a new frontier for wound healing? *Burns Trauma* **3** 9

Masini B D *et al* 2009 Resource utilization and disability outcome assessment of combat casualties from Operation Iraqi Freedom and Operation Enduring Freedom *J. Orthop Trauma* **23** 261–6

Mathison M *et al* 2012 *In vivo* cardiac cellular reprogramming efficacy is enhanced by angiogenic preconditioning of the infarcted myocardium with vascular endothelial growth factor *J. Am. Heart Assoc.* **1** e005652

McLennan S V *et al* 2008 The anti-inflammatory agent propolis improves wound healing in a rodent model of experimental diabetes *Wound Rep. Reg.* **16** 706–13

Mendes J J *et al* 2013 Wound healing potential of topical bacteriophage therapy on diabetic cutaneous wounds *Wound Rep. Reg.* **21** 595–603

Mirastschijski U *et al* 2004 Effects of a topical enamel matrix derivative on skin wound healing *Wound Rep. Reg.* **12** 100–8

Mudge E, Price P, Neal W and Harding K G 2013 A randomized controlled trial of larval therapy for the debridement of leg ulcers: results of a multicenter, randomized, controlled, open, observer blind, parallel group study *Wound Rep. Reg.* **22** 43–51

Murakami K *et al* 2010 Enhanced healing of mitomycin C-treated healing-impaired wounds in rats with hydrosheets composed of chitin/chitosan, fucoidan, and alginate as wound dressings *Wound Rep. Reg.* **18** 478–85

Naska S *et al* 2016 Identification of drugs that regulate dermal stem cells and enhance skin repair *Stem Cell Rep.* **6** 74–84

Nicholls J and Saunders N 1996 Regeneration of immature mammalian spinal cord after injury *Trends Neurosci.* **19** 229–34

Niederost B, Zimmerman D R, Schwab M E and Bandtlow C E 1999 Bovine CNS myelin contains neurite growth-inhibitory activity associated with chondroitin sulfate proteoglycans *J. Neurosci.* **19** 8979–9889

Nikulina E *et al* 2004 The phosphodiesterase inhibitor rolipram delivered after a spinal cord lesion promotes axonal regeneration and functional recovery *Proc. Natl Acad. Sci. USA* **101** 8786–90

Occleston N L, O'Kane S, Goldspink N and Ferguson M 2008 New therapeutics for the prevention and reduction of scarring *Drug Discov. Today* **13** 973–81

O'Neill P, Whalley K and Ferretti P 2004 Nogo and Nogo-66 receptor in human and chick: implications for development and regeneration *Dev. Dyn.* **231** 109–21

Oppenheim J S, Spitzer D E and Winfree C J 2009 Spinal cord bypass surgery using peripheral nerve transfers: review of translational studies and a case report on its use following complete spinal cord injury in a human *Neurosurg. Focus* **26** E6

Owens B D *et al* 2007 Characterization of extremity wounds in Operation Iraqi Freedom and Operation Enduring Freedom *J. Orthop. Trauma* **21** 254–7

Owens B D *et al* 2008 Combat wounds in Operation Iraqi Freedom and Operation Enduring Freedom *J. Trauma* **64** 295–9

Owino V, Yang S Y and Goldspink G 2001 Age-related loss of skeletal muscle function and the inability to express the autocrine form of insulin-like growth factor-1 (MGF) in response to mechanical overload *FEBS Lett.* **505** 259–63

Pelled G *et al* 2010 Direct gene therapy for bone regeneration: gene delivery, animal models, and outcome measures *Tissue Eng.* B **16** 13–20

Pertusi G *et al* 2012 Selective release of cytokines, chemokines, and growth factors by minced skin micrografts technique for chronic ulcer repair *Wound Rep. Reg.* **20** 178–84

Philp D *et al* 2003 Thymosin β4 and a synthetic peptide containing its actin-binding domain promote dermal wound repair in db/db diabetic mice and in aged mice *Wound Rep. Reg.* **11** 19–24

Picard F, Hersant B, Bosc R and Meningaud J-P 2015 The growing evidence for the of platelet-rich plasma on diabetic chronic wounds: a review and a proposal for a new standard care *Wound Rep. Reg.* **23** 638–843

Plemel J R and Craig J J 2012 Motor axonal regeneration following cord transection *J. Neurosci.* **32** 15645–6

Priya K S *et al* 2004 Celosia argentea Linn leaf extract improves wound healing in a rat burn model *Wound Rep. Reg.* **12** 618–25

Qian L *et al* 2012 *In vivo* reprogramming of murine cardiac fibroblasts into induced cardiomyocytes *Nature* **485** 593–604

Ramer L M, Ramer M S and Steeves J D 2005 Setting the stage for functional repair of spinal cord injuries: a cast of thousands *Spinal Cord* **43** 134–61

Reddy G, Stehno-Bittel L and Enwemeka C S 2001 Laser photostimulation accelerates wound healing in diabetic rats *Wound Rep. Reg.* **9** 248–55

Ribeiro M P *et al* 2009 Development of a new chitosan hydrogel for wound dressing *Wound Rep. Reg.* **17** 817–24

Ripamonte U 2010 Soluble and insoluble signals sculpt osteogenesis in angiogenesis *World J. Biol. Chem.* **1** 109–32

Roberts A B 1995 Transforming growth factor-β: activity and efficacy in animal models of wound healing *Wound Rep. Reg.* **3** 408–18

Ross E V and Domankevitz Y 2005 Laser treatment of leg veins: physical mechanisms and theoretical considerations *Lasers Surg. Med.* **36** 105–16

Roy S *et al* 2011 Platelet-rich fibrin matrix improves wound angiogenesis via inducing endothelial cell proliferation *Wound Rep. Reg.* **19** 753–66

Rowland J W, Hawryluk G W J, Kwon B and Fehlings M G 2008 Current status of acute spinal cord injury pathophysiology and emerging therapies: promise on the horizon *Neurosurg. Focus* **25** 1–17

Ruas J L *et al* 2012 A PGC-1α isoform induced by resistance training regulates skeletal muscle hypertrophy *Cell* **151** 1319–31

Ruschel J *et al* 2015 Systemic administration of epothilone B promotes axon regeneration after spinal cord injury *Science* **348** 347–52

Samantaray S *et al* 2010 Neuroprotective drugs in traumatic CNS injury *Open Drug Discov. J.* **2** 174–80

Sarkar A *et al* 2010 Combination of stromal cell-derived factor-1 and collagen-glycosaminoglycan scaffold delays contraction and accelerates reepithelialization of dermal wounds in wild-type mice *Wound Rep. Reg.* **19** 71–9

Sasaki J *et al* 2010 The PtdIns (3,4)P_2 phosphatase INPP4A is a suppressor of excitotoxic neuronal death *Nature* **465** 497–501

Satoh A, Cummings G M C, Bryant S V and Gardiner D M 2010 Neurotrophic regulation of fibroblast dedifferentiation during limb skeletal regeneration in the axolotl (*Ambystoma mexicanum*) *Dev. Biol.* **337** 444–57

Saunders N R *et al* 1995 Repair and recovery following spinal cord injury in a neonatal marsupial (*Monodelphus domestica*) *Clin. Exp. Pharmacol. Physiol.* **22** 518–26

Sazani P *et al* 2015 *In vitro* pharmacokinetic evaluation of Eteplirsen, SRP-4045, and SRP-4053; three phosphorodiamidate morpholino oligomers (PMO) for the treatment of patients with Duchenne muscular dystrophy (DMD) *Neurology* **84** P5.061

Schnell L *et al* 1994 Neurotrophin-3 enhances sprouting of corticospinal tract during development and after adult spinal cord lesion *Nature* **367** 170–3

Seeherman H, Wozney J and Li R 2002 Bone morphogenetic protein delivery systems *Spine* **27** S16–23

Selig H F *et al* 2013 The use of a polylactide-based copolymer as a temporary skin substitute in deep dermal burns: 1-year follow-up results of a prospective clinical noninferiority trial *Wound Rep. Reg.* **21** 402–9

Sen C K *et al* 2009 Human skin wounds a major and snowballing threat to public health and the economy *Wound Rep. Reg.* **17** 763–71

Shah N J *et al* 2014 Adaptive growth factor delivery from a polyelectroyte coating promotes synergistic bone tissue repair and reconstruction *Proc. Natl Acad. Sci. USA* **111** 12847–52

Sharokhi S, Arno A and Jeschke M 2014 The use of dermal substitutes in burn injury: acute phase *Wound Rep. Reg.* **22** 14–22

Shen Y *et al* 2009 PTPσ is a receptor for chondroitin sulfate proteoglycan, an inhibitor of neural regeneration *Science* **326** 592–6

Sheridan R L and Choucair R J 1997 Acellular allogeneic dermis does not hinder initial engraftment in burn wound resurfacing and reconstruction *J. Burn Care Rehabil.* **18** 496–9

Shive M S *et al* 2015 BST-CarGel® treatment maintains cartilage repair superiority over microfracture at 5 years in a multicenter randomized controlled trial *Cartilage* **6** 62–72

Sierpinski P *et al* 2008 The use of keratin biomaterials derived from human hair for the promotion of rapid regeneration of peripheral nerves *Biomaterials* **29** 118–28

Singh S 2007 From exotic spice to modern drug? *Cell* **130** 765–8

Smart N *et al* 2007 Thymosin β4 induces adult epicardial progenitor mobilization and neovascularization *Nature* **445** 177–82

Sofroniew M V 2018 Dissecting spinal cord regeneration *Nature* **557** 343–50

Song K *et al* 2012 Heart repair by reprogramming non-myocytes with cardiac transcription factors *Nature* **485** 599–604

Steeves J, Fawcett J and Tuszynski M 2004 Report of international clinical trials workshop on spinal cord injury *Spinal Cord* **42** 591–7

Stevenson S 1999 Biology of bone grafts *Orthop. Clin. North Am.* **30** 543–52

Steward O *et al* 1999 Genetic approaches to neurotrauma research: opportunities and potential pitfalls of murine models *Exp. Neurol.* **157** 19–42

Suckow M A, Voytik-Harbin S L, Terril L A and Badylak S F 1999 Enhanced bone regeneration using porcine small intestinal submucosa *J. Invest. Surg.* **12** 277–87

Sumitra M, Panchatcharam M, Subramani V and Lonchin S 2009 *Emblica officinalis* exerts wound healing action through up-regulation of collagen and extracellular signal-regulated kinases (ERK ½) *Wound Rep. Reg.* **17** 99–107

Sun G and Mal J J 2011 Engineering dextran-based scaffolds for drug delivery and tissue repair *Nanomedicine* **7** 1771–84

Sumitra M *et al* 2009 *Emblica officinalis* exerts wound healing action through up-regulation of collagen and extracellular signal-regulated kinases (ERK ½) *Wound Rep. Reg.* **17** 99–107

Tabakow P *et al* 2014 Functional regeneration of supraspinal connections in a patient with transected spinal cord following transplantation of bulbar olfactory ensheathing cells with peripheral nerve bridging *Cell Transplant.* **23** 1631–55

Teuschi A H *et al* 2015 A new preparation method for anisotropic silk fibroin nerve guidance conduits and its evaluation *in vitro* and in a rat sciatic nerve defect model *Tissue Eng.* C **21** 945–57

Thom S R *et al* 2011 Vasculogenic stem cell mobilization and wound recruitment in diabetic patients: Increased cell number and intracellular regulatory protein content associated with hyperbaric oxygen therapy *Wound Rep. Reg.* **19** 149–61

Toma J G, McKenzie J A, Bagli D and Miller E F 2005 Isolation and characterization of multipotent skin-derived precursors from human skin *Stem Cells* **23** 727–37

Treherne J M *et al* 1992 Restoration of conduction and growth of axons through injured spinal cord of the neonatal opossum in culture *Proc. Natl Acad. Sci. USA* **89** 431–4

Tsintou M, Dalamagkas K and Seifalian A M 2015 Advances in regenerative therapies for spinal cord injury: a biomaterials approach *Neural Regen. Res.* **10** 726–42 .

Tysseling-Mattiace V M *et al* 2008 Self-assembling nanofibers inhibit glial scar formation and promote axon elongation after spinal cord injury *J. Neurosci.* **28** 3814–23

Van der Veen V C, Boekema B, Ulrich M M W and Middelkoop E 2011 New dermal substitutes *Wound Rep. Reg.* **19** S59–65

Viateau V *et al* 2007 Long-bone critical-size defects treated with tissue-engineered grafts: a study on sheep *J. Orthopaed. Res.* **25** 741–9

Voight J and Driver V R 2012 Hyaluronic acid derivatives and their healing effect on burns, epithelial surgical wounds, and chronic wounds: a systematic review and meta-analysis of randomized controlled trials *Wound Rep. Reg.* **20** 317–31

Wainwright D J *et al* 1996 Clinical evaluation of an acellular allograft dermal matrix in full thickness skin burns *J. Burn. Care Rehabil.* **17** 124–36

Wakabayashi Y *et al* 2001 Functional recovery and regeneration of descending tracts in rats after spinal cord transection in infancy *Spine* **26** 1215–22

Wang X-Q *et al* 2008 Conservative surgical debridement as a burn treatment: Supporting evidence from a porcine burn model *Wound Rep. Reg.* **16** 774–83

Wilgus R *et al* 2003 Reduction of scar formation in full-thickness wounds with topical celecoxib treatment *Wound Rep. Reg.* **11** 15–34

Wojtowicz A M *et al* 2011 Coating of biomaterial scaffolds with the collagen-mimetic peptide GFOGER for bone defect repair *Biomaterials* **31** 2574–82

Wu K Y *et al* 2005 Local translation of RhoA regulates growth cone collapse *Nature* **436** 1020–4

Wu X *et al* 2015 Organic light emitting diode improves diabetic cutaneous wound healing in rats *Wound Rep. Reg.* **23** 104–14

Xie J L *et al* 2008 Basic fibroblast growth factor (bFGF) alleviates the scar of the rabbit ear model in wound healing *Wound Rep. Reg.* **16** 576–81

Yannas I V 2001 *Tissue and Organ Regeneration in Adults* (New York: Springer)

Yiu G and He Z 2006 Glial inhibition of CNS axon regeneration *Nat. Rev. Neurosci.* **7** 617–27

Zhang L, Chen J and Han C 2009 A multicenter clinical trial of recombinant human GM-CSF hydrogel for the treatment of deep second-degree burns *Wound Rep. Reg.* **17** 685–9

Zhao H *et al* 2006 SBD.4 stimulates regenerative processes in vitro, and wound healing in genetically diabetic mice and in human skin/severe-combined immunodeficiency mouse chimera *Wound Rep. Reg.* **14** 593–601

Zukor K and He Z 2011 Regenerative medicine: drawing breath after spinal injury *Nature* **475** 178–9

Chapter 8

Cell transplants as regenerative therapy

Summary

This chapter explores the use of cell transplants for skin, neural tissue, liver and pancreas, musculoskeletal tissue, myocardium and the hematopoietic system. Cultured keratinocytes can be sprayed onto skin wounds and burns, where they proliferate to cover the wound. MSCs and adipose-derived stem cells differentiate into cells of the granulation tissue and/or exert paracrine effects on host tissue. Schwann cells, neural stem cells and other cell types have been tested for their ability to promote the regeneration of peripheral nerve and spinal cord in animal and clinical studies of SCI, but so far with little positive result. On the other hand, progress is being made on dissecting the molecular biology of neurodegenerative disease initiation and progression, using iPSCs derived from the fibroblasts of patients, and clinical trials in which RPE progenitors derived from human ESCs improved vision in cases of macular degeneration. Beta cells have now been produced in enough numbers from human ESCs and iPSCs to replace β-cells destroyed by disease. Satellite cells derived from iPSCs with a genetically corrected dystrophin gene have been successfully transplanted into muscle of a dystrophic mouse model. Animal or human MSCs are capable of bridging CSDs in bones of animal models, and clinical protocols are in use for autogeneic chondrocyte transplant to repair chondral defects. Some success has been had in transplanting iPSC-derived cardiomyocytes to improve cardiac function in monkeys, and several hematopoietic deficiency disorders have been treated by transplants of genetically corrected HSCs.

8.1 Skin

Acute excisional wounds, chronic wounds, burns and genetic skin diseases have been treated with keratinocytes applied as sheets of cultured cells, known as cultured epithelial autografts or allografts (CEAs). Autografts of cultured epidermal cells have proven successful in resurfacing and closing burn wounds (Kym *et al* 2015).

One drawback of autografts is that their culture from an epidermal biopsy takes two weeks or more, whereas burns may require them immediately. Allogeneic keratinocyte sheets can be banked and used off the shelf, but these are subject to immunorejection, though this may be slow enough to allow sufficient protection until host epidermis can cover the wound. Either way, the epidermis formed by resurfacing deep burns or excisional wounds with keratinocyte sheets does not form hair follicles due to lack of BMP signaling from dermal papillae. Dermal papilla cells derived from human ESC cultures and transplanted under the skin of nude mice were able to induce hair follicle formation (Gnedeva *et al* 2015), suggesting a possible way to provide enough of these cells for follicle formation in patients who have suffered extensive excisional wounds or burns.

Genetically modified keratinocytes expressing growth factors that enhance healing or corrected for disease-causing mutations represent an opportunity to improve repair in epidermal genetic skin diseases (Sun *et al* 2014). A good example is junctional epidermolysis bullosa (JEB), an epidermal blistering disease caused by mutations in the gene encoding laminin B3 (LAMB3) of the basement membrane (Christiano and Uitto 1996). Keratinocytes taken from an unaffected skin area of a patient with JEB were retrovirally transfected with the normal gene and expanded in culture to form a sheet. The sheet of corrected keratinocytes was transplanted back onto the legs of the patient, where it produced a stable epidermis with normal functional levels of laminin B3 (Mavilio *et al* 2006). The epidermis had been maintained for six years at last report (De Rosa *et al* 2014). More recently, 80% of the epidermis of a 7-year-old child who had lost almost his whole epidermis to JEB, was replaced in the same way with several sheets of genetically corrected keratinocytes derived from a small area of unaffected skin. After 21 months, there appeared to be full recovery of the epidermis, with no blistering (Hirsch *et al* 2017).

Mesenchymal stem cells have positive effects on wound repair. Umbilical cord MSCs accelerated the repair of fresh cutaneous wounds in mice (Luo *et al* 2010), and spheroids of bone marrow MSCs induced rapid healing of scratch wounds in cultured monolayers of keratinocytes (Peura *et al* 2009). Culture medium conditioned by umbilical cord MSCs promoted the migration and proliferation of human skin fibroblasts *in vitro* (Jeon *et al* 2010). The effect of the MSCs may be mediated by paracrine HGF, since antibody to HGF or inhibitors of the HGF receptor c-Met and its intracellular signaling pathway impeded healing in the scratch wounds. MSCs tend to have poor engraftment after transplant to skin. Delivering bone marrow-derived MSCs to excisional wounds of mice in a pullulan–collagen hydrogel improved their survival and engraftment (Rustad *et al* 2011). The MSCs differentiated into fibroblasts, pericytes and endothelial cells, which was associated with increased VEGF levels, angiogenesis and improved wound closure. The effect appeared to be mediated by structural and paracrine contributions of the cells.

Transplants of adipose tissue-derived stem cells (ADSCs) are receiving increased attention as therapies for repair of skin wounds. ADSCs were reported to enhance the repair of rat skin wounds better than transplanted fibroblasts (Sheng *et al* 2013). ADSCs from streptozotocin-induced diabetic mice, however, had less therapeutic effect, indicating that diabetes impairs their function (Cianfarani *et al* 2013).

Treatment of foot ulcers in diabetic patients with processed lipoaspirate (PLA), a heterogeneous population of mast cells, pericytes, adipocytes, fibroblasts and endothelial cells, achieved complete healing in 100% of the treatment cohort (Han *et al* 2010). Healing was most likely due to paracrine factors produced by the aspirate, since *in vitro* analysis of the effects of PLA cells on diabetic fibroblasts showed better survival and higher cell proliferation and collagen synthesis of these fibroblasts. In another study, GFP-labeled adipose stem cells accelerated wound closure in an excisional wound model of normal and diabetic rats (Nie *et al* 2011). The cells were reported to differentiate into epithelial and endothelial lineages, but also secreted pro-angiogenic VEGF, HGF and FGF-2, suggesting both structural and paracrine contributions.

A novel technique designed to bypass the need for culture *in vitro* of keratinocyte sheets has proven successful in healing second-degree burns (Wood 2011) and venous ulcers (Kirsner *et al* 2013) (figure 8.1). A spray gun is used to deposit fresh, enzymatically dissociated autologous keratinocytes onto the burn wounds after debridement of charred epidermis (Gerlach *et al* 2011). The wound surface serves as the substrate for clonal expansion of individual keratinocytes, which resurface the wound. A number of second-degree burn patients and patients with venous ulcers have been treated successfully with this procedure, which takes little more than two hours. To assure survival of the keratinocytes, dressings with circulating nutrient solutions are placed on the sprayed burn. The mean time required for re-epithelialization of burns that covered an average 5% of the total body surface was 12.6 days. The technique has not yet been applied to third degree burns or deep excisional wounds because epidermal regeneration is poor in the absence of dermis and restoration of anchoring fibrils is slow, leading to blistering of the epidermis. In these cases, the spray technique might be used to provide dermal fibroblasts and keratinocytes in succession. Or a mixture of keratinocytes and dermal fibroblasts could be applied that would then sort out into dermal and epidermal layers.

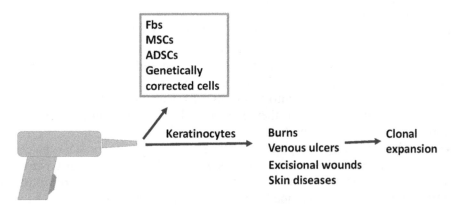

Figure 8.1. Spray gun method of distributing a cell suspension onto damaged skin. The method has been used thus far to distribute keratinocytes onto second-degree burns and venous ulcers, but theoretically could be used to distribute keratinocytes onto excisional wounds and skin damaged by epidermal disease. Other cell types (box) might also be delivered by this method.

Genetically corrected cells could be delivered by this method to skin or other tissues compromised by disease.

The future promises new sources of autogeneic keratinocytes and other skin cell types by the directed differentiation of iPSCs. Skin cells differentiated from iPSCs derived from fibroblasts of patients with skin diseases will also be used to investigate the cellular and molecular nature of the diseases and to screen for interventional drugs.

8.2 Neural tissues

8.2.1 Peripheral nerve

Schwann cells harvested from peripheral nerves and added to biomaterial conduits have been shown to improve peripheral nerve regeneration in experimental animals (Yannas 2001). Harvest of Schwann cells from patients, however, is not practical, so other cells have been tested for their ability to enhance peripheral axon regeneration. Olfactory epithelial cells (OECs) are the counterparts of Schwann cells that sheath the regenerating axons of olfactory epithelial stem cells and guide and promote their extension to the olfactory lobe. OECs injected on both sides of the suture line of a microsurgically repaired rat sciatic nerve survived and formed a type of myelin around the regenerating axons (Radtke et al 2011). The conduction velocity of the regenerated axons was improved and the rats displayed improved stepping movements, suggesting enhanced axon regeneration. Bone marrow cells seeded into polyhydroxybutyrate conduits or chitosan gel sponges were reported to enhance axon regeneration across gaps (Tohill et al 2004, Ishikawa et al 2009).

A variety of cell types have been reported able to transdifferentiate into Schwann cells, including bone marrow cells, skin derived precursor (SKP) cells (cells related to embryonic neural crest cells), ADSCs, hair follicle stem cells and olfactory ensheathing cells (OECs) (Walsh and Midha 2009, for a review). ADSC-derived Schwann cells stimulated neurite outgrowth from dorsal root ganglion explants (Radtke et al 2011) and accelerated the functional motor and sensory recovery from a nerve crush injury (Lopatina et al 2011). SKP-derived Schwann cells re-myelinated injured nerves of wild-type or *shiverer* mutant mice, which are genetically deficient in myelin basic protein (McKenzie et al 2006). Mouse and human hair follicle stem cells, which like Schwann cells are derived embryonically from neural crest cells, were reported to differentiate into Schwann cells when grafted into a 2 mm gap in the mouse sciatic nerve, where they enhanced the rate of axon regeneration and restoration of nerve function (Amoh et al 2009).

Pluripotent stem cells are a potential source of large numbers of Schwann cells for transplant. However, of the many attempts to use Schwann and other types of cells to improve the regeneration of peripheral nerves, none have resulted in improvements greater than those obtained with autografts of peripheral nerve. The use of PSC-derived Schwann cells is thus unlikely to give better results within the context of the types of experiments performed.

8.2.2 Spinal cord

spinal cord injury

Injured spinal cord has been a major target of cell transplant therapy (Barnabe-Heider and Frisen 2008, for a review). Both neural and non-neural cell transplants have been reported to have beneficial effects on SCI in animal studies, and several clinical trials have been conducted (figure 8.2). While there has been some success at promoting axon regeneration, any therapeutic effect on preventing or reversing paralysis has been modest at best.

8.2.2.1 Non-neural cells

Bone marrow MSCs (Akiyama *et al* 2002), adipose-derived stem cells (Kang *et al* 2006), human umbilical cord blood cells (Kuh *et al* 2005), and other MSCs (Sheth *et al* 2008) have all been reported to support some functional recovery after SCI in rats. Although these cells were claimed to differentiate into glial and neural cells, the evidence for this is not strong and they likely exerted their effect via paracrine and juxtacrine factors. Evidence for such effects is that splenic dendritic cells transplanted into mouse SCIs activated the proliferation and differentiation of host NSCs into new neurons and induced axon sprouting accompanied by partial recovery from hind limb paralysis (Mikami *et al* 2004). Co-culture of spinal cord NSCs with dendritic cells significantly enhanced the survival and proliferation of the NSCs. Medium conditioned by dendritic cells had only one-tenth the enhancing activity seen in co-cultures, indicating that the major effect of the dendritic cells on neurons is mediated by juxtacrine interaction.

Mesenenchymal stem cells exposed to an inflammatory environment have been shown to down regulate immune responses, and to have promise in combating rejection and graft versus host disease after organ transplant (English *et al* 2010, for a review). Bone marrow MSCs (BMSCs) have been heavily investigated as modulators of SCI injury effects. Neirinckx *et al* (2014) have reviewed the effects of BMSCs and neural crest stem cells (NCSCs) that reside in the bone marrow on the primary and secondary phases of SCI in animal experiments (see also Holmes 2017). During the primary damage phase, these cells support neuron survival and axon

SPINAL CORD TRAUMA		RETINAL DISEASE	NEURODEGENERATIVE DISEASES
NON-NEURAL	*NEURAL*	Night blindness (*Gnat -/-*	Parkinson's disease: iPSC-
ADSCs,	Schwann cells	Mouse) : PSC-derived rod	derived neural progenitors;
		Precursors	fetal mesencephalic cells*
Splenic	OECs		
Dendritic cells		Dry AMD: hESC-derived	Huntington's disease: iPSC-
	NSCs	RPE progenitors*	Derived GABA striatal neurons
Bone marrow			
MSCs*	PSC-derived neurons		
hUBCs*			
Macrophages			

Figure 8.2. Summary of cell types that have been found to have positive effects on spinal cord trauma, retinal disease, and neurodegenerative disease. Asterisks indicate cells that have been used in human trials.

regeneration via neuroprotective neurotrophic factors, immunomodulation and regulation of inflammation, antioxidative actions and antiapoptotic actions. During the secondary damage phase, they support remyelination, reduce glial scar formation by degradation of extracellular matrix, and promote angiogenesis.

MSCs also enhance axon regeneration by paracrine action. Rat bone marrow MSCs genetically modified to express the multineurotrophin MNT1, which binds to TRKA, B, and C receptors and p75^{NTR} receptors, enhanced axon regeneration after SCI (Kumagi *et al* 2013). Transplants to rat SCIs of human umbilical cord blood MSCs transfected with a *Wnt3a* gene construct resulted in greater expression of axonal regeneration marker genes and significantly higher scores on motor behavior tests (Seo *et al* 2017). Intravenous delivery of MSCs has been reported to improve functional outcome in rats (Osaka *et al* 2010).

8.2.2.2 Schwann and olfactory ensheathing cells

Rat and human Schwann cells expanded *in vitro* support axon regeneration across rat SCI lesions with modest functional recovery, via secretion of neurotrophic factors (Bunge 2002). A common approach has been to replace a resected segment of spinal cord with a PVC tube filled with a Schwann cell suspension. Significant axon regeneration was reported in SCI rats after transplanting collagen-embedded Schwann cells genetically modified to hypersecrete NGF (Weidner *et al* 1999). However, supraspinal axons typically fail to re-enter the cord after crossing the lesion in these experiments (Plant *et al* 2001).

OECs injected directly into, or rostral and caudal to, SCI lesions in rodents promoted axon regeneration across the lesions and beyond, with partial functional recovery (Santos-Benito and Ramon-Cueto 2003, Richter and Roskams 2008). OECs appear to be neuroprotective through their secretion of neurotrophic factors while also promoting axon extension through their expression of adhesion molecules, preventing cavitation, enhancing vascularization, and promoting the branching of neighboring intact axons (Ramer *et al* 2005, Richter and Roskams 2008). Immortalized OECs transplanted into thoracic hemisection lesions supported the partial functional recovery of tactile sense and proprioceptive functions required to walk a grid without missing the wires (DeLucia *et al* 2003). The inclusion of other cell types, such as Schwann cells, olfactory nerve fibroblasts, and neural precursor cells increased the effectiveness of OECs (Ramon-Cueto *et al* 1998, Barnett and Chang 2004, Wang *et al* 2010).

8.2.2.3 Neural stem and progenitor cells

Partial restoration of function in rodent spinal cords has been reported many times after injecting neural/glial precursors isolated from the cord or differentiated from PSCs *in vitro* (Mothe and Tator 2012). Fetal rat neural precursors expanded *in vitro* enhanced functional recovery of rats after transplantation into SCIs (Ogawa *et al* 2002). Fetal rat neural precursors engineered to express the neural transcription factors HB9, Nkx6.1 and neurogenin proliferated, colonized the ventral horn, expressed motor neuron differentiation markers and projected cholinergic axons into the ventral root (Bohl *et al* 2008). ESC-derived neurons were reported to survive, integrate into surrounding tissue and restore some function in SCI rats (Deshpande *et al* 2006). Rat embryonic glial progenitors injected into lesioned adult rat spinal cord were reported to differentiate into oligodendrocytes and astrocytes and to reduce scarring and expression of inhibitory CSPGs by host cells. Transplantation of astrocytes differentiated from glial-restricted precursors *in vitro* significantly suppressed scarring and improved axon regeneration and functional recovery in SCI rats (Davies *et al* 2006), a result consistent with a subset of glial scar astrocytes being an aid to axon regeneration (chapter 7).

Remyelination of spared axons by oligodedrocyte precursors has been an important goal of SCI cell transplant therapies. Proof of principle was provided by experiments in which glial precursors derived from mouse or human ESCs were shown to remyelinate axons in rodent genetic models of demyelinating disorders (Brustle *et al* 1999, Nistor *et al* 2005). Remyelination of axons by neurospheres derived from NSCs of the lateral ventricles of the brain was also demonstrated in a mouse model of multiple sclerosis (EAE) (Pluchino *et al* 2005). Fifty percent of adult brain-derived neural precursors differentiated into oligodendrocytes or oligoden-drocyte precursors when grafted into SCI lesions of mice, resulting in partial functional recovery (Karimi-Abdolrezaee *et al* 2006). Highly pure populations of oligodendrocyte precursors derived from human ESCs differentiated into oligoden-drocytes that enhanced remyelination and improved motor function when injected into adult rat spinal cord lesions seven days after injury (Kierstead *et al* 2005).

GFP-marked NSCs embedded in fibrin matrices containing a growth factor cocktail and grafted into rat SCIs at the T3 level did not migrate beyond the lesion borders, but extended axons for distances of 25 mm rostrally (to C4) and caudally (to the upper lumbar cord), as well as into host ventral roots (Lu *et al* 2012). Host supraspinal axons regenerated into the graft to make new neuronal relays that restored electrophysiological activity and some movement about each joint of the hind limb; a human NSC line and NSCs derived from hESCs gave similar results when transplanted into rat SCIs using the same protocol. The degree of recovery, however, was indistinguishable from spontaneous recovery. Human fetal spinal cord stem cells transplanted into rat SCIs were reported to improve hind limb paw placement, normalized thermal and tactile pain/escape thresholds, and ameliorated muscle spasticity (van Gorp *et al* 2013).

The clinical usefulness of cell transplants for SCI requires a source of autogeneic NSCs (to avoid immunorejection) that can be easily expanded. Thus, the derivation of neural precursors from PSCs *in vitro* and the direct transdifferentiation of non-neural cells to neural cells *in vitro* and *in vivo* is an area of intense research (Amamoto and Arlotta 2014, Southwell *et al* 2014). NSCs derived from mouse and human iPSCs transplanted into the developing chick neural tube engrafted into the ventral horn at the exact position where endogenous motor neurons reside at stage 31 (Son *et al* 2011). NSCs derived from hiPSCs and embedded in a fibrin matrix with neurotrophic factors differentiated into new neurons with extensive axon growth and synapse formation with host neurons after grafting into immunodeficient rat SCIs created by C5 lateral hemisections (Lu *et al* 2014). These hiPSCs were derived from dermal fibroblasts of an 86 year old man, showing that neurons derived from aged human fibroblasts retain their functional capabilities. Transdifferentiation *in vitro* of human fibroblasts to neurons has been achieved by transduction with genes for neural transcription factors and by microRNAs (Son *et al* 2011, Caiazzo *et al* 2011, Yoo *et al* 2011). Oligodendrocyte precursors capable of myelinating axons in hypomyelinated mice have likewise been made by transdifferentiation with oligo-dendroglia genes (Najm *et al* 2013, Yang *et al* 2013), and pericytes from the human brain have been transdifferentiated into neurons (Karow *et al* 2012). These cells could potentially be expanded for transplant.

Despite a few modestly positive results, the vast majority of NSC transplants has proven no more successful in repairing spinal cord injuries than the use of other cell types (Popovich 2012). A major problem is the difficulty in comparing studies that have used experimental protocols that vary in terms of animal species and strains used, type and extent of lesion inflicted, and controls. Another problem is that the locomotor rating scale of Basso *et al* (1995) used to measure recovery of motor function in rats does not accurately reflect motor recovery in mice. To get around this problem, Basso *et al* (2006) developed a new scale to assess mouse locomotor function, the Basso Mouse Scale (BMS). This scale was found to be a much more sensitive and reliable locomotor measure in SCI mice. Human iPSCs transplanted to contusive SCIs made in NOD-SCID mice survived, migrated and differentiated into neurons, astrocytes and oligodendrocytes (Nori *et al* 2011). The grafted neurons extended axons that formed synapses with host neurons, and increased myelination took place with improvement of motor function as judged by Rotorod test, treadmill gait and BMS score.

8.2.2.4 Clinical trials of cell transplants for SCI

Clinical trials of cell transplant therapies for SCI have been, or are being, conducted worldwide, but in general have been disappointing (Hawryluk *et al* 2008). Transplantation of bone marrow cells has generally been shown to be safe, and at least three studies suggested modest improvements in motor and sensory functions after six months (Park *et al* 2005, Yoon *et al* 2007, Saito *et al* 2008). Umbilical cord blood transplants were reported to partially restore some sensory and motor function (Kang *et al* 2005). Cells of the fetal olfactory bulb have been used in China on a large number of SCI patients, but with little hard evidence for any functional improvement (Dobkin *et al* 2006). Likewise, SCI has been treated in Portugal by transplantation of autologous olfactory mucosal implants with reported motor and sensory improvement (Lima *et al* 2006), but a similar study in Australia found no motor improvement (Feron *et al* 2005).

Macrophages activated by exposure to myelin to induce scavenging of myelin breakdown products have been reported to promote axon regeneration and partial functional recovery after implantation into SCI lesions of adult rats (Rapalino *et al* 1998). In small phase I clinical trials, activated autologous macrophages transplanted within two weeks of SCI, or bone marrow cells transplanted 2–12 years after complete SCI were reported to confer modest functional improvement (Steeves *et al* 2004). However, a study of SCI dogs implanted with autologous myelin activated macrophages showed no evidence of axon regeneration or recovery of motor function (Assina *et al* 2008). None of these trials has resulted in an FDA-approved treatment for SCI.

In a phase I clinical trial, patients suffering from the demyelinating disorder Pelizaeus–Merzbacher disease received transplants of oligodendrocyte precursors. MRI scans taken over 12 months indicated that the transplanted cells were engaged in myelination (Gupta *et al* 2012). Geron conducted the first clinical trials of ESC-derived human oligodendrocyte precursors for remyelination in acute SCI patients (Bretzner *et al* 2011). The trial was subsequently abandoned for financial reasons. At the same time, StemCells derived human adult NSCs from fetal tissue and began clinical trials for chronic spinal cord injury based on positive results in mouse experiments. This trial has since been abandoned as well due to lack of efficacy (Mack 2011). Recently, a pre-clinical trial was reported in which neural progenitor cells, derived from an eight-week old human embryonic spinal cord and maintained as a stable GFP+ cell line, were grafted into cervical SCIs (hemisections)of rhesus monkeys with subsequent immunosuppression to prevent rejection (Rosenzweig *et al* 2018). The grafted cells extended up to hundreds of thousands of axons into the host spinal cord, where they formed reciprocal synaptic connections with host circuits. Likewise, host axons regenerated into the grafts and formed synapses. The grafts matured over nine months, were well-tolerated and improved forelimb function beginning several months after grafting. These results suggest that human SCIs could potentially be treated with grafts of neural progenitor cells derived from autogeneic human iPSCs.

8.2.3 Retinal regeneration

The application of stem cell therapy to retina regeneration perhaps holds the most promise currently for restoring the structure and function of a neural tissue by cell transplantation (Roska and Sahel 2018, for a review). The eye is ideal for treatment with stem cells because it is externally visible, requires fewer cells due to its small size, and is an immune-privileged site, meaning that allogeneic cells can be used. In addition, transplanted cells can be tracked non-invasively by a technique called optical coherence tomography (OCT), which gives high-resolution detail.

The $Gnat1^{-/-}$ mouse is a model of stationary night blindness caused by a lack of rod photoreceptors. $Gnat1^{-/-}$ mice regained night vision after transplant of rod precursors from normal mouse eyes. Human iPSCs directed to a retinal progenitor fate were able to integrate into a normal mouse retina and express photoreceptor markers (Lamba *et al* 2010). When rod precursors derived from mouse ESCs were injected subretinally into the $Gnat1^{-/-}$ mouse, they migrated and integrated into the recipient retina (Gonzalez-Cordero *et al* 2013). Here, they formed mature rods that connected into the existing retinal circuitry and responded to pharmacological stimuli, like endogenous rods, thus showing that PSCs are potential sources for photoreceptors and perhaps other types of retina cells to be used as regenerative therapy.

Although there are literally hundreds of diseases that affect the eye, a prime therapeutic target of eye disease is age-related macular degeneration (AMD), the leading cause of vision impairment and blindness in the developed world. There are two forms of AMD, wet and dry, and both cause loss of central vision by destruction of photoreceptors. Wet AMD is the less common (10% of cases), but is the more serious of the two. Photoreceptor destruction is due to leakage from blood vessels into the central retina. The dry form of AMD (90% of cases), often referred to as atrophic AMD, is due to the degeneration and death of RPE cells in the macular region of the retina and is characterized by macular thinning and the formation of yellow crystalline deposits called drusen. The RPE nourishes the photoreceptor layer of the retina and its compromise thus causes a loss of photoreceptors. An inherited form of juvenile macular degeneration similar to the dry form of AMD is Stargardt's disease (SD).

Two clinical trials for safety and tolerability have been conducted to determine the efficacy of RPE progenitors derived from hESCs to improve dry AMD and SD (Schwartz *et al* 2015, Song *et al* 2015). These trials involved 18 Caucasian patients followed for a median time of 22 months and four Asian patients followed for one year, respectively. Cells were transplanted into one eye, leaving the other as an untreated control. No significant adverse post-operative effects were encountered in either trial. Most encouraging was the finding that visual acuity improved in the treated eye, sometimes quite substantially, in a high percentage of patients. None of the treated eyes lost visual acuity, whereas in most of the untreated eyes visual acuity continued to decline. Further clinical trials are underway in the US, Britain and Japan to treat AMD and other eye diseases by photoreceptor and RPE replacement using hiPSCs derivatives, as well as paracrine treatment of eye tissues via transplantation of human umbilical cord tissue (Perkel 2014).

8.2.4 Pluripotent stem cells and neurodegenerative disease

Neurodegenerative diseases such as Parkinson's, Huntington's, Alzheimer's and amyotrophic lateral sclerosis (ALS, Lou Gehrig's disease) are increasing as the world population ages, and are costly in terms of the health care required. These diseases are characterized by loss of neurons from different parts of the brain and spinal cord that rob an individual of his or her physical and/or mental abilities. The feature common to all of them is the misfolding of proteins to a pathological state in which they can induce their normal counterparts to misfold and form aggregates that compromise cell function, a new form of infectivity. Such proteins are called 'prion-like' because they mimic the action of the prion protein, which has been shown to cause encephalopathies such as the wasting disease of deer, mad cow disease, kuru and Creutzfeld–Jacob disease (Prusiner 2014, for a fascinating review). The prion protein is transmissible through the food chain, and there may be other means of transmission of prion-like proteins through cell interactions (Polymenidou and Cleveland 2011, Prusiner 2014, Walker and Jucker 2015, Jaunmuktane *et al* 2015, Frontzek *et al* 2016, Abbott 2016, for reviews).

Pluripotent stem cells are viewed as sources of neurons to replace cells lost to neurodegenerative disease, or as 'drugstores' that secrete paracrine neuroprotective factors to inhibit disease progression. Human iPSCs have been made from the fibroblasts of patients with Parkinson's (Wernig *et al* 2008), Alzheimer's (Israel *et al*

2012), ALS (Hedges *et al* 2016) and Huntington's (Carter *et al* 2014) to study the cellular and molecular basis for initiation and progression of these diseases and to screen drugs for therapeutic effects (disease in a dish model).

Autologous neural progenitors derived from iPSCs of Parkinsonian rats and monkeys showed long-term survival after transplant into their brains with little sign of inflammatory cells or reactive glia (Emborg *et al* 2013, Hallet *et al* 2015). Experiments transplanting human iPSC-derived neurons have begun to explore their therapeutic effects on neurodegenerative disease. Jeon *et al* (2012) directed the differentiation of iPSCs from a Huntington's patient to GABAergic striatal neurons, the neurons most susceptible to this disease. When transplanted into the striatum of a rat model of HD, significant behavioral recovery was observed.

Cell transplants for neurodegenerative diseases must remain healthy to function over the long term. There is some evidence from autopsies of patients who had received transplants of fetal mesencephalic cells suggesting that misfolded α-synuclein in the host tissue had been transmitted to healthy graft cells. Histological examination of the brain revealed the presence of α-synuclein positive Lewy bodies in some graft cells (Kordower *et al* 2008). Consistent with transmissibility of infective protein, injection of brain extracts from mice containing aggregated α-synuclein or synthetic human or mouse α-synuclein stimulated Lewy body formation, pathology and neurodegeneration in normal mouse brain (Mougenot *et al* 2012, Luk *et al* 2012a, 2012b, Masuda-Zuzukake *et al* 2013). More evidence is needed, however, to definitively establish the probability of transmission of misfolded proteins from grafted cells.

8.3 Liver and pancreas

8.3.1 Liver

There is a great interest in cell transplant therapies for liver failure and type 1 diabetes. Protocols have been developed to direct the differentiation of mouse iPSCs to hepatocytes (Zaret and Grompe 2008, Si-Tayeb *et al* 2010). Hepatocytes derived from human iPSCs repopulate mouse livers compromised by dimethylnitrosamine to mimic liver cirrhosis, and repair the tissue (Liu *et al* 2012). Mouse fibroblasts have been transdifferentiated directly to hepatocytes (Sekiya and Suzuki 2011, Guo *et al* 2017). These hepatocytes were able to rescue the livers of tyrosinemic mice after transplant, although not as efficiently as transplants of primary hepatocytes.

Hepatocyte transplants are not yet in the clinic. A major issue for clinical application is the availability of human hepatocytes. High-throughput screening has identified two classes of small molecules that have the potential for generating renewable sources of functional human hepatocytes (Shan *et al* 2013). The first class stimulated the proliferation of primary human hepatocytes *in vitro*, while the second class promoted the differentiation of hiPSC-derived hepatocytes toward a more mature phenotype than previously obtainable. Human fibroblasts have been trans-differentiated to a multipotential progenitor cell (hiMPC) state from which endoderm cells and subsequently hepatocytes can be differentiated (Zhu *et al*

2014). These hepatocytes underwent further maturation after transplantation and could repopulate and rescue tyrosinemic mouse livers.

8.3.2 Pancreas

Remarkable advances have been made in developing cell transplant therapies for type 1 and type 2 diabetes. The types of therapies under consideration for clinical trials have been reviewed recently by Zhou and Melton (2018). Type 1 diabetes is an autoimmune disorder that attacks β-cells. To replace these cells in diabetic mice, fibroblasts have been reprogrammed into endoderm-like cells (Li *et al* 2014). Using a differentiation protocol that included retinoic acid and the small synthetic molecules A83-01 (a TGF-β receptor inhibitor), LDE225 (a Hedgehog pathway inhibitor) and 2-phospho-L-ascorbic acid (pVc), these cells were differentiated to immature insulin-secreting cells that matured when transplanted under the kidney capsule and reduced diabetic hyperglycemia. Human ESC cultures have been derived by SCNT from somatic cells of a newborn and an adult woman with type 1 diabetes (Yamada *et al* 2014). These cells provide the potential to generate β-cells for transplant, drug screening and study of the cell and molecular biology of diabetes (figure 8.3). iPSCs have also been induced from human pancreatic β-cells and will be useful to study the genesis and progression of diabetes (Mayhew and Wells 2010).

Roughly 340–750 million islet cells are required to resolve type 1 diabetes in islet transplantation protocols (McCall and Shapiro 2012). To be clinically useful, we need to be able to derive from human PSCs this number of mature β-cells capable of sensing blood glucose concentration. This requirement has been met by a protocol enabling the production of 300 million mature β-cells per 1000 ml of culture medium from either human ESCs or iPSCs (Pagliuca *et al* 2014). Combinations of growth factors and other molecules important to pancreatic development are applied sequentially to the PSCs to modulate multiple developmental signaling pathways

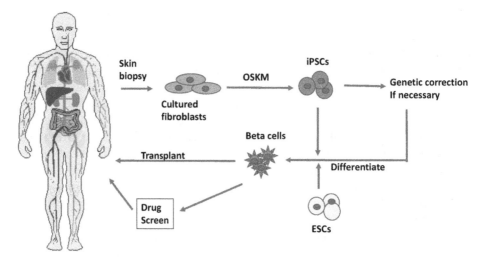

Figure 8.3. Use of ESCs and iPSCs to derive differentiated cell types for transplant and drug screening. The example here is β-cells of the pancreas.

leading to β-cell differentiation. The morphology and insulin expression of β-cells derived in this way are identical to native β-cells, they are glucose-sensitive, and they ameliorate hyperglycemia when transplanted into diabetic mice. They represent a means of replacing β-cells destroyed by disease, as well as opportunities to study the genesis of type 1 diabetes. Start-up companies for β-cell production have been formed and clinical trials of PSC-derived β-cells have been initiated.

Type 2 diabetes is characterized by peripheral insulin resistance that leads progressively to the same end point as type 1 diabetes, loss of β-cells. Type 2 diabetes may also be amenable to treatment by cell transplant. A high-fat diet (HFD) model of type 2 diabetes has been developed (Bruin *et al* 2015). Pancreatic progenitor cells derived from human ESCs partially improved glucose tolerance after transplantation to HFD mice. Full glucose tolerance was achieved, however, by combining the cell transplants with either sitagliptin or metformin, two drugs used to lower glucose production by the liver and decrease insulin resistance.

8.4 Musculoskeletal tissues

8.4.1 Duchenne muscular dystrophy

Cell transplant therapy for DMD has focused on the replacement of dystrophic myofibers with self-renewing wild-type SCs delivered by injection or in a biodegradable scaffold (Skuk and Tremblay 2011). Early trials injecting wild-type SCs into mouse *mdx* muscle or into the muscles of DMD patients were minimally effective because only a small percentage of the grafted cells were self-renewing SCs. When higher percentages of self-renewing skeletal muscle precursors were isolated and injected into *mdx* muscles they differentiated into myofibers and also self-renewed. Dystrophin expression was restored and muscle histology and contractile function was significantly improved (Cerletti *et al* 2008).

A major problem with injection of even self-renewing satellite cells directly into muscles is that the cells do not migrate uniformly from individual injection sites and dystrophin does not diffuse throughout myofibers after stem cell fusion (Skuk and Tremblay 2011). The regenerative process is thus restricted to a small volume of muscle. This problem might be solved by a combination genetic correction and transplant approach using a mouse model lacking both dystrophin and utrophin that approximates human DMD more closely than the *mdx* model. iPSCs derived from fibroblasts of this mouse and genetically corrected for the dystrophic genotype were differentiated into Pax7+ skeletal muscle precursors. When transplanted into the tibialis anterior muscle these precursors seeded the satellite cell compartment and regenerated new myofibers that responded to injury by regeneration and self-renewal (Filareto *et al* 2013). More importantly, corrected cells could engraft over a large muscle volume after intravenous injection, which would carry the cells into the muscle by vascular routes. To improve engraftment potential, human iPSC-derived DMD muscle progenitor cells were corrected by CRISPR/Cas9 and cells bearing the muscle-specific receptors ERBB3 and nerve growth factor receptor (NGR) were selected. These cells efficiently restored dystrophin levels after transplant to dystrophin-deficient muscles of the *mdx* mouse (Hicks *et al* 2018).

8.4.2 Critical size gaps in bone

Injection of osteogenic cell suspensions is insufficient to repair CSDs because the cells are not constrained to remain in the gap, so the cells are seeded into a biodegradable scaffold, with or without growth factors, that is then implanted into the gap. The scaffold does not necessarily offer mechanical support, but is osteoconductive, allowing the osteogenic cells to migrate, and osteoinductive, promoting proliferation and differentiation to produce new bone (figure 8.4). The most commonly used osteogenic cells are MSCs of one type or another, but particularly bone marrow MSCs.

Osteoconductive and inductive scaffolds seeded with culture-expanded bone marrow MSCs were developed in the 1990s and the first decade of the 21st century. The delivery scaffolds were hydroxyapatite/β-tricalcium phosphate cylinders, or synthetic polymers alone or in combination with calcium or hydroxyapatite that can be formed as pastes and pressed into CSDs. Other scaffolds have been made from synthetic poly (lactic-co-glycolic acid) and poly(ε-caprolactone). These studies showed that animal or human MSCs were capable of making new bone in CSDs of both endochondral or intramembranous bones of rodent models (Cowan *et al* 2004, Dupont *et al* 2010). However, the volume and strength of the bone formed, though improved over controls, does not reach the original level, even when exogenous growth factors are delivered with the MSCs.

Vascularization is essential for bone formation in CSDs and its importance in regeneration of bone across a CSD has long been recognized (Hankenson *et al* 2011). One of the most important growth factors to stimulate vascularization is

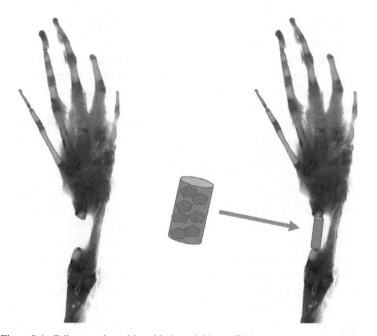

Figure 8.4. Cells transplanted in a biodegradable scaffold to regenerate gaps in bone.

VEGF (Keramaris *et al* 2008). The discovery that endothelial colony forming cells are the actual endothelial progenitor cells that participate in angiogenesis (Yoder *et al* 2007) led to the use of ECFCs as a means to improve the vascularization of regenerating bone in rat femur CSDs (Chandrasakar *et al* 2011). Incorporation of ECFCs into hydroxyapatite/tricalcium phosphate (HA/TCP) scaffold implanted into a CSD promoted significantly more vascularization and bone formation than control scaffolds lacking ECFCs. While partially successful in experimental animals, none of these cell transplant strategies have yet attained clinical status due to less than optimal bone filling and/or integration with the remaining bone ends.

A new type of MSC has been identified in fragments of cortical bone that has the same multipotent differentiation capability as bone marrow MSCs, but which are larger in size, respond more readily to pro-inflammatory cytokines, have lower proliferation kinetics and display greater commitment toward the osteogenic lineage (Corradetti *et al* 2015). These differences appear to be the result of specialized niche factors that may make these cells a superior candidate for use in cell transplants to repair CSDs.

Electroacupuncture (EA) is a form of acupuncture where a small electric current is passed between pairs of acupuncture needle to reduce electrical resistance and increase conductivity. Two sets of acupuncture points on the hind limb of rats and limbs of humans were stimulated that promoted the release of MSCs into the circulation via a hypothalamic/sympathetic nerve pathway (Salazar 2017). This treatment induced repair processes in the partially ruptured Achilles tendon. When stimulated in human subjects at these points, there was a similar release of MSCs into the circulation from adipose tissue. These results suggest that EA could be a means to mobilize MSCs from adipose tissue to heal various types of injured musculoskeletal tissues.

8.4.3 Articular cartilage defects

8.4.3.1 Therapy for chondral and osteochondral defects

Cell transplants are a rapidly developing approach to the repair of articular cartilage (Orth *et al* 2014, Makris *et al* 2015, Mardones *et al* 2015). In the 1980s, the successful repair of osteochondral defects in animals by chondrocyte transplants led to a clinical therapy called autologous chondrocyte implantation (ACI) (Brittberg 2008). This therapy is in wide use today for traumatic injury to articular cartilage under the name Carticel, a division of Genzyme Corporation. A small piece of autologous cartilage is removed by arthroscopy from healthy non-weight bearing cartilage of the injured joint and the chondrocytes enzymatically disassociated (figure 8.5). The chondrocytes are then expanded *in vitro* where they dedifferentiate, and are then injected as a concentrated suspension of 3–5 million cells under a flap of periosteum stitched over the lesion, where they redifferentiate into new hyaline cartilage. Transplanting the dedifferentiated cells in a collagen hydrogel enhanced redifferentiation (Ochi *et al* 2001). Good clinical outcomes over 2–9 years have been reported for ACI treatment of patients with femoral condylar lesions (Kon *et al* 2009), although a five-year study comparing ACI with microfracture for femoral

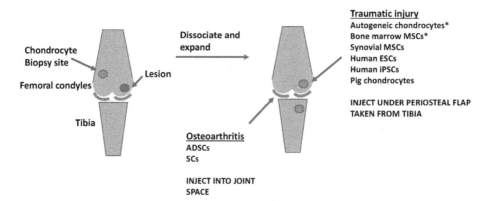

Figure 8.5. The Brittberg technique of repairing chondral injuries (blue) due to trauma. A biopsy (green) is taken from a non-weight-bearing region of articular cartilage. The chondrocytes are dissociated, expanded *in vitro* and injected as a pellet under a periosteal flap (red) taken from the tibia. Chondrocytes derived from other cell types have also been tested in animal experiments and in human trials (asterisks). More recently, adipose stem cells and satellite cells have been tested for their effects on damage due to osteoarthritis. Purple = menisci.

condylar defects revealed no significant difference in outcome (Knutsen *et al* 2007). Physical activity has been shown to improve the results of ACI even more (Kreuz *et al* 2007).

The use of other cell sources to supply chondrocytes for ACI would avoid the need to harvest a piece of the patient's knee cartilage. Bone marrow MSCs are able to regenerate chondrocytes in defects of animal articular cartilage (Ajibade *et al* 2014, Grassel and Lorenz 2014, Yamasaki *et al* 2014). Clinical use of bone marrow MSCs to repair chondral defects has been limited. Injection of a patient's knee with autologous bone marrow MSCs and platelet lysate, along with dexamethasone, was reported to reduce the size of a medial femoral chondral defect (Centeno *et al* 2008). MRI data showed a significant decrease in defect size and improvement in pain and function measures three months later.

Synovial MSCs can substitute for articular chondrocytes in rabbit chondral defects (Jones and Pei 2012), and ectopic cartilage for ACI has been induced from synovial cells *in situ* by stitching collagen/ fibrin patches impregnated with BMP-2 to the synovial membrane (Hunziger *et al* 2015). Human ESCs and iPSCs have been differentiated into MSCs for use in the production of chondrocytes (Lian *et al* 2012a) and hESCs have been differentiated directly into chondrocytes *in vitro* (Oldershaw *et al* 2010, Cheng *et al* 2013). Clinical use of MSCs in osteochondral defects has been limited so far, but there are reports that implants of circulating MSCs and bone marrow MSCs reduced pain and improved cartilage repair for up to six years (Lee *et al* 2012, Saw *et al* 2013). In addition to the often poor redifferentiation of dedifferentiated chondrocytes and MSCs into articular cartilage that mimics native articular cartilage, a major issue with cell transplants to heal chondral defects is the less than satisfactory integration of the transplant-derived cartilage with the cartilage that surrounds the defect space.

The development of protocols for high quality chondrocyte differentiation from hESCs and hiPSCs, and xenogeneic sources holds the promise of unlimited cell sources for repair of articular cartilage. Human iPSCs as pellets or incorporated into an alginate hydrogel formed articular cartilage of substantially higher quality in rat osteochondral defects than defects filled with alginate alone (Ko *et al* 2014). Xenogeneic chondrocytes derived from pig femoral condyle and cultured for 14–21 days formed tissue with the characteristics of articular cartilage when implanted into chondral defects in rabbit femoral condyles under a periosteal flap (Ramallal *et al* 2004). Interestingly, the new tissue was composed of only 27.5% pig cells. Since the periosteal flap does not appear to contribute to new cartilage (Kajatani *et al* 2004), this suggests that the xenografted cells might have stimulated host chondrocytes to divide by paracrine or juxtacrine action. Similar results were obtained in a model consisting of pig chondrocytes implanted into defects drilled into cores of human articular cartilage maintained *in vitro* (Fuentes-Boquete *et al* 2004).

8.4.3.2 Therapy for osteoarthritis

A major orthopedic problem is the destruction of articular cartilage with age and/or overuse by osteoarthritis that afflicts a large percentage of the population. Tearing and maceration of the menisci is the cause of many knee problems and often leads to osteoarthritis. Several case reports of patients treated with adipose-derived stem cells in a differentiation-promoting mixture of hyaluronic acid, platelet-rich plasma (PRP) to provide multiple growth factors, calcium chloride to activate PRP, and low-dose dexamethasone indicate that this protocol enhances the healing of meniscal fibrocartilage (Pak 2011, Pak *et al* 2014). However, multiple treatments were required over a period of months to achieve maximum pain reduction, as opposed to one-time surgical knee replacement. Mesenchymal stem cells injected into osteoarthritic knees (figure 8.5) have been reported to improve symptoms, and several clinics are now offering MSC treatments for osteoarthritis (Orozco *et al* 2013, Koh *et al* 2013, Grassel and Lorenz 2014, Kim *et al* 2014). A large number of clinical trials are underway to test the efficacy of MSC and ADSC therapies for osteoarthritis (Orth *et al* 2014).

8.5 Cardiac tissues

8.5.1 Satellite and bone marrow cell infusions

A variety of cell types have been transplanted to myoinfarcted hearts in human patients and experimental animals (figure 8.6). Adult stem cell transplants have been extensively explored as a means of rescuing the structure and function of the failing heart in patients (Segers and Lee 2008, Hansson *et al* 2009). Satellite cells and bone marrow cells have been tested in clinical trials for their ability to enhance left ventricular function after infusion or injection into myocardial infarct regions. There is no evidence that these cells transdifferentiate into cardiomyocytes (Reinecke *et al* 2002). A summary of the results of 28 clinical trials from 2002–20007 in which bone marrow cells were delivered to infarcted myocardium indicated that the effect on left ventricular ejection fraction (LVEF) was highly variable, ranging from −3.0% to +12%, with most values ranging from +2.8% to +9% (Segers and Lee 2008). Improvements have been attributed to paracrine effects that promote angiogenesis,

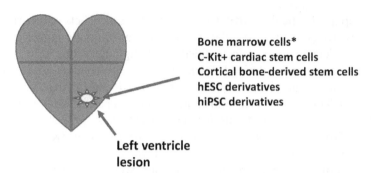

Bone marrow cells*
C-Kit+ cardiac stem cells
Cortical bone-derived stem cells
hESC derivatives
hiPSC derivatives

Left ventricle
lesion

Figure 8.6. Delivery of stem cells or stem cell derivatives to myocardial infarcts. Asterisk indicates human trials.

resulting in the delivery of more oxygen to the heart muscle. This is consistent with evidence that bone marrow cells injected into infarcted pig hearts up regulated expression of the angiogenesis-promoting factors FGF-2, VEGF, IL-1β and TNF-α (Kamahita *et al* 2001), and that injection of VEGF into myocardial infarcts of sheep significantly increased angiogenesis in the peri-infarct region (Chachques *et al* 2004). In addition, genetic marking experiments suggested that bone marrow hemato-poietic c-Kit cells induce the differentiation of unknown progenitor cells of the heart into new cardiomyocytes (Loffredo *et al* 2011).

To further analyze the variability of the therapeutic effects of bone marrow cell transplants on infarcted hearts, 49 clinical trials were examined for discrepancies (dissimilarities) in study design, methods and baseline characteristics, as well as the effect of sample size on therapeutic results (Nowbar *et al* 2014). Over 600 discrepancies were found in these studies. Only five trials were highly similar in design and these showed essentially zero effect on ejection fraction. In the other trials, increasing numbers of discrepancies were correlated with increasing positive effect on ejection fraction. This data implies that bone marrow cell infusion or injection does not have even a paracrine therapeutic effect on left ventricular function. Despite the lack of evidence for beneficial effects of bone marrow cells, even more ambitious clinical trials are underway by cardiologists who are convinced of the paracrine potential for bone marrow cell transplants to ameliorate damage to, or regenerate, myocardial tissue after myocardial infarction (Doss 2015).

8.5.2 Cardiac c-Kit cells

A population of c-Kit+ stem cells (c-Kit cardiac stem cells, CSCs) identified in the adult rat heart has been reported to differentiate into cardiomyocytes and endothe-lial cells when injected into the borders of cardiac infarcts in adult rats (Beltrami *et al* 2003, Bearzi *et al* 2007). Other studies, however, have suggested that these c-Kit+ cells do not contribute differentiated cardiomyocytes to the developing or adult heart, or after injury, and do not develop into cardiomyocytes after transplantation into injured heart muscle (Zaruba *et al* 2010), but do generate cardiac endothelial cells (van Berlo *et al* 2014). A clinical trial is underway to test the safety and feasibility of using autologous c-Kit+ cardiac cells as an adjunct therapy for coronary bypass patients.

Cortical bone-derived stem cells (CBSCs) have been compared with CSCs for their ability to improve cardiac function in infarcted hearts of mice (Ong and Wu 2013). CBSCs were reported to improved survival, cardiac function and angiogenesis, and to reduce infarct size better than CSCs, through both paracrine factors (FGF-2 and VEGF) and differentiation into immature cardiomyocytes, endothelium and vascular smooth muscle.

8.5.3 Cardiomyocytes derived from PSCs

Human cardiomyocytes have been derived in high yield from ESC cultures by temporal modulation of Wnt signaling under defined growth factor-free conditions (Lian *et al* 2012b). ESC-derived human cardiomyocytes have been shown to improve the function of infarcted mouse and rat hearts (Caspi *et al* 2007, van Laake *et al* 2007). Human ESC-derived cardiomyocytes (hESC-CMs) transplanted to injured guinea pig or monkey hearts contracted synchronously with host myocardium and improved heart function (Shiba *et al* 2016). The grafts protected against arrhythmia, but showed both electrically coupled and uncoupled regions compared to the uniformly coupled cardiomyocytes of the normal heart. A preclinical study of hESC-CMs transplanted to infarcted hearts of four monkeys showed extensive remuscularization of myocardium compared to three control infarcted hearts. Functional improvement was variable, however, yielding no statistically significant data due to the small sample size (Chong *et al* 2014).

Transplantation of cardiomyocytes derived from iPSCs in several animal models of myocardial infarct has had only limited effect due to insufficient optimization (maturation) of the injected cells, resulting in poor engraftment. By contrast, hiPSC-derived cardiomyocytes matured for 20 days after initial differentiation engrafted well into infarcted mouse hearts, where they proliferated, underwent further maturation and provided significant functional improvement. In another study, cardiomyocytes derived from a monkey iPSC culture and grafted to the infarcted hearts of five other monkeys survived for 12 weeks with minimal immunosuppression, regenerated myocardial tissue, and improved contractile function (Funakoshi *et al* 2016).

There are still some problems to be solved before cardiomyocytes derived from pluripotent stem cells can be used clinically. Ventricular arrythmias were observed in monkeys after transplant of PSC-derived cardiomyocytes (Chong *et al* 2014), suggesting that this could be a problem for similar transplants in human hearts. An *in vitro* model of right ventricular arrhythmia has been developed using cardiomyocytes derived from hiPSCs of patients with the genetic disease arrhythmogenic right ventricular dysplasia/cardiomyopathy (ARVD/C) to study ways of dealing with transplant arrhythmia (Kim *et al* 2013). Another problem is that the derived cardiomyocytes do not achieve an adult level of maturation. To address this problem, a 3D culture system combined with electrical stimulation has been developed that markedly improves the maturation of derived cardiomyocytes (Nunes *et al* 2013).

Cardiomyocytes derived by transdifferentiation of cardiac fibroblasts offer a potential regenerative therapy for injured myocardium. So far, cardiac fibroblasts of mice have been transdifferentiated *in vitro* to cardiomyocytes (Efe *et al* 2011, Zwi-Dantsis *et al* 2012), but since yields of these cells are low they have not yet been used for transplant studies. There have been no reports of successful transdifferentiation of cardiac fibroblasts *in situ,* but we may expect success in the near future.

8.6 Hematopoietic deficiency disorders

Transplants of genetically corrected cells have been used to treat genetic hematopoietic disorders caused by faulty genes. Several gene therapy trials have been conducted to cure severe combined immunodeficiency disease in children due to adenosine deaminase deficiency (ADA-SCID) or SCID-X1, an X-linked inherited disorder caused by γc cytokine deficiency leading to defective T and NK lymphocyte differentiation or function. Blaese *et al* (1995) transduced the peripheral T-cells of two patients *ex vivo* with a retroviral construct containing the gene for ADA and infused the transfected cells back into the patients. In a study in which ADA replacement enzyme therapy was not available, $CD34^+$ HSCs from two patients were transduced with a retroviral vector-ADA gene and infused back into the patients (Aiuti *et al* 2002). The infusions were preceded by low-intensity non-ablative chemotherapy to provide a developmental advantage to the transduced cells. The patients responded well to the therapy, showing normal growth and development. There was sustained engraftment and differentiation of the transduced HSCs into multiple blood cell lineages and improved immune function.

SCID-X1 is a disease of males caused by inactivating mutations in the gene encoding the interleukin-2 receptor common gamma chain (IL2RG), located on the X-chromosome. Affected boys fail to develop T-cells and NK cells and suffer from a functional B cell defect. While the disease can be treated by a matched bone marrow transplant, such a transplant is available only to a third of boys. SCID-X1 has been treated with autologous $CD34^+$ HSCs transduced *ex vivo* with a correct copy of IL2RG-cDNA in a retroviral vector (Cavazzana-Calvo *et al* 2000, Baum *et al* 2004). Counts of B, T and NK cells were comparable to age-matched normal children, as were the functions of these cells, including antigen-specific responses and the recipients had normal growth and development without side effects.

Despite largely positive results, several of these patients developed leukemia, which has been attributed to the insertion of the retrovirus near the cancer-promoting gene LMO-2 (Cavazzana-Calvo *et al* 2004). Further trials were subsequently put on hold for a short time, pending a review of the retroviral vector problem. Trials were renewed with lentiviral vectors, which proved not to cause tumors when introduced into mouse HSCs (Montini *et al* 2006). Lentiviral vectors have since been used to treat several kinds of genetic disorders. A lentiviral gene construct has been used to treat X-linked adrenoleukodyastrophy (ALD) in two boys (Cartier *et al* 2009). ALD is specific to males and is due to a mutation in the *ABCD* gene encoding the ALD protein, causing a deficiency that disrupts myelin maintenance in oligodendrocytes and microglia. Autogeneic HSCs were genetically corrected *ex vivo* by lentiviral introduction of a wild-type *ABCD* gene and re-infused after myeloablative treatment. Starting at 14–16 months post-transplant, demyelination halted. Many cell types of the hematopoietic system expressed the ALD protein, but the effect was probably mediated by microglia, which have a monocytic origin. Another hematopoietic disorder treated by reinfusions of autogeneic HSCs transfected with normal genes in lentiviral vectors is HbE/β-thalassemia, a disorder in which a mutation in the hemoglobin gene compromises hemoglobin production (Cavazzana-Calvo *et al* 2010).

Other genetic diseases that have been treated by transplantation of cells genetically corrected with lentiviral constructs are: Wiscott–Aldrich syndrome, a primary immunodeficiency (Aiuti *et al* 2013); metachromatic leukodystrophy (MLD), a neurodegenerative lysosomal disorder caused by aryl sulfatase A deficiency (Biffi *et al* 2013), in which rescue is provided by elevation of aryl sulfatase A activity in cerebrospinal fluid; and SCID-X1, where a lentiviral vector was used to introduce a recombination DNA template of the normal gene (Genovese *et al* 2014). Hurler syndrome is a genetic disease in which there is abnormal buildup of glycosaminoglycans due to mutation of the gene for α-L-iduronidase. The disease can be treated by allogeneic transplant of HSCs from a family member (Souillet *et al* 2003), but the long-term outcome is highly variable (Aldenhoven *et al* 2015). It has been possible to genetically correct iPSCs from Hurler patients *in vitro* with the virally delivered α-L-iduronidase gene and to then differentiate these corrected cells into hematopoietic cells. Clinical trials of these cells have not yet been published (Tolar *et al* 2011).

References

Abbott A 2016 The red-hot debate about transmissible Alzheimer's *Nature* **531** 294–7

Aiuti A *et al* 2002 Correction of ADA-SCID by stem cell gene therapy combined with nonmyeloablative conditioning *Science* **296** 2410–3

Aiuti A *et al* 2013 Lentiviral-based gene therapy of hematopoietic stem cells in Wiskott–Aldrich syndrome *Science* **341** 865

Ajibade D A *et al* 2014 Emerging applications of stem cell and regenerative medicine to sports injuries *Orthoped. J. Sports Med.* **2** 2325967113519935

Akiyama Y, Radtke C and Kocsis J D 2002 Remyelination of the rat spinal cord by transplantation of identified bone marrow stromal cells *J. Neurosci.* **22** 6623–30

Amoh Y *et al* 2009 Human hair follicle pluripotent stem (hfPS) cells promote regeneration of peripheral-nerve injury: an advantageous alternative to ES and iPS cells *J. Cell Biochem.* **107** 1016–20

Aldenhoven M *et al* 2015 Long-term outcome of Hurler syndrome patients after hematopoietic cell transplantation: an international multicenter study *Blood* **125** 2164–72

Amamoto R and Arlotta P 2014 Development-inspired reprogramming of the mammalian central nervous system *Science* **343** 1239882

Amoh Y *et al* 2009 Human hair follicle pluripotent stem (hfPS) cells promote regeneration of peripheral-nerve injury: an advantageous alternative to ES and IPS cells *J. Cell. Biochem.* **107** 1016–20

Assina R *et al* 2008 Activated autologous macrophage implantation in a large-animal model of spinal cord injury *Neurosurg. Focus* **25**(5) E3

Barnabe-Heider F and Frisen J 2008 Stem cells for spinal cord repair *Cell* **3** 16–24

Barnett S C and Chang L 2004 Olfactory ensheathing cells and CNS repair: going solo or in need of a friend? *Trends Neurosci.* **27** 54–60

Basso D M, Beattie M A and Bresnahan J C 1995 A sensitive and reliable locomotor rating scale for for open field testing in rats *J. Neurotrauma* **12** 1–21

Basso D M *et al* 2006 Basso mouse scale for locomotion detects differences in recovery after spinal cord injury in five common mouse strains *J. Neurotrauma* **23** 635–59

Baum C *et al* 2004 Chance or necessity? Insertional mutagenesis in gene therapy and its consequences *Mol. Therapy* **9** 5–13

Bearzi C *et al* 2007 Human cardiac stem cells *Proc. Natl Acad. Sci. USA* **104** 14068–73

Beltrami A P *et al* 2003 Adult cardiac stem cells are multipotent and support myocardial regeneration *Cell* **114** 763–76

Biffi A *et al* 2013 Lentiviral hematopoietic stem cell gene therapy benefits metachromatic leukodystrophy *Science* **341** 864

Blaese R M *et al* 1995 T lymphocyte-directed gene therapy for ADA⁻ SCI: Initial trial results after four years *Science* **270** 475–80

Bohl D *et al* 2008 Directed evolution of motor neurons from genetically engineered neural precursors *Stem Cells* **26** 2564–75

Bretzner F, Gilbert F, Baylis F and Brownstone R M 2011 Target populations for first-in-human embryonic stem cell research in spinal cord injury *Cell Stem Cell* **8** 468–75

Brittberg M 2008 Autologous cartilage implantation-technique and long-term follow-up *Injury* **39** S40–9

Bruin J E *et al* 2015 Diet-induced diabetes and obesity with human embryonic stem cell-derived pancreatic progenitor cells and antidiabetic drugs *Stem Cell Rep.* **4** 605–20

Brustle O *et al* 1999 Embryonic stem cell-derived glial precursors. A source of myelinating transplants *Science* **285** 754–6

Bunge M B 2002 Bridging the transected or contused adult rat spinal cord with Schwann cell and olfactory ensheathing glia transplants *Prog. Brain Res.* **137** 275–82

Caiazzo M *et al* 2011 Direct generation of functional dopaminergic neurons from mouse and human fibroblasts *Nature* **476** 224–7

Carter R *et al* 2014 Reversal of cellular phenotypes in neural cells derived from Huntington's disease monkey-induced pluripotent stem cells *Stem Cell Rep.* **3** 1–9

Cartier N *et al* 2009 Hematopoietic stem cell gene therapy with a lentiviral vector in X-linked adrenoleukodystrophy *Science* **326** 818–23

Caspi O *et al* 2007 Transplantation of human embryonic stem cell-derived cardiomyocytes improves myocardial performance in infarcted rat hearts *J. Am. Coll. Cardiol.* **50** 1884–93

Cavazzana-Calvo M *et al* 2000 Gene therapy of human severe combined immunodeficiency (SCID-X1) disease *Science* **288** 669–72

Cavazzana-Calvo M, Thrasher A and Mavilio F 2004 The future of gene therapy *Nature* **427** 779–81

Cavazzana-Calvo M *et al* 2010 Transfusion independence and *HMGA2* activation after gene therapy of human β-thalassemia *Nature* **467** 318–22

Centeno C J, Busse D, Keohan C and Freeman M 2008 Increased knee cartilage volume in degenerative joint disease using percutaneously implanted, autologous mesenchymal stem cells, platelysate and dexamethasone *Am. J. Case Rep.* **9** 201–6

Cerletti M *et al* 2008 Highly efficient, functional engraftment of skeletal muscle stem cells in dystrophic muscles *Cell* **134** 37–47

Chachques J C *et al* 2004 Angiogenic growth factors and/or cellular therapy for myocardial regeneration: a comparative study *J. Thor. Cardiovasc. Surg.* **128** 245–53

Chandrasakar K S *et al* 2011 Blood vessel wall-derived endothelial colony-forming cells enhance fracture repair and bone regeneration *Calcif. Tissue Int.* **89** 347–57

Cheng A, Hardingham T E and Kimber S J 2013 Generating cartilage repair from pluripotent stem cells *Tissue Eng.* B **20** 257–66

Chong J J H *et al* 2014 Human embryonic-stem-cell-derived cardiomyocytes regenerate non-human primate hearts *Nature* **510** 273–7

Christiano A M and Uitto J 1996 Molecular complexity of the cutaneous basement membrane zone: revelations from the paradigms of epidermolysis bullosa *Exp. Dermatol.* **5** 1–11

Cianfarani F *et al* 2013 Diabetes impairs adipose tissue-derived stem cell function and efficiency in promoting wound healing *Wound Rep. Reg.* **21** 545–53

Corradetti B *et al* 2015 Osteoprogenitor cells from bone marrow and cortical bone: Understanding how the environment affects their fate *Stem Cells Dev.* **24** 1112–23

Cowan C M *et al* 2004 Adipose-drived adult stromal cells heal critical-size mouse calvarial defects *Nat. Biotechnol.* **22** 560–7

Davies J E *et al* 2006 Astrocytes derived from glial-restricted precursors promote spinal cord repair *J. Biol.* **5** 7

DeLucia T A *et al* 2003 Use of a cell line to investigate olfactory ensheathing cell-enhanced axonal regeneration *Anat. Rec.* B **271B** 61–70

De Rosa L *et al* 2014 Long-term stability and safety of transgenic cultured epidermal stem cells in gene therapy of junctional epidermolysis bullosa *Stem Cell Rep.* **26** 1–8

Deshpande D M *et al* 2006 Recovery from paralysis in adult rats using embryonic stem cells *Ann. Neurol.* **60** 32–44

Dobkin B H, Curt A and Guest J 2006 Cellular transplants in China: observational study from the largest human experiment in chronic spinal cord injury *Neurorehabil. Neural Repair* **20** 5–13

Doss M X 2015 Cell based regenerative therapy of heart failure: where do we stand? *Int. J. Stem Cell Res.* **1** e101

Dupont K M *et al* 2010 Human stem cell delivery for treatment of large segmental bone defects *Proc. Natl Acad. Sci. USA* **107** 3305–10

Efe J A *et al* 2011 Conversion of mouse fibroblasts into cardiomyocytes using a direct reprogramming strategy *Nat. Cell Biol.* **13** 215–24

Emborg M E *et al* 2013 Induced pluripotent stem cell-derived neural cells survive and mature in the nonhuman primate brain *Cell Rep.* **3** 646–50

English K, French A and Wood K J 2010 Mesenchymal stromal cells: facilitators of successful transplantation? *Cell Stem Cell* **7** 431–42

Feron F *et al* 2005 Autologous olfactory ensheathing cell transplantation in human spinal cord injury *Brain* **128** 2951–60

Filareto A *et al* 2013 An *ex vivo* gene therapy approach to treat muscular dystrophy using inducible pluripotent stem cells *Nat. Commun.* **4** 1549

Frontzek K *et al* 2016 Amyloid-beta pathology and cerebral amyloid angioplasty are frequent in iatrogenic Creuzfeldt–Jacob disease after dural grafting *Swiss Med. Wkly.* **146** w14287

Fuentes-Boquete I *et al* 2004 Pig chondrocyte xenoimplants for human chondral defect repair: an *in vitro* model *Wound Rep. Reg.* **12** 444–52

Funakoshi S *et al* 2016 Enhanced engraftment, proliferation, and therapeutic potential in heart using optimized human iPSC-derived cardiomyocytes *Sci. Rep.* **6** 19111

Genovese P *et al* 2014 Targeted genome editing in human repopulating haematopoietic stem cells *Nature* **510** 235–40

Gerlach J C *et al* 2011 Method for autologous single skin cell isolation for regenerative cell spray transplantation with non-cultured cells *Int. J. Artif. Organs* **34** 271–9

Gnedeva K *et al* 2015 Derivation of hair-inducing cell from human pluripotent stem cells *PLoS One* **10** e0116892

Gonzalez-Cordero A *et al* 2013 Photoreceptor precursors derived from three-dimensional embryonic stem cell cultures integrate and mature within adult degenerate retina *Nat. Biotechnol.* **8** 741–52

Grassel S and Lorenz J 2014 Tissue engineering strategies to repair chondral and osteochondral tissue in osteoarthritis: use of mesenchymal stem cells *Curr. Rheumatol. Rep.* **16** 452

Guo R *et al* 2017 Chemical cocktails enable hepatic reprogramming of mouse fibroblasts with a single transcription factor *Stem Cell Rep.* **9** 499–512

Gupta N, Henry R G and Strober J *et al* 2012 Neural stem cell engraftment and myelination in the human brain *Sci. Transl. Med.* **4** 155ra137

Hallet P J *et al* 2015 Successful function of autologous iPSC-derived dopamine neurons following transplantation in a non-human primate model of Parkinson's disease *Cell Stem Cell* **16** 269–74

Han S-K, Kim H-R and Kim W-K 2010 The treatment of diabetic foot ulcers with uncultured, processed lipoaspirate cells: a pilot study *Wound Rep. Reg.* **18** 342–8

Hankenson K D, Dishowitz M, Gray C and Schenker M 2011 Angiogenesis in bone regeneration *Injury* **42** 556–61

Hansson E M, Lindsay M E and Chien K R 2009 Regeneration next: Toward Heart stem cell therapeutics *Cell Stem Cell* **5** 364–77

Hawryluk G W J, Rowland J, Kwon B K and Fehlings M G 2008 Protection and repair of the injured spinal cord: a review of completed, ongoing, and planned clinical trials for acute spinal cord injury *Neurosurg. Focus* **25** E14

Hedges E C, Mehler V J and Nishimura A 2016 The use of stem cells to model amyotrophic lateral sclerosis and frontotemporal dementia: from basic research to regenerative medicine *Stem Cells Int.* **2016** 9279516

Hicks M R *et al* 2018 ERBB3 and NGFR mark a distinct skeletal muscle progenitor cell in human development and hPSCs *Nat. Cell Biol.* **20** 46–57

Hirsch T *et al* 2017 Regeneration of the entire human epidermis using transgenic stem cells *Nature* **551** 327–32

Holmes D 2017 Spinal cord injury: spurring regrowth *Nature* **552** 549–51

Hunziger E B, Lippuner K, Keel M and Shintani N 2015 Age-independent cartilage generation for synovium-based aurologous chondrocyte implantation *Tissue Eng.* A **21** 2089–98

Ishikawa N *et al* 2009 Peripheral nerve regeneration by transplantation of BMSC-derived Schwann cells as chitosan gel sponge scaffolds *J. Biomed. Mater. Res.* A **89** 1118–24

Israel M A *et al* 2012 Probing sporadic and familial Alzheimer's disease using induced pluripotent stem cells *Nature* **482** 216–20

Jaunmuktane Z *et al* 2015 Evidence for human transmission of amyloid-beta pathology and cerebral amyloid angiopathy *Nature* **525** 247–50

Jeon Y K *et al* 2010 Mesenchymal stem cells' interaction with skin: wound-healing effect on fibroblast cells and skin tissue *Wound Rep. Reg.* **18** 655–61

Jeon I *et al* 2012 Neuronal properties, *in vivo* effects, and pathology of a Huntington's disease patient-derived induced pluripotent stem cells *Stem Cells* **30** 2054–62

Jones B A and Pei M 2012 Synovium-derived stem cells: a tissue-specific stem cell for cartilage engineering and regeneration *Tissue Eng.* B **18** 301–11

Kajatani K *et al* 2004 Role of the periosteal flap in chondrocyte transplantation: an experimental study in rabbits *Tissue Eng.* **10** 331–42

Kamahita H *et al* 2001 Implantation of bone marrow mononuclear cells into ischemic myocardium enhances collateral perfusion and regional function via side supply of angioblasts, angiogenic ligands and cytokines *Circulation* **104** 1046–56

Kang K S *et al* 2005 A 37-year-old spinal cord-injured female patient, transplanted of multipotent stem cells from human UC blood, with improved sensory perception and mobility, both functionally and morphologically: a case study *Cytotherapy* **7** 368–73

Kang S K *et al* 2006 Autologous adipose tissue-derived stromal cells for treatment of spinal cord injury *Stem Cells Dev.* **15** 583–94

Karimi-Abdolrezaee S *et al* 2006 Delayed transplantation of adult neural precursor cells promotes remyelination and functional recovery after spinal cord injury *J. Neurosci.* **26** 3377–89

Karow M, Schichor C, Beckervordersndforth R and Berninger B 2012 Lineage-reprogramming of pericyte-derived cells of the adult human brain into induced neurons *J. Vis. Exp.* **87** PMC4181648

Keramaris N C *et al* 2008 Fracture vascularity and bone healing: a systematic review of the role of VGEF *Injury* **39S2** S45–57

Kierstead H S *et al* 2005 Human embryonic stem cell-derived oligodendrocyte progenitor cell transplants remyelinate and restore locomotion after spinal cord injury *J. Neurosci.* **25** 4694–705

Kim C *et al* 2013 Studying arrhythmogenic right ventricular dysplasia with patient-specific iPSCs *Nature* **494** 105–10

Kim J D *et al* 2014 Clinical outcome of autologous bone marrow aspirates concentrate (BMAC) injection in degenerative arthritis of the knee *Eur. J. Orthop. Surg. Traumatol.* **24** 1505–11

Kirsner R A *et al* 2013 Durability of healing from spray-applied cell therapy with human allogeneic fibroblasts and keratinocytes for the treatment of chronic venous leg ulcers: a 6-month follow-up *Wound Rep. Reg.* **21** 682–87

Knutsen G *et al* 2007 A randomized trial comparing autologous chondrocyte implantation with microfracture *J. Bone Joint Surg.* **89A** 2105–12

Ko J-Y, Kim K-I, Park S and Im G-I 2014 *In vitro* chondrogenesis and *in vivo* repair of osteochondral defect with human induced pluripotent stem cells *Biomaterials* **35** 3571–81

Koh Y G *et al* 2013 Infrapatellar fat pad-derived mesenchymal stem cell therapy for knee osteoarthritis *Knee* **19** 902–7

Kon E *et al* 2009 Arthroscopic second-generation autologous chondrocyte implantation compared with microfracture for chondral lesions of the knee: prospective non-randomized study at 5 years *Am. J. Sports Med.* **37** 33–41

Kordower J H *et al* 2008 Lewy body-like pathology in long term embryonic nigral transplants in Parkinson's disease *Nat. Med.* **14** 504–6

Kreuz P C *et al* 2007 Importance of sports in cartilage regeneration after autologous chondrocyte implantation *Am. J. Sports Med.* **35** 1261–68

Kuh S U *et al* 2005 Functional recovery after human umbilical cord blood cells transplantation with brain-derived neurotrophic factor into the spinal cord injured rat *Acta Neurochir.* **147** 985–92

Kumagi G *et al* 2013 Genetically modified mesenchymal stem cells (MSCs) promote axonal regeneration and prevent hypersensitivity after spinal cord injury *Exp. Neurol.* **248** 369–80

Kym D *et al* 2015 The application of cultured epithelial autografts improves survival in burns *Wound Rep. Reg.* **23** 340–4

Lamba D A *et al* 2010 Generation, purification and transplantation of photoreceptors derived from human induced pluripotent stem cells *PLoS One* **5** e8763

Lee D H *et al* 2012 Synovial fluid CD34(-)CD44(+)CD90(+) mesenchymal stem cell levels are associated with the severity of primary knee osteoarthritis *Osteoarth. Cart.* **20** 106–9

Li K and Xu S *et al* 2014 Small molecules facilitate the reprogramming of mouse fibroblasts into pancreatic lineages *Cell Stem Cell* **14** 228–36

Lian X *et al* 2012a Robust cardiomyocyte differentiation from human pluripotent stem cells via temporal modulation of canonical Wnt signaling *Proc. Natl Acad. Sci. USA* **109** E1848–57

Lian Q *et al* 2012b Functional mesenchymal stem cells derived from human induced pluripotent stem cells attenuate limb ischemia in mice *Circulation* **121** 1113–23

Lima C *et al* 2006 Olfactory mucosa autografts in human spinal cord injury: a pilot clinical study *J. Spinal Cord Med.* **29** 191–203

Liu H, Chaudhari P, Choi S M and Jang Y-Y 2012 Applications of human induced pluripotent stem cell-derived hepatocytes *Stem Cells and Cancer Stem Cells* vol 3 ed M Hyat (Dordrecht: Springer), pp 213–20

Loffredo F S, Steinhauser M L, Gannon J and Lee R T 2011 Bone marrow-derived cell therapy stimulates endogenous cardiomyocyte progenitors and promotes cardiac repair *Cell Stem Cell* **8** 389–98

Lopatina T *et al* 2011 Adipose-derived stem cells stimulate regeneration of peripheral nerves: BDNF secreted by these cells promotes nerve healing and axon growth *de novo PLoS One* **6** e17899

Lu P *et al* 2012 Long-distance growth and connectivity of neural stem cells after severe spinal cord injury *Cell* **150** 1264–73

Lu P *et al* 2014 Long-distance axonal growth from human induced pluripotent stem cells after spinal cord injury *Neuron* **83** 789–96

Luk K C *et al* 2012a Intracerebral inoculation of pathological α-synuclein initiates a rapidly progressive neurodegenerative α-syncucleinopathy in mice *J. Exp. Med.* **209** 975–86

Luk K C *et al* 2012b Pathological α-synuclein transmission initiates Parkinson-like neuro-degeneration in nontransgenic mice *Science* **338** 949–53

Luo G *et al* 2010 Promotion of cutaneous wound healing by local application of esenchymal stem cells derived from human umbilical cord blood *Wound Rep. Reg.* **18** 506–13

Mack G S 2011 ReNeuron and StemCells get green light for neural stem cell trials *Nat. Biotechnol.* **29** 95–7

Makris E A *et al* 2015 Repair and tissue engineering techniques for articular cartilage *Nat. Rev. Rheumatol.* **11** 21–34

Mardones R, Jofre C M and Minguell J J 2015 Cell therapy and tissue engineering approaches for cartilage repair and/or regeneration *Int. J. Stem Cells* **8** 48–53

Masuda-Zuzukake M *et al* 2013 Prion-like spreading of pathological α-synuclen in brain *Brain* **136** 1128–38

Mavilio F *et al* 2006 Correction of junctional epidermolysis bullosa by transplantation of genetically modified epidermal stem cells *Nat. Med.* **12** 1397–402

Mayhew C N and Wells J M 2010 Converting human pluripotent stem cells into beta cells. Recent advances and future challenges *Curr. Opin. Organ Transplant.* **15** 54–60

McCall M and Shapiro A M J 2012 Update on islet transplantation *Cold Spring Harbor Perspect. Med.* **2** a007823

Mc Kenzie I A *et al* 2006 Skin-derived precursors generate myelinating Schwann cells for the injured and dysmyelinated nervous system *J. Neurosci.* **26** 6651–60

Mikami Y *et al* 2004 Implantation of dendritic cells in injured adult spinal cord results in activation of endogenous neural stem/progenitor cells leading to *de novo* neurogenesis and functional recovery *J. Neurosci. Res.* **76** 453–65

Montini E *et al* 2006 Hematopoietic stem cell gene transfer in a tumor-prone mouse model uncovers low genotoxicity of lentiviral vector integration *Nat. Biotechnol.* **24** 687–96

Mothe A J and Tator C H 2012 Advances in stem cell therapy for spinal cord injury *J. Clin. Invest.* **122** 3824–34

Mougenot A L *et al* 2012 Prion-like acceleration of a synucleinopathy in a transgenic mouse model *Neurobiol. Aging* **33** 2225–28

Najm F J *et al* 2013 Transcription factor-mediated reprogramming of fibroblasts to expandable, myelogenic oligodendrocyte progenitor cells *Nat. Biotechnol.* **31** 426–33

Neirinckx V *et al* 2014 Concise review: Spinal cord injuries: how could adult mesenchymal and neural crest stem cells take up the challenge? *Stem Cells* **32** 829–43

Nie C *et al* 2011 Locally administered adipose-derived stem cells accelerate wound healing through differentiation *Cell Transplant.* **20** 205–16

Nistor G I *et al* 2005 Human embryonic stem cells differentiate into oligodendrocytes in high purity and myelinate after spinal cord transplantation *Glia* **49** 385–96

Nori S *et al* 2011 Grafted human-induced pluripotent stem-cell-derived neurospheres promote motor functional recovery after spinal cord injury in mice *Proc. Natl Acad. Sci. USA* **108** 16825–30

Nowbar A N *et al* 2014 Discrepancies in autologous bone marrow stem cell trials and enhancement of ejection fraction (DAMASCENE): weighted regression and meta-analysis *BMJ* **348** g2688

Nunes S S *et al* 2013 Biowire: a platform for maturation of human pluripotent stem cell-derived cardiomyocytes *Nat. Methods* **10** 781–6

Ochi M, Uchio Y, Tobita M and Kuriwaka M 2001 Current concepts in tissue engineering technique for repair of cartilage defect *Artif. Organs* **25** 172–9

Ogawa Y *et al* 2002 Transplantation of *in vitro*-expanded fetal neural progenitor cells results in neurogenesis and functional recovery after spinal cord contusion injury in adult rats *J. Neurosci. Res.* **69** 925–33

Ong S-G and Wu J C 2013 Cortical bone-derived stem cells: a novel class of cells for myocardial protection *Circ. Res.* **113** 480–3

Oldershaw R A, Baxter M A, Lowe E T, Bates N, Grady L M, Soncin F, Brison D, Hardingham T E and Kimber S J 2010 Directed differentiation of human embryonic stem cells toward chondrocytes *Nat. Biotechnol.* **28** 1187–94

Orozco L *et al* 2013 Treatment of knee osteoarthritis with autologous mesenchymal stem cells: two-year follow-up results *Transplantation* **97** e66–8

Orth P *et al* 2014 Current prospects in stem cell research for knee cartilage repair *Stem Cells Cloning* **7** 1–17

Osaka M *et al* 2010 Intravenous administration of mesenchymal stem cells derived from bone marrow after contusive spinal cord injury improves functional outcome *Brain Res.* **1343** 226–35

Pagliuca F W *et al* 2014 Generation of functional human pancreatic β-cells *in vitro Cell* **159** 428–39

Pak J 2011 Regeneration of human bones in hip ostronecrosis and human cartilage in knee osteoarthritis with autologous adipose-tissue-derived stem cells: a case series *J. Med. Case Rep.* **5** 296

Pak J, Lee J H and Lee S H 2014 Regenerative repair of damaged meniscus with autologous adipose tissue-derived stem cells *Biomed. Res. Int.* **2014** 436029

Park H C, Shim Y S, Ha Y, Yoon S H, Park S R, Choi B H and Park H S 2005 Treatment of complete spinal cord injury patients by autologous bone marrow cell transplantation and administration of granulocyte-macrophage colony stimulating factor *Tissue Eng* **11** 913–22

Perkel J M 2014 Eyes on the prize *Scientist* **28** October

Peura M *et al* 2009 Bone marrow mesenchymal stem cells undergo nemosis and induce keratinocyte wound healing utilizing the HGF/c-Met PI3K pathway *Wound Rep. Reg.* **17** 569–77

Plant G W, Bates M L and Bunge M B 2001 Inhibitory proteoglycan immunoreactivity is higher at the caudal than the rostral Schwann cell graft-transected spinal cord interface *Mol. Cell. Neurosci.* **17** 471–87

Pluchino S *et al* 2005 Neurosphere-derived multipotent precursors promote neuroprotection by an immunomodulatory mechanism *Nature* **436** 266–71

Polymenidou M and Cleveland D W 2011 The seeds of neurodegeneration: prion-like spreading in ALS *Cell* **147** 498–508

Popovich P G 2012 Building bridges for spinal cord repair *Cell* **150** 1105–6

Prusiner S 2014 *Madness and Memory-The Discovery of Prions-A New Biological Principle of Disease* (New Haven, CT: Yale University Press)

Qi M 2014 Transplantation of encapsulated pancreatic islets as a treatment for patients with type 1 diabetes mellitus *Adv. Med.* **2014** 429710

Radtke C, Wewetzer K, Reimers K and Vogt P M 2011 Transplantation of olfactory ensheathing cells as adjunct cell therapy for peripheral nerve injury *Cell Transplant.* **20** 145–52

Ramallal M *et al* 2004 Xeno-implantation of pig chondrocytes into rabbit to treat localized articular cartilage defects: an animal model *Wound Rep. Reg.* **12** 337–45

Ramer L M, Ramer M S and Steeves J D 2005 Setting the stage for functional repair of spinal cord injuries: a cast of thousands *Spinal Cord* **43** 134–61

Ramon-Cueto A, Plant G W, Avila J and Bunge M B 1998 Long-distance axonal regeneration in the transected adult rat spinal cord is promoted by olfactory ensheathing glia transplants *J. Neurosci.* **18** 3803–15

Rapalino O *et al* 1998 Implantation of stimulated homologous macrophages results in partial recovery of paraplegic rats *Nat. Med.* **4** 814–21

Reinecke H, Poppa V and Murry C E 2002 Skeletal muscle stem cells do not transdifferentiate into cardiomyocytes after cardiac grafting *J. Mol. Cell Cardiol.* **34** 241–9

Richter M W and Roskams A J 2008 Olfactory ensheathing cell transplantation following spinal cord injury: hype or hope? *Exp. Neurol.* **209** 353–67

Rosenzweig E S *et al* 2018 Restorative effects of human neural stem cell grafts on the primate spinal cord *Nat. Med.* **24** 484–90

Roska B and Sahel J-A 2018 Restoring vision *Nature* **557** 359–67

Rustad K C *et al* 2011 Enhancement of mesenchymal stem cell angiogenic capacity and stemness by a biomimetic hydrogel scaffold *Biomaterials* **33** 80–90

Saito F *et al* 2008 Spinal cord injury treatment with intrathecal autologous bone marrow stromal cell transplantation: the first clinical case report *J. Trauma* **64** 53–9

Salazar T E 2017 Electroacupuncture promotes CNS-dependent release of mesenchymal stem cells *Stem Cells* **35** 1303–15

Santos-Benito F F and Ramon-Cueto A 2003 Olfactory ensheathing glia transplantation: a therapy to promote repair in the mammalian central nervous system *Anat. Rec. Part B New Anat.* **271B** 77–85

Saw K Y *et al* 2013 Articular cartilage regeneration with autologous peripheral blood stem cells versus hyaluronic acid: a randomized controlled trial *Arthroscopy* **29** 684–94

Schwartz S D *et al* 2015 Human embryonic stem cell-derived retinal pigment epithelium in patients with age-related macular degeneration and Stargardt's macular dystrophy: follow-up of two open-label phase ½ studies *Lancet* **385** 509–16

Segers V F M and Lee R T 2008 Stem-cell therapy for cardiovascular disease *Nature* **451** 937–42

Sekiya S and Suzuki A 2011 Direct conversion of mouse fibroblasts to hepatocyte-like cells by defined factors *Nature* **475** 390–93

Seo D K *et al* 2017 Enhanced axonal regeneration by transplanted Wnt3a-secreting human mesenchymal stem cells in a rat model of spinal cord injury *Acta Neurochir.* **159** 947–57

Shan R E *et al* 2013 Identification of small molecules for human hepatocyte expansion and iPS differentiation *Nat. Chem. Biol.* **9** 514–20

Sheng L, Yang M, Liang Y and Li Q 2013 Adipose-derived stem cells (ADSCs) transplantation promotes regeneration of expanded skin using a tissue expansion model *Wound Rep. Reg.* **21** 746–54

Sheth R N, Manzano G, Li X and Levi A D 2008 Transplantation of human bone marrow-derived stomal cells into the contused spinal cord of nude rats *J. Neurosurg. Spine* **8** 153–62

Shiba Y *et al* 2016 Allogeneic transplantation of iPS cell-derived cardiomyocytes regenerates primate hearts *Nature* **538** 399–396

Si-Tayeb K *et al* 2010 Highly efficient generation of human hepatocyte-like cells from induced pluripotent stem cells *Hepatology* **51** 297–305

Skuk D and Tremblay J P 2011 Myoblast transplantation in skeletal muscles *Principles of Regenerative Medicine* 2nd edn ed A Atala, R Lanza, J A Thompson and R Nerem (San Diego, CA: Elsevier/Academic), pp 779–93

Son E Y *et al* 2011 Conversion of mouse and human fibroblasts into functional spinal motor neurons *Cell Stem Cell* **9** 205–18

Song W K *et al* 2015 Treatment of macular degeneration using embryonic stem cell-derived retinal pigment epithelium: preliminary results in Asian patients *Stem Cell Rep.* **4** 860–72

Souillet G *et al* 2003 Outcome of 27 patients wuth Hurler's syndrome transplanted from either related or unrelated haemattopoietic stem cell sources *Bone Marrow Transplant.* **31** 1105–17

Southwell D G *et al* 2014 Interneurons from embryonic development to cell-based therapy *Science* **344** 167

Steeves J, Fawcett J and Tuszynski M 2004 Report of international clinical trials workshop on spinal cord injury *Spinal Cord* **42** 591–7

Sun B K, Siprashvili Z and Khavari P A 2014 Advances in skin grafting and treatment of cutaneous wounds *Science* **346** 941–5

Tohill M, Mantovani C, Wiberg M and Terenghi G 2004 Rat bone marrow mesenchymal stem cells express glial markers and stimulate nerve regeneration *Neurosci. Lett.* **362** 200–3

Tolar J and Lees C J *et al* 2011 Hematopoietic differentiation of pluripotent stem cells from patients with mucopolysaccharidosis type I (Hurler syndrome) *Blood* **117** 839–47

van Berlo J H *et al* 2014 c-Kit+ cells minimally contribute cardiomyocytes to the heart *Nature* **509** 347–341

van Gorp S *et al* 2013 Amelioration of motor/sensory dysfunction and spasticity in a rat model of acute lumbar spinal cord injury by human neural stem cell transplantation *Stem Cell Res. Therapy* **4** 57

van Laake L W *et al* 2007 Human embryonic stem cells-derived cardiomyocytes survive and mature in the mouse heart and transiently improve function after myocardial infarction *Stem Cell Res.* **1** 9–24

Walker L C and Jucker 2015 Neurodegenerative diseases: expanding the prion concept *Annu. Rev. Neurosci.* **38** 87–103

Walsh S and Midha R 2009 Practical considerations concerning the use of stem cells for peripheral nerve repair *Neurosurg. Focus* **26** 1–8

Wang G, Ao Q, Gong K, Zuo H, Gong Y and Zhang X 2010 Synergistic effect of neural stem cells and olfactory ensheathing cells on repair of adult rat spinal cord injury *Cell Transplant.* **19** 1325–37

Weidner N, Blesch A, Grill R J and Tuszynski M H 1999 Nerve growth factor-hypersecreting Schwann cell grafts augment and guide spinal cord axonal growth and remyelinate central nervous system axons in a phenotypically appropriate manner that correlates with expression of L1 *J. Comput. Neurol.* **413** 495–506

Wernig M *et al* 2008 Neurons derived from reprogrammed fibroblasts functionally integrate into the fetal brain and improve symptoms of rats with Parkinson's disease *Proc. Natl Acad. Sci. USA* **105** 5856–61

Wood F 2011 Tissue engineering of skin *Principles of Regenerative Medicine* 2nd edn ed A Atala, R Lanza, J A Thomson and R Nerem (San Diego, CA: Elsevier/Academic), pp 1063–78

Yamada M *et al* 2014 Human oocytes reprogram adult somatic nuclei of a type 1 diabetic to diploid pluripotent stem cells *Nature* **510** 533–6

Yamasaki S *et al* 2014 Cartilage repair with autologous bone marrow mesenchymal stem cell transplantation: review of preclinical and clinical studies *Cartilage* **5** 196–202

Yang N *et al* 2013 Generation of oligodendrogial cells by direct lineage conversion *Nat. Biotechnol.* **31** 434–9

Yannas I V 2001 *Tissue and Organ Regeneration in Adults* (New York: Springer)

Yoder M C *et al* 2007 Redefining endothelial progenitor cells via clonal analysis and hematopoietic stem/progenitor principles *Blood* **109** 1801–9

Yoo A S *et al* 2011 MicroRNA-mediated conversion of human fibroblasts to neurons *Nature* **476** 228–31

Yoon S H *et al* 2007 Complete spinal cord injury treatment using autoologous bone marrow cell transplantation and bone marrow stimulation with granulocyte macrophage-colony stimulating factor: phase I/II clinical trial *Stem Cells* **25** 2066–73

Zaret K S and Grompe M 2008 Generation and regeneration of cells of the liver and pancreas *Science* **322** 1490–4

Zaruba M-M, Soonpa M, Reuter S and Field L J 2010 Cardiomyogenic potential of c-kit$^+$-expressing cells derived from neonatal and adult mouse hearts *Circulation* **121** 1992–2000

Zhou Q and Melton D A 2018 Pancreas regeneration *Nature* **557** 351–8

Zhu S, Wang H and Ding S 2014 Mouse liver repopulation with hepatocytes generated from human fibroblasts *Nature* **508** 93–7

Zwi-Dantsis L *et al* 2012 Derivation and cardiomyocyte differentiation of induced pluripotent stem cells from heart failure patients *Eur. Heart J.* **34** 1575–86

IOP Publishing

Foundations of Regenerative Biology and Medicine

David L Stocum

Chapter 9

Biomimetic tissues and organs

Summary

Biomimetic tissues and organs are under development to alleviate the shortage of organ and tissue transplants, and to provide platforms for drug testing and disease modeling. Skin equivalents are in clinical use or in development to replace damaged skin. Biomimetic muscle derived by culturing myogenic cells under tension in molds, or made of elastic polymers or twisted nanofiber yarns, is under development. Biomimetic bone consisting of scaffolds seeded with osteogenic cells have been successfully implanted in patients to bridge critical size defects in long bones. A biomimetic phalange has been constructed *in vitro* and the body has been used as a bioreactor to mature human biomimetic mandibular bone for transplant. Human articular cartilage can be constructed either by seeding a scaffold with MSCs or forcing MSCs to initiate chondrogenesis in the absence of a scaffold. Transplants of biomimetic myocardial tissue made *in vitro* have had success in improving the function of infarcted rat hearts. Biomimetic blood vessels with high burst strength have been built by seeding scaffolds seeded with bone marrow cells; grafts of such vessels to replace the peripheral pulmonary arteries of patients were functioning normally over a mean of 5.8 years. Devices that mimic liver and pancreatic function have been tested successfully in animals, and organic versions are under development. Tracheal mucosa has been successfully patched with biomimetic tissue, and some progress has been made in the development of biomimetic alveolar lung tissue. An implantable biomimetic kidney has been developed to the point where it is ready for clinical trial, and the development of organic biomimetic bladders is well along.

9.1 Skin

A standard treatment for an excisional wound or burn is the application of an autogeneic meshed split-thickness skin graft (MSTSG). The mesh makes it possible to stretch a skin graft over an area larger than the dimensions of the unmeshed graft (figure 9.1(A)). Biomimetic skin equivalents (BSEs) are engineered to mimic normal

(A) MSTSG   **(B) BSE**

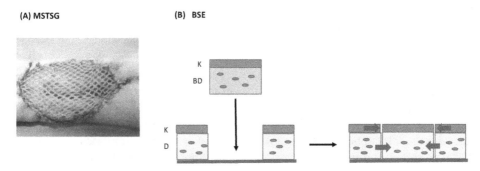

Figure 9.1. (A) Meshed split-thickness skin graft (MSTSG). Reproduced with permission from (Stocum 2012). Copyright 2012 Elsevier. (B) Biomimetic skin equivalent (BSE). A biomimetic dermis consisting of a matrix containing allogeneic fibroblasts is overlaid with keratinocytes. Host fibroblasts invade the dermal matrix to replace the door fibroblasts and the keratinocytes of the construct are replaced by host keratinocytes.

skin, thus avoiding the necessity for taking skin from one place to graft to another. A skin equivalent consists of a biomimetic dermis consisting of a natural or synthetic biodegradable scaffold seeded with allogeneic fibroblasts, surfaced with keratinocytes (figure 9.1(B)). The use of allogeneic fibroblasts allows BSEs to be banked for use 'off the shelf'. BSEs are essentially living wound dressings used to cover excisional wounds, burns and chronic wounds (MacNeil 2007). Allogeneic keratinocytes can be stored and added to the dermal equivalent, or autogeneic keratinocytes might be added by spray gun (chapter 8). Allogeneic fibroblasts and keratinocytes are immunorejected by the host, but secrete growth factors and cytokines that encourage migration and proliferation of host fibroblasts and keratinocytes into the wound to replace the dressing (Ehrlich 2004, Jimenez and Jimenez 2004).

Several skin equivalents are in current clinical use. Transcyte™ and Orcel® are two FDA-approved constructs for excisional wounds and burns. The Transcyte™ dermis consists of human dermal fibroblasts grown on nylon mesh; the Orcel® dermis is made of bovine collagen containing human dermal cells. The collagen-based Apligraf® and the polyester-based Dermagraft® are bioartificial skins approved for use on diabetic and venous ulcers. Apligraf® was reported to result in a significantly higher frequency of healed wounds and a reduced median time to complete wound closure of venous ulcers (Falanga *et al* 1998) and diabetic ulcers (Boulton *et al* 2004). When compared in a randomly controlled clinical trial to a collagenous acellular template of dehydrated human amnion for effectiveness in healing diabetic foot ulcers, Apligraf® increased the probability of healing by 97% (Kirsner *et al* 2015).

The fibroblasts of Dermagraft® synthesize a matrix containing dermal type I, III and VI collagens, elastin, tenascin, fibronectin, hyaluronic acid, chondroitin sulfates, and the major dermal proteoglycan core protein, decorin, and also synthesize IGF-1 and 2, FGF-2, PDGF, HGF and VEGF (Landeen *et al* 1992). Re-epithelialization and vascularization of the wound is rapid, and there is minimal inflammatory response. Immunostaining for laminin and type IV collagen revealed the presence of a continuous basement membrane at the epidermal-dermal junction. Newer skin equivalents are under development or in clinical trials (Greaves *et al* 2013, for a review).

There are several problems with skin equivalents that have not been completely resolved. They do not promote the development of hair follicles and skin glands, and because they are cryopreserved before use, the viability of their fibroblasts may be reduced. Immunorejection of allogeneic cells can destroy the capacity of a BSE to provide paracrine factors and signal for the in-migration of host cells. They are more expensive than MSTSGs or acellular dermal templates and have been reported to be no better for healing excisional wounds and burns and restoring cosmetic appearance (Boyce 2001). These facts suggest that acellular dermal templates overlaid with MSTSGs or keratinocytes might be a better option for promoting host skin tissue. However, a statistical analysis comparing Apligraf® versus acellular pig SIS in healing venous leg ulcers found that Apligraf® reduced the median time to wound closure by 44%, achieved healing 19 weeks sooner, and increased the probability of healing by 29% over that of pig SIS (Marston et al 2014). To address the problem of allogeneic fibroblast rejection, a dermal equivalent was made of bovine type I collagen/ chondroitin-6-sulfate seeded with fibroblasts genetically modified to express the immunosuppressive molecule, indolamine 2,3-dioxygenase (IDO). This construct suppressed T-cell infiltration in rat skin wounds, enhanced angiogenesis by a factor of four, and significantly accelerated repair of the wound over seven days (Forouzandeh et al 2010). It is likely that immune rejection would be minimized if skin equivalents could be rapidly made using fresh autogeneic fibroblasts or allogeneic fibroblasts from a cell bank with a wide range of histocompatibility antigens. Perhaps a protocol could be developed like that for keratinocytes in which autogeneic dermal fibroblasts could be obtained by skin biopsy and injected into a scaffold that is applied to a burn or wound, or sprayed directly onto the wound, as for keratinocytes.

9.2 Muscle

Biomimetic muscle constructed *in vitro* theoretically could be grafted to restore volumetric muscle loss or even restore whole muscles (Vigodarzeve and Mantero 2014). For clinical purposes, the muscle-forming cells need to be autogeneic satellite cells isolated by biopsy, or derived from hiPSCs. Initial experimental attempts to make biomimetic muscle combined rat satellite cells and fibroblasts to engineer myooids *in vitro* (Dennis et al 2001). The satellite cells fused to form myotubes that remodeled themselves into a cylinder of myofibers surrounded by interstitial fibroblasts organized into fascicles surrounded by perimysium. The isometric force generated after electrical stimulation was only about 1% of the control value for adult muscles. These myooids reached only a very small diameter (0.3–0.4 mm) because of their dependence on diffusion for gas exchange and nutrients. Myofiber formation with good histological structure has been attained by culturing human myoblasts in silicon molds with anchors at the ends for contraction (figure 9.2). These myofibers respond to drugs (acetylcholine, caffeine) in the same way as normal human muscle (Madden et al 2015).

Myooids have been implanted into experimental animals, where they were vascularized and innervated by the host, enhanced by inclusion of endothelial cells

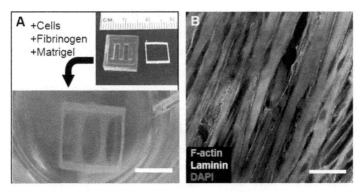

Figure 9.2. (A) Culture system for developing muscle bundles from myoblasts in a fibrinogen/Matrigel matrix and anchored on Velcro at either end. (B) Differentiated muscle fibers stained to reveal F-actin (green), laminin (white) and nuclei (blue). Reproduced from (Madden *et al* 2015). Open Access.

and nerve segments in the implant (Levenberg *et al* 2005, Kang *et al* 2012). Rat satellite cells mixed into a fibrinogen-containing hydrogel and cultured in a cylindrical mold differentiated into mature myofibers that were quickly vascularized and achieved significant contractile function when implanted into a chamber under the back skin (Juhas *et al* 2014). The myofibers were able to regenerate after injury *in vitro*, indicating the presence of a satellite cell compartment in the cultured muscle; presumably, the same response to injury would be observed in the implanted muscle. In another study, rat muscle was enzymatically digested and the cells grown on strips of decellularized bladder matrix where they formed immature myotubes (Corona *et al* 2012). In this case, the myotubes were 'exercised' by slowly expanding and contracting the strips. After implantation into a gap in the latissimus dorsi muscle the myotubes matured into myofibers that integrated with host muscle.

The construction of acellular synthetic muscles is a novel approach to muscle replacement. An artificial polyprotein made from small GB1 proteins (immunoglobulin G-binding protein) derived from *Streptococcus* bacteria was shown capable of spring-like rapid and reversible high fidelity folding with low mechanical fatigue during repeated stretching-relaxation cycles (Cao and Li 2007). Lv *et al* (2010) made an elastic polymer by inserting the genes for GBI and resilin, an insect elastomeric protein that confers the elasticity necessary for insects such as fleas to jump up to 150 times their length, into *Escherichia coli* to express the proteins. The proteins were then cross-linked photochemically to produce the elastic polymer. The polymer mimicked the structure and function of titin, the large protein responsible for the passive elastic properties of the myofiber contractile apparatus, suggesting the potential for assembly of other artificial muscle proteins that could be combined into higher orders of assembly to form biomimetic muscle.

Substitutes for organic muscle have also been created from twisted (torsional) carbon nanofiber yarns (Foroughi *et al* 2011, Lima *et al* 2012). The yarns can be actuated to contract by ionic swelling in a liquid electrolyte, electromagnetic effects, or thermal expansion of infiltrated paraffin wax, but they are expensive and difficult to make (Yuan and Poulin 2014). Much less expensive torsional artificial muscle has been made from sewing thread and fishing line that can contract by 49% and lift loads over 100 times that of human muscle of the same length and weight (Haines *et al* 2014). Further development of these technologies could be incorporated into robotic designs, as well as integrated with organic structure.

9.3 Bone

Several approaches have been designed for the construction of biomimetic bone (Ng *et al* 2017, for a review). Biomimetic bones are designed to bridge defects of critical size or greater with new tissue while providing mechanical support (figure 9.3). They have been constructed of porous scaffolds such as calcium phosphate-chitosan-RGD (Chen *et al* 2013), bioactive glass microspheres (Perez *et al* 2014), or apatite-coated poly (lactic-*co*-glycolic acid) (PLGA) (Cowan *et al* 2004, Cancedda *et al* 2007) into which are seeded MSCs isolated from the bone marrow or derived from PSCs, or adipose-derived donor adult stromal (ADAS) cells. A poly(ε-caprolactone) scaffold seeded with human MSCs formed significantly more bone than unseeded scaffold when implanted into rat femur CSDs (Dupont *et al* 2010). Exogenous growth factors can be added to biomimetic bone constructs to promote the differentiation of seeded cells. For example, blocks of Gelfoam impregnated with culture-expanded human MSCs and a solution containing 10 μg ml^{-1} each of BMP-2 and FGF-2 completely healed 4 mm defects in the mouse parietal bone with a quality superior to that formed by MSCs in Gelfoam without the growth factors (Akita *et al* 2008).

Biomimetic bones have already been put to clinical use. In 2001, it was reported that three patients recovered limb function after receiving implants of a shaped macroporous hydroxyapatite scaffold seeded with autologous bone marrow cells to bridge 4.0–7.0 cm CSDs (Quarto *et al* 2001). The patients had experienced no problems 6–7 years after the surgery (Marcacci *et al* 2006). Two cases of partial mandible replacement were reported using constructs consisting of a PLLA scaffold

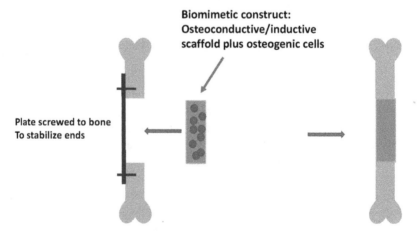

Figure 9.3. Biomimetic bone construct derived by seeding an osteoconductive/osteoinductive scaffold with osteogenic cells and its insertion to bridge a large gap in a bone that has been stabilized with a metal plate. The new bone integrates with the host bone.

containing particulate cancellous bone and marrow from the ilium (Kinoshita *et al* 1996). In one case, the right side of the mandible was partially resected. The biomimetic replacement formed bone with an alveolar ridge that allowed the patient to wear a denture. In the other case, the whole front of the mandible from the left to right molar region had been lost, creating a cosmetic deformity that made speaking and chewing difficult. The biomimetic bone construct restored the continuity of the mandible, markedly improving the cosmetic appearance and speaking and chewing abilities of the patient.

Replacement of large volumes of tissue loss requires constructs to be pre-vascularized. To solve this problem for large constructs, the body has been used as a bioreactor to make total mandible replacements (Tatara *et al* 2014, for a review). A biomimetic whole mandible was constructed for a patient who had lost most of his mandible to oral cancer nine years earlier (Warnke *et al* 2004, 2006). The construct was composed of a titanium micromesh scaffold seeded with autologous bone marrow cells, and small blocks of bovine bone matrix plus BMP-2 (figure 9.4). It was implanted into a pocket under the patient's latissimus dorsi muscle, where it was infiltrated by blood vessels. Seven weeks later the implant was removed and joined to the remainder of the patient's mandible. Blood vessels of the construct were connected to blood vessels in the patient's neck and a skin graft used to cover the surface. The neomandible continued to develop new bone and the patient was able to chew solid food for the first time in nearly a decade, but died 15 months later.

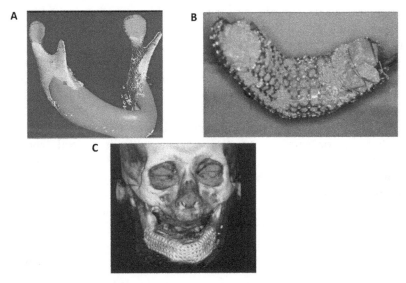

Figure 9.4. Use of the body as a bioreactor to construct a biomimetic mandible. (A) Digitally generated computer model of missing mandibular bone lost to cancer. (B) The model was used to construct a titanium mesh cage, which was filled with bone mineral blocks infiltrated with a mixture of recombinant human BMP-7 and bone marrow. This construct was vascularized by incubating it in a pocket made under the latissimus dorsi muscle of the patient. (C) Three-dimensional CT scan of the biomimetic mandible after removing it from the pocket and joining it to the patient's remaining mandible. Reproduced with permission from (Warnke *et al* 2004). Copyright 2004 Elsevier.

A similar construct composed of a coral-derived hydroxyapatite scaffold loaded with bone marrow cells and BMP-7 was used to replace a segment of mandible lost to surgery for a mandibular tumor (Heliotis *et al* 2006). The construct was pre-vascularized by implanting it into the pectoralis muscle. Clinical trials are now underway to replace maxillofacial deficiencies with biomimetic bone constructed with autogeneic fat cells.

In much the same way, ectopic bone for transplant has been manufactured by using the body as a bioreactor (Stevens *et al* 2005). A sub-periosteal space made on the tibial surface of rabbits was injected with ~200 μl of alginate gel cross-linked by $CaCl_2$. MSCs from the periosteum migrated into the gel and formed compact bone in an intramembranous fashion, accompanied by neovascularization that attained normal levels by 12 weeks and dissolution of the alginate. The mechanical properties of the bone were comparable to those of normal bone. Radiographic and histologic evaluation six weeks after harvesting the neobone and transplanting it to a defect in the contralateral tibia revealed that the transplanted bone had remodeled and become integrated with the surrounding bone. Delivery of liposome encapsulated Suramin, which inhibits vascular invasion, into the bioreactor space resulted in neocartilage formation, suggesting the possibility of creating new cartilage for transplant into endochondral or chondral defects by this method.

An experimental phalange was constructed *in vitro* by seeding pig bone marrow MSCs into alginate or fibrin hydrogels molded from digitized images into the shape of a human distal phalanx (Sheehy *et al* 2015). A cap of self-assembled chondrocytes was adhered to one end of the construct, which was then cultured in cartilage-promoting medium. The constructs underwent extensive endochondral ossification on their periphery and articular cartilage formation when implanted subcutaneously into nude mice. Phalanges with a central channel in the mold showed osteogenesis around the channel and on the periphery, imitating a bone marrow cavity. Constructs of the temporomandibular condyle have also been successfully differentiated *in vitro*.

These kinds of studies illustrate the potential for creating a wide variety of biomimetic bones or cartilage templates using scaffolds shaped from digitized medical images and seeded with human MSCs or chondrocytes derived from ADSCs or iPSCs. These could be cultured at suitable body sites to be vascularized before transplant. The fact that cartilage differentiation is favored by low oxygen concentrations might allow a considerable scale-up in the size of endochondral constructs. The addition of periosteal fibroblasts and MSCs to the construct would enable angiogenic signals from the hypertrophied chondrocytes of the construct to initiate blood vessel invasion with subsequent replacement of cartilage with bone.

9.4 Articular cartilage

Several types of biodegradable polymers have been used as scaffolds to prepare biomimetic articular cartilage *in vitro*. Three-dimensional scaffolds of polyglycolic or polyglactic acid, collagen gels, hybrid scaffolds of collagen plus polylactic acid or poly (DL-lactic-*co*-glycolic acid) (PLGA), SIS, and β-chitin from squid, all supported the differentiation of allogeneic cultured human or animal chondrocytes, or bone marrow MSCs into tissue that histologically and biochemically resembled hyaline cartilage (Freed and Vunjak-Novakovic 1995, Wakatani *et al* 1998, Chen *et al* 2004). When this cartilage was implanted into chondral and osteochondral defects *in vivo*, however, repair was not optimal. The constructs either did not differentiate the zones of normal articular cartilage, and/or did not integrate well with host articular cartilage. Integration could be enhanced in osteochondral defects by coating the defects with a high-strength adhesive prior to implantation (Wang *et al* 2007). New scaffolds such as polyvinyl alcohol nanofibers (Coburn *et al* 2012), as well as hydrogels for controlled release of bioactive factors (Rey-Rico *et al* 2016) support chondrocyte differentiation *in vitro* and in osteochondral defects *in vivo*. Seeding chondrocytes into hyaluronate-based or a chemically cross-linked dextran-poly (ethylene glycol) hydrogel was reported to result in articular cartilage-like tissue when implanted into rodent osteochondral defects (Toh *et al* 2010, Jukes *et al* 2008).

Well-developed, mechanically strong pieces of human articular cartilage have been constructed *in vitro* without scaffolds by forcing MSCs to go through the initial step of cartilage formation, mesenchymal condensation (Bhumiratana *et al* 2014). Aggregates of condensed cells (condensed mesenchymal bodies, CMBs) were capable of filling a chondral defect made in the cartilaginous condyle of an explant of bone. The cells formed articular-like cartilage that integrated with the surrounding articular cartilage. This method might be used to make neocartilage of any shape (figure 9.5) without need for a scaffold. Biomimetic articular cartilage has not yet been tested in clinical trials and its long-term biochemical and structural integrity in animal models is largely unknown.

Closely associated with the articular cartilage of the knee are the medial and lateral menisci, C-shaped fibrocartilages with an avascular center and a vascularized

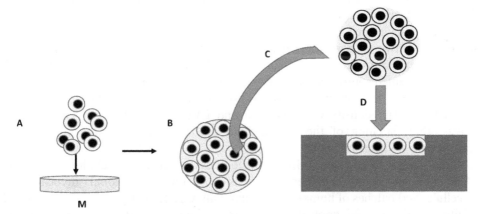

Figure 9.5. Scaffold-free production of biomimetic articular cartilage. (A) MSCs or chondroblasts (yellow) are cultured in a mold (M) shaped in the form of a chondral or osteochondral defect, with appropriate proliferation and differentiation factors. (B) The precursors differentiate into chondrocytes that secrete cartilage matrix (tan). (C) The differentiated slab of cartilage is removed from the mold, and trimmed to size if necessary. (D) The cartilage is fitted into the defect to mature further and integrate with the surrounding cartilage (purple).

perimeter. The menisci function as distributors of mechanical load and as shock absorbers during locomotion, and are a frequently injured tissue. Meniscus-like tissue has been created by seeding multilayered silk scaffolds with human bone marrow stem cells, followed by differentiation in chondrogenic medium containing TGF-β3 (Mandal *et al* 2011). A plastic meniscus composed of a polycarbonate-urethane (PCU) center reinforced circumferentially by ultra-high molecular weight polyethylene is currently in a phase III clinical trial in the US, Europe and Israel. The PCU is pliable, conforms to the site of implant, and does not require fixation to the bone or adjacent soft tissue.

9.5 Cardiovascular tissue

9.5.1 Cardiac muscle

A major problem with the injection of cell suspensions into infarcted myocardium (chapter 8) is loss of at least 50% of the cells by extrusion and death of 90% of the remaining cells by one week post-transplant, leaving less than 5% of the original number of injected cells (Laflamme and Murry 2011). Organizing the cells within a biodegradable scaffold appears to improve cell survival and cardiac function; thus, patching ischemic myocardium with pieces of biomimetic myocardium might be the most immediately useful way to reverse heart failure. The patches can be pre-vascularized *in vitro* by the addition of endothelial cells, or implanted and vascularized by host capillaries. Pre-vascularization of alginate/neonatal cardio-myocyte patches by transplantation onto the rat omentum gave improved cardiac function when the patch was removed and transplanted onto infarcted rat hearts (Dvir *et al* 2009). The patch showed structural and electrical integration into the host myocardium when examined 28 days after transplant.

The stiffness of the scaffolds used to construct bioartificial myocardium distorts the geometry of the host ventricular wall. To circumvent this problem, sheets of myocardium have been grown *in vitro* without scaffolds. For example, fibroblast cell sheets were grown in combination with spheres of cardiac progenitor cells derived from atrial explants of rat and human heart (Zakharova *et al* 2010). When the sheets were transplanted onto the infarcts of rat hearts, about 15% of the cardiac progenitor cells differentiated into cardiomyocytes. Cardiac wall thinning was reduced, capillary density was increased and LVEF was increased. These results suggest that patching of the myocardium can be achieved with prefabricated scaffold-free myocardial wall.

A new idea is to add stem cells or cardiomyocytes back to decellularized cardiac ECM and determine whether the cells will form functional myocardium. Decellularized patches of human ventricular wall were seeded with MSCs suspended in fibrin hydrogel and applied to myocardial infarcts made in nude rat hearts (Godier-Furnemont *et al* 2011). The MSCs migrated into the ischemic myocardium and greatly enhanced vascular network formation by their secretion of paracrine factors such as SDF-1, and baseline levels of left ventricular systolic dimensions and contractility were recovered. This technology has also been applied to whole rat hearts (Ott *et al* 2008). The heart matrix, which included blood vessel matrix, was installed in a bioreactor and seeded with either freshly isolated rat neonatal cardiomyocytes or aortic endothelial cells and the construct perfused with medium by pulsatile flow to mimic systole and diastole. Constructs seeded with cardiomyocytes formed immature cross-striated fibers by days 8–10, but the hearts could pump at a capacity of only about 2% of the adult heart. Constructs seeded with endothelial cells formed endothelial monolayers in large and small coronary vessels. Whether simultaneous seeding with cardiomyocytes and endothelial cells would have improved function is not known. The technology has also been applied to re-seeding a decellularized (except for skeleton) rat limb with myoblasts, fibroblasts and endothelial cells (Jank *et al* 2015). The seeded cells engrafted in their normal compartments. The myofibers formed by the myoblasts generated force of about 80% of neonatal myofibers. The construct filled with blood when transplanted to a recipient rat. These results hint at the future possibility of mimicking decellularized matrix on a human scale with a synthetic scaffold bioprinted from digitized medical images and seeding it with the appropriate cells in the presence of growth factors to create a complete bioartificial heart or limb.

9.5.2 Blood vessels

Biomimetic blood vessels would fill a need for the replacement of damaged or diseased blood vessels. The ideal biomimetic blood vessel should mimic the normal vessel as closely as possible in terms of thickness, structure and function so that it will have the requisite flexibility (compliance) and burst strength, and be mechanically stable enough to be easily sutured. This goal has not yet been attained, but progress has been made in constructing bioartificial vessels using synthetic polymers and autologous vascular cells that approximate the composition, architecture and function of natural vessels. The various approaches to biomimetic blood vessel construction have been reviewed in detail by Pashneh-Tala *et al* (2016).

Small-caliber biomimetic arteries have been constructed by seeding non-crosslinked polyglycolic acid tubes with bovine aortic smooth muscle cells (Niklason *et al* 1999). After culture for eight weeks in bioreactors under pulsatile flow to mimic the hydrodynamic conditions of fetal development, an endothelial cell suspension was injected into the lumen and allowed to adhere. The constructs were then cultured under luminal flow for an additional three days. Pre-seeding of scaffolds with smooth muscle cells and fibroblasts significantly increased the attachment of endothelial cells over constructs in which endothelial cells were seeded directly onto the scaffold wall. Histological and biochemical evaluation of the constructs revealed an endothelium-lined wall containing collagen and smooth muscle actin that was similar to that of native artery. The vessel exhibited burst strength greater than that of human saphenous vein and within the range suitable for arterial grafting.

Bioartificial arteries made in the same way with autologous carotid artery cells and grown under pulsatile flow were implanted into the right saphenous artery of miniature pigs (Jain 2003). The grafts remained open for the four-week duration of the experiment, whereas those grown in the absence of pulsatile flow thrombosed at three weeks, attesting to the importance of physical forces in vascular development. Zhang *et al* (2008) made tubular scaffolds by electrospinning reconstituted silk fibroin from silkworm cocoons and coating the tubes with Matrigel on their inner diameter. The tubes were then seeded with human aortic endothelial cells and smooth muscle cells. In a bioreactor under pulsatile flow the cells attached and proliferated, suggesting that silk has promise as vascular graft scaffolding. Autologous endothelial stem cells of bone marrow or peripheral blood were shown by Noishiki *et al* (1996) and Kaushal *et al* (2001) to form a non-thrombogenic endothelial surface when seeded onto small-diameter decellularized arteries or synthetic vascular grafts implanted into dogs or sheep. The grafts remained patent for four months when grafted into the aorta or carotid artery and their contractility and relaxation responses were similar to those of native arteries.

Other experiments have focused on creating blood vessels *in vitro* from cells only, without any natural or synthetic scaffolds. To do this, sheets of human endothelial and smooth muscle cells from umbilical veins and fibroblasts from dermis were grown in culture (L'Heureux *et al* 1998). The construct was made in a series of steps, starting with the production of an inner membrane (IM) by dehydrating a tubular fibroblast sheet. The IM was slipped over a perforated tubular mandrel (outside diameter 3 mm), wrapped with a sheet of smooth muscle and cultured in a bioreactor under luminal flow for a week. A sheet of adventitial fibroblasts was then applied to the outside of the construct and it was cultured for another eight weeks before removing the mandrel and seeding the luminal surface with endothelial cells. The mature construct bore a striking histological and ultrastructural resemblance to a native human artery, with burst strength significantly higher than that of human saphenous vein. The endothelial cells did not adhere platelets of human blood, which is clinically essential to prevent thrombosis. These vessels (5 cm length) were xenografted without endothelial cells (to avoid rejection) into the femoral arteries of dogs. At the end of one week, the patency rate was only 50%, but the histological architecture of the graft was retained. Such grafts have been shown to survive for up to 8 months in rat, dog and non-human primate models (L'Heureux *et al* 2006).

Microvessels (capillaries and venules) will be the most difficult small-caliber vessels to construct. They can be made by bioprinting endothelial cells or aggregates, but to form patterned vessels the cells needed to be printed on patterned surfaces. Human retinal endothelial cells have been patterned by seeding them through one end of a silicon wafer with surface grooves 150 μm in length and 50 μm in diameter (Kulkarni *et al* 2004). The other end of the wafer contained reservoirs of VEGF in Matrigel released into the grooves to stimulate microvessel formation by the endothelial cells migrating along the grooves. If patterned microvessels could be established on biodegradable chips seeded with endothelial cells and pericytes, and supplemented with angiogenic growth factors, it might be possible to provide biomimetic organ constructs or regenerating tissues with a microvascularization that aids in their survival.

An acellular blood vessel has been developed for implantation into the arms of end-stage renal disease patients to replace current polytetrafluoroethylene (PTFE) grafts that convey blood into the dialysis machine (Lawson *et al* 2016). PTFE grafts are prone to infection and stenosis due to intimal hyperplasia and thrombosis. The new blood vessel was constructed by seeding a biodegradable polymer with human vascular smooth muscle cells and incubated in a bioreactor (figure 9.6). The vessel was then decellularized, leaving a tube of ECM proteins. In a clinical phase-2 trial involving 60 patients followed for a mean of 16 months, these vessel grafts proved to be safe and functional, warranting further clinical study.

The construction of large-diameter vessels is more difficult, due to the scale-up in the cellular components required. A large-diameter biomimetic artery has been made by seeding a poly-4-hydroxybutane scaffold with sheep autologous carotid artery smooth muscle cells and cultured under pulsatile flow (Opitz *et al* 2004). Two days after the addition of endothelial cells, the constructs were wrapped in sheep SIS to increase mechanical strength, and implanted into the descending aorta. Up to three months after implantation, the grafts were fully patent, with no signs of occlusion, intimal thickening, or dilatation. Scanning electron microscopy revealed a confluent luminal endothelium. By six months, the grafts exhibited partial thrombus formation and significant dilatation. These abnormalities were associated with a deficiency in the number and distribution of elastic fibers compared to native artery. Elastin fibers are found throughout the wall of native aorta, whereas in the constructs, elastin fibers were located only in the luminal part of the wall. This work thus demonstrates the importance of elastin synthesis and distribution to achieving normal mechanical properties of biomimetic vessels.

In a clinical study, 12–24 mm diameter vascular grafts composed of biodegradable PGA/PLLA seeded with autologous bone marrow cells were used to replace the peripheral pulmonary arteries of 23 patients (Hibino *et al* 2010). These patients were followed for a mean of 5.8 years, during which time the grafts were reported to have shown no complications such as thrombosis, stenosis or obstructions.

9.6 Liver and pancreas

9.6.1 Liver

Extracorporeal liver devices are meant as a bridge to sustain liver function after acute liver failure until the patient can receive a liver transplant or their own liver regenerates. There are two types of devices, one a filtration system for liver dialysis that removes waste molecules, the other a hepatocyte-based bioartificial liver (BAL) that can also function in metabolic detoxification and synthetic activity. The focus here is on the latter device.

Extracorporeal biomimetic livers have been developed in large animal studies. The most compact and successful LAD designs have been hollow fiber bioreactors containing either allogeneic or xenogeneic hepatocytes (Wolfe *et al* 2002, Tilles *et al* 2002). The number of hepatocytes in the bioreactor needed to achieve maximum efficacy is estimated to be at least 10^{10}, and preferably the full human liver-equivalent of 2.8×10^{11} (Strain and Neuberger 2002). This number is difficult to achieve, but the ability to differentiate liver cells from hiPSCs may make it feasible. One successful design is called the HepatAssist system (Arbios Systems, Inc). The bioreactor is a cartridge containing artificial capillaries made of polysulfone hollow

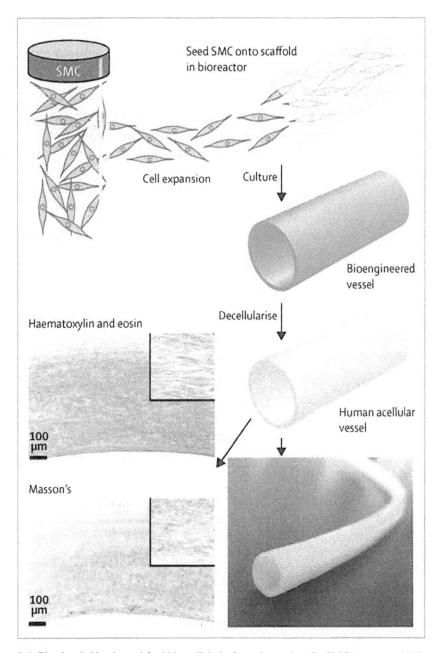

Figure 9.6. Biomimetic blood vessel for kidney dialysis. Smooth muscle cells (SMC) are expanded in culture and seeded onto a scaffold in a bioreactor. The bioengineered vessel (orange) is then decellularized to leave the ECM and create an acellular vessel (lower right inset). Panels at left show stained sections of vessel wall. Reproduced with permission from (Lawson *et al* 2016). Copyright 2016 Elsevier.

fibers with a nominal 0.15 μm pore size (figure 9.7). The spaces between the fibers contain 7×10^9 pig hepatocytes attached to collagen-coated dextran beads. Blood flows from the femoral artery and is first passed through a plasmapheresis machine to separate the plasma from the cells. The blood cells are returned and the plasma goes through two charcoal columns and through the hollow fibers of the bioreactor, where it bathes the hepatocytes through the fiber pores and is detoxified. The plasma is then returned to the circulation via the femoral vein. A phase I/II clinical study of this system on 39 patients found that the procedure was well tolerated by patients and that the bioreactor performed well. Six patients survived without transplantation, indicating that their liver had regenerated, and the remainder received liver transplants. The overall 30[+] days survival rate for all patients was 90%, compared to 50%–60% for acute liver failure patients, including those who received transplants that failed (Mullon and Solomon 2000).

Hepatocytes function better and last longer in bioreactors if they are in the form of spheroids rather than monolayers (Strain and Neuberger 2002). The hepatocytes re-acquire normal polarization (asymmetric localization of surface proteins that confer different functions on different parts of the cell) when aggregated into spheroids. Rat hepatocyte spheroids embedded in collagen gel within the spaces between the polysulfone fibers of a bioreactor provided a four-fold improvement in albumin synthesis and P-450 enzyme activity over that of hepatocytes dispersed in the collagen (Wu *et al* 1995). An advanced system called the Spheroid Reservoir Bioartificial Liver (SRBAL) has been developed in which blood is passed through a reservoir of spherical hepatocyte aggregates (Erro *et al* 2013, Glorioso *et al* 2015). When tested on pigs with induced acute liver failure, this system achieved 83% survival of the animals over 90 h. Possibly even more improvement in BAL function might be achieved by seeding bioreactors with tiny organoids (see chapter 10) composed of more than one liver cell type.

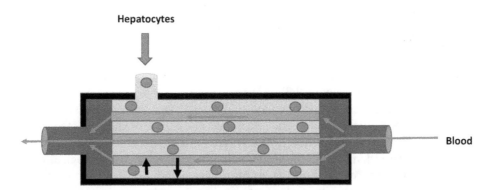

Figure 9.7. Diagram of bioartificial liver (BAL) device. The device consists of a housing (black) containing hollow fibers (blue) through which flow blood or medium (red arrows). Hepatocytes (green) are delivered into the device through a port, where they occupy the spaces between the hollow fibers. Metabolites made by the hepatocytes diffuse into the fibers, while toxic molecules in the blood diffuse outward into the spaces between the fibers to be detoxified by the hepatocytes (black arrows).

Ultimately, we would like to have an implantable biomimetic liver, and research in this direction has begun. A culture system has been developed consisting of human hepatocytes, human endothelial cells, and mouse fibroblasts grown in a three-dimensional plastic scaffold (Chen *et al* 2011). The endothelial cells and fibroblasts provide factors necessary for the survival and vascularization of the hepatocytes. After one week of culture, the cells formed a 20 mm long miniature liver organoid. The scaffold provided protection against the immune system, allowing the liver to be implanted into the abdominal cavity of recipient mice. The implant connected to the host circulation and was shown to express most of the human liver enzymes and to metabolize drugs. Such a liver organoid theoretically could be scaled up to human size.

In another experiment rat livers were decellularized, leaving the liver ECM and vessel matrix (Uygun *et al* 2010). The matrix was then reseeded with rat hepatocytes and endothelial cells to recreate a liver, which was cultured for two weeks in a perfusion chamber. During this time, the hepatocytes separated from the endothelial cells and distributed themselves around the vessels made by the endothelial cells. The hepatocytes were shown to express drug-metabolizing enzymes. When transplanted in place of one rat kidney and attached to the circulation by the renal artery and vein, the reseeded livers became perfused with blood. Livers removed after 8 h for histological examination showed only slight apoptosis. Histological examination indicated normal hepatocyte morphology, and immunohistochemical staining for liver proteins indicated that normal function had been maintained after transplant.

9.6.2 Pancreas

Biomimetic pancreata are systems of islets or β-cells sequestered from the immune system by encapsulating them in implantable polymer microcapsules or encasing them in an implantable device that interfaces with the vascular system.

9.6.2.1 Encapsulation of β-cells
Beta cells can be encapsulated within microspheres, fibers, or sheets made of materials with a pore size that allows O_2/CO_2 and nutrient/metabolite exchange, prevents immunorejection by denying access to T-cells, and does not stimulate inflammation and scar tissue formation that would block nutrient and gas exchange (Qi 2014, for a review) (figure 9.8). Alginate has been the principal microcapsule material, and when used to encapsulate allogeneic islets has proved successful in normalizing blood glucose in diabetic mice, dogs and humans when injected into the portal vein of the liver or intraperitoneally. Xenogeneic islets from dogs, pigs and cows have reversed hyperglycemia in diabetic mice. Long-term histological examination (30–130 days) revealed normal appearing α, β and δ cells. Diabetic monkeys treated by intraperitoneal transplant of pig islets in alginate-polylysine-alginate microcapsules were rendered insulin-independent for 120 days to over two years with fasting blood glucose levels and glucose clearance in the normal range and experiments on diabetic pigs using encapsulated human islets gave similar results (Sun *et al* 1996, Barnett *et al* 2007). In one clinical study, alginate microcapsules

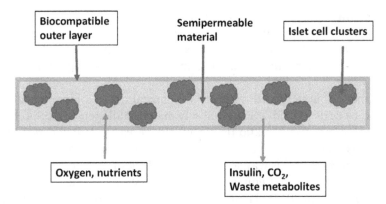

Figure 9.8. Islet cell clusters encapsulated in a semipermeable sheet biomaterial bounded by a biocompatible outer membrane. The capsule material protects the islets from immune attack, while allowing oxygen and nutrients to diffuse in and the insulin, CO_2, and waste metabolites to diffuse out.

loaded with islets harvested from cadavers were implanted into the peritoneal cavity of a patient who was a type I diabetic for 30 years. The patient was able to discontinue insulin injections nine months after implant and diabetic symptoms such as peripheral neuropathy and foot ulcers had resolved (Soon-Shiong *et al* 1994).

Encapsulated islets depend on diffusion for survival, and this appears to be sufficient as long as the microcapsules remain intact. However, the capsules eventually are walled off by a foreign body response that results in inflammation and fibrosis, leading to islet lysis and necessitating a new implant. Delivery site is also problematic. Both intraperitoneal injection and injection into the portal vein of the liver are invasive and portal injection carries some risk of bleeding and thrombosis. To find a biomaterial that inhibits fibrosis a library of alginate derivatives was generated by attaching different small molecules to the alginate polymer chain. In tests on mice and non-human primates, three triazole (a ring with two carbon atoms and three nitrogen atoms) derivatives were found to lower the foreign body response (Vegas *et al* 2016a). The best of these, triazole-thiomorpholine dioxide (TMTD) alginate, was used to encapsulate human islet cells generated from human PSCs (Vegas *et al* 2016b). The islets produced insulin in proportion to blood glucose levels for about six months (the length of the experiment) when implanted into the intraperitoneal cavity of mice. Furthermore, microcapsules not carrying islet cells survived in the intraperitoneal space of non-human primates for six months without fibrosis.

9.6.2.2 Implantable pancreatic devices

Implantable bioartificial pancreatic devices have been under development for more than three decades. The devices consist of a polymer housing enclosing an islet-filled chamber. Earlier versions contained allogeneic or xenogeneic (pig) islets embedded in alginate hydrogel and were tested on type 1 diabetic dogs (Maki *et al* 1996, Lum *et al* 1992, Lanza *et al* 1999). The device was implanted intraperitoneally and connected to the iliac vessels at each end to form an arteriovenous shunt. Blood glucose concentrations and other metabolic indicators were normalized for several months in these dogs and the device remained patent for up to 3.5 years in normal dogs, demonstrating its excellent biocompatibility (Sullivan *et al* 1991, Maki *et al* 1996). Late vascular thrombosis was encountered in the majority of dogs used in the studies, however.

The number of human β-cells required for clinical use of a biomimetic pancreatic device could be derived from PSCs or closely matched allogeneic β-cells chosen from a bank. Implanting a pancreatic device would be easier if it did not have to be surgically anastomosed to the circulation, but research has shown that the islets of devices not connected in this way soon fail due to lack of oxygen. New versions of an implantable free device that enhances oxygen supply have been designed (Ludwig *et al* 2012, Barkai *et al* 2013). These versions have a densely packed islet chamber in a semipermeable membrane sitting directly above a gas chamber connected to inflow and outflow ports (figure 9.9). A gas mixture high in oxygen is injected once a day into this chamber. The devices proved to have good functionality and maintained normoglycemia for six months in STZ-induced diabetic rats.

Another experimental pancreatic device has been designed that consists of a 2 cm-long stainless steel cylinder with 450 μm pores, stoppered on one end and containing a plunger that fills the interior of the cylinder (Pileggi *et al* 2006). The device was implanted subcutaneously in STZ-induced diabetic rats for 40 days to become invested on the inside and outside of the cylinder wall with vascularized connective tissue. The plunger was then removed, the cylinder filled with a solution of islets and the end stoppered. The device restored normoglycemia and histological examination of the islet tissue showed good preservation of islet structure and extensive vascular networks. The construct did not, however, confer immunoisolation, but could be modified to do so by encapsulating the islets.

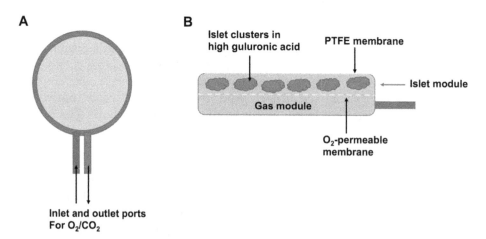

Figure 9.9. Biomimetic pancreatic device. (A) Top view. (B) Internal schematic.

9.7 Airways

9.7.1 Trachea

The trachea is a thin tube stiffened by a series of cartilage rings on the outside. Biomimetic tracheas or bronchial tubes can be created *in vitro* and sutured into tracheal defects, where they become vascularized and re-epithelialized. They have been made as a single construct of polypropylene mesh cylinders with polypropylene rings attached to the external surface and covered with a scaffold material that supports the differentiation of chondrocytes or MSCs seeded over the rings (Naito *et al* 2010). Alternatively, the rings have been made separately in 3D hydrogel chondrocyte cultures and then attached to the polypropylene cylinder (Nomoto *et al* 2013). The constructs become vascularized when implanted back into recipients with tracheal defects. Clinical implantation of bioartificial airways is not approved for routine use, though it has been tried on a number of patients. Unfortunately, several deaths resulted due to the collapse of the construct after implantation and there were a number of bioethical issues revealed that were related to the surgical protocols. No further bioartificial trachea implants have been allowed in patients since 2014. Further animal studies are needed to thoroughly assess the technology, and a protocol for this purpose has been established for a rat model (Jungebluth *et al* 2014).

The tracheal mucosa has been successfully patched with biomimetic tissue in a patient with an extensive chemical burn (Steinke *et al* 2015). The patch was made from a piece of pig decellularized jejunal wall reseeded with skeletal muscle and endothelial cells derived from a sample of the patient's thigh muscle and skin. When examined by bronchoscopy and biopsy 2.5 years later, the repaired mucosal region was indistinguishable from native airway mucosa.

9.7.2 Lung

The only treatment for end-stage interstitial pulmonary fibrosis (IPF) and chronic pulmonary obstructive disease (COPD), which includes chronic bronchitis and emphysema, is a lung transplant. Very few donor lungs are available, however, and the transplant survival rates are far less than for kidney or heart transplants. A readily available supply of biomimetic lung tissue has the potential to alleviate these problems, and research to provide such tissue has begun.

Gelfoam is a scaffold material that supports the development of alveolar-like epithelia from fetal lung cells *in vitro* (Birchall *et al* 2011). Gelfoam containing fetal rat lung cells was injected into normal rat lungs, where it supported the development of the cells into alveolar-like structures before it degraded (Anrade *et al* 2007). There was some evidence that recipient alveolar cells contributed to these structures, suggesting that the donor cells provided paracrine factors that stimulated growth and differentiation of host alveolar stem cells. In another experiment, adipose-derived stromal cells expressing HGF were incorporated into a polyglycolic acid scaffold and the construct was implanted into rats that had undergone resection of lung tissue (Shigemura *et al* 2006). Alveolar and vascular regeneration were substantially accelerated after one week, and the rats showed improved gas exchange and exercise tolerance.

The decellularization approach has been applied to making biomimetic rat lungs (Petersen *et al* 2010, Ott *et al* 2010). Decellularized lungs were repopulated with neonatal lung epithelial cells delivered through the trachea, and lung endothelial or human umbilical cord vein cells through the pulmonary artery. The constructs were cultured in a bioreactor where they regenerated region-specific tissue with the characteristics of normal alveolar tissue. When transplanted in place of a normal lung, this lung construct functioned with efficient gas exchange and vital capacity over a 2–6 h period. Human lung segments obtained from a tissue bank were decellularized and seeded with A549 lung carcinoma cells and endothelial cells. The A549 cells adhered well to the alveolar matrix and the endothelial cells to the vascular channels, indicating the feasibility of constructing a human bioartificial lung in this way. A critical need in this approach is the development of advanced bioreactors scaled to clinical application. Such a bioreactor has been developed that provides a mechanical environment mimicking the chest cavity in terms of fluid suspension, negative pressure ventilation, and physiologic perfusion (Raredon *et al* 2016). This bioreactor could potentially be adapted for the construction of other biomimetic organs as well.

9.8 Kidney and bladder

9.8.1 Kidney

The only cure for end-stage renal disease is a kidney transplant. According to 2013 numbers from the National Transplantation Data Report, US Department of Health and Human Services, 2013, the ratio of donor organ availability to patients needing a transplant is over 5:1. Implantable bioartificial kidneys would constitute a major advance in alleviating the donor shortage and avoiding the inconveniences and complications of dialysis.

A promising strategy to this end has been the construction of an extracorporeal kidney device that couples a dialysis hemofilter to a biomimetic proximal tubule for reabsorbing water, forming the equivalent of a giant nephron (Kim *et al* 2015, for a review). The hemofilter cartridge (glomerulus) is connected in series to a renal assist device (RAD) seeded with proximal tubule cells for reabsorption of water from ultrafiltrate (Johnston and Humes 2011, Oo *et al* 2011). Clinical studies carried out on patients with renal failure demonstrated that this device is effective.

A multidisciplinary effort among several university medical centers called the Kidney Project has the goal of miniaturizing this device to an implantable size. Miniaturization has been made possible by advances in biomaterial membranes that enhance the filtration capacity of the hemocartridge. The device consists of two units coupled to one another that are connected to a renal artery on one end (blood in) and a renal vein on the other (blood out). Blood flows into a hemocartridge to produce an ultrafiltrate that is sent to the adjacent unit, a biocartridge containing proximal tubule cells. The biocartridge reabsorbs water and electrolytes from the ultrafiltrate to be returned to the blood, while the remainder of the ultrafiltrate is eliminated as urine through a connection tó the bladder. Proof of efficacy of the miniature unit has been shown in pigs (Kim *et al* 2016). A clinical trial of the miniaturized device was scheduled for 2017, but has been delayed.

A second strategy is to build completely organic biomimetic kidneys. The subset of parietal epithelial cells isolated from human Bowman's capsule described in chapter 3 might be a source of cells that could be used in a bioartificial kidney. Yoo *et al* (1996) reported that functional glomeruli and proximal tubules self-organized into functional nephric units from disassociated mouse renal cells seeded into the lumen of tubular scaffolds. The scaffolds were made from silastic catheters surrounded by a polycarbonate membrane and coated with collagen. When the nephric units were connected to a reservoir and the whole construct implanted into athymic mice, they filtered waste from the blood and produced urine that collected in the reservoir. Decellularized cadaver kidneys, which preserve the ECM of the blood vessels, tubules and glomerular membranes, are being explored as scaffolds for rebuilding human kidneys. Renal cells seeded into decellularized porcine kidney ECM and grafted into athymic mice were able to form tubule and glomerular-like structures (Amiel *et al* 2000). Human umbilical venous endothelial cells and rat neonatal kidney cells seeded into decellularized rat kidneys repopulated the scaffold and formed glomerular and tubule tissue, but the function of the neotissue was poor (Song 2013).

9.8.2 Bladder

The wall of the urinary bladder consists of three layers, an inner uroepithelium, an intermediate connective tissue layer and an outer smooth muscle layer. Construction of bioartificial bladder tissue is an alternative preferable to surgical reconstruction of damaged bladder from intestinal tissue, or diversion of urine into the ileum or colon. Bioartificial bladder wall has been fabricated from cultured pig or human urothelial and bladder smooth muscle cells seeded onto either pig SIS or bladder wall matrix (Gabouev *et al* 2003, Zhang *et al* 2004). Both matrices supported development of a double layer of urothelial and smooth muscle cells. A full-thickness urinary wall has also been made *in vitro* without using any scaffold material (Fossum *et al* 2004). Urothelial cells, fibroblasts and smooth muscle cells were plated on top of one another, in that order. After three weeks of co-culture, a well-organized three-layered tissue was produced that could be mechanically handled, although the tissue was thin and the urothelial component was only a monolayer.

Total bioartificial bladders have been constructed by seeding bladder epithelial cells into either synthetic scaffolds or decellularized bladder matrix. Neobladders were constructed by molding a biodegradable polyglycolic acid mesh coated with poly-DL-lactide-*co*-glycolide 50:50 into the shape of a bladder (Oberpennig *et al* 1999). Autologous urothelial cells and smooth muscle cells were derived from biopsies of individual dogs who were to receive the neobladders. The luminal side of the scaffold was seeded with the cultured urothelial cells and the exterior side was seeded with the cultured smooth muscle cells. The constructs were transplanted to each donor dog after partial cystectomy that spared the top (ureteral side) of the bladder. Control dogs received either no construct or acellular scaffold. Urodynamic studies, radiographic cystograms, gross anatomical inspection and histological and immunocytochemical evaluation showed that, compared to controls, animals that received the bioartificial bladder had excellent bladder capacity and a histologically normal trilayered structure with ingrowth of neural tissue. A similar experiment was performed using allogeneic bladder submucosa (UBS) as the scaffold and autogeneic urothelial cells (Yoo et al 1998). Partial cystectomies were performed and the defects closed by grafting either seeded or unseeded (control) UBS. Cells migrated into the matrix in the controls. Histologically, the constructs in both groups had a normal trilayered cellular organization with nerve fibers, but bladders augmented with seeded UBS showed a 99% increase in capacity compared to only a 30% increase in the unseeded group.

Anthony Atala at the Wake Forest Institute for Regenerative Medicine has implanted bioartificial bladders into seven patients with bladder damage or malfunction (Atala et al 2006, Atala 2008). Scaffolds of collagen or a composite of collagen and polyglycolic acid were seeded with autogeneic urothelial and smooth muscle cells expanded from a biopsy and the bladders were implanted with or without an omental wrap. Follow-up urodynamic studies (22–63 months) indicated that the function of the bladders, particularly those implanted with an omentum wrap, was satisfactory with regard to compliance, urine capacity and continence. Further clinical trials of neobladders are now underway.

References

Akita S, Akino K, Imaizumi T and Hirano A 2008 Basic fibroblast growth factor accelerates and improves second-degree burn wound healing *Wound Rep. Reg.* **16** 635–41

Amiel G E, Yoo J J and Atala A 2000 Renal tissue engineering using a collagen-based kidney matrix *Tissue Eng.* **6** 685

Anrade C F et al 2007 Cell-based tissue engineering for lung regeneration *Am. J. Physiol. Lung Cell Mol. Physiol.* **292** L510–8

Amiel G E, Yoo J J and Atala A 2000 Renal tissue engineering using a collagen-based kidney matrix *Tissue Eng.* **6** 685

Atala A et al 2006 Tissue-engineered autologous bladders for patients needing cystoplasty *Lancet* **367** 1241–6

Atala A 2008 Genitourinary system *Principles of Regenerative Medicine* ed A Atala, R Lanza, J A Thompson and R Nerem (San Diego, CA: Elsevier/Academic), pp 1126–37

Barkai U et al 2013 Enhanced oxygen supply improves islet viability in a new bioartificial pancreas *Cell Transplant.* **22** 1463–76

Barnett B P et al 2007 Magnetic resonance-guided, real-time targeted delivery and imaging of magnetocapsules immunoprotecting pancreatic islet cells *Nat. Med.* **13** 986–91

Bhumiratana S et al 2014 Large, stratified, and mechanically functional human cartilage grown *in vitro* by mesenchymal condensation *Proc. Natl Acad. Sci. USA* **111** 6940–5

Birchall M A, Janes S and Macchiarini P 2011 Regenerative medicine of the respiratory tract *Principles of Regenerative Medicine* 2nd edn ed A Atala, R Lanza, J A Thompson and R Nerem (San Diego, CA: Elsevier/Academic), pp 1079–90

Boulton A J M, Kirsner R S and Vileikyte L 2004 Neuropathic diabetic foot ulcers *New Eng. J. Med.* **351** 48–55

Boyce S T 2001 Design principles for composition and performance of cultured skin substitutes *Burns* **27** 523–33

Cancedda R, Giannoni P and Mastrogiacomo M 2007 A tissue engineering approach to bone repair in large animal models and in clinical practice *Biomaterials* **28** 4240–50

Cao Y and Li H 2007 Polyprotein of GB1 is an ideal artificial elastomeric protein *Nat. Mater.* **6** 109–14

Chen G, Sato T, Ushida T, Ochiai N and Tateishi T 2004 Tissue engineering of cartilage using a hybrid scaffold of synthetic polymer and collagen *Tissue Eng.* **10** 323–30

Chen A A, Thomas L L, Ong R E, Schwartz R E, Golub T R and Bhatia S N 2011 Humanized mice with ectopic artificial liver tissues *Proc. Natl Acad. Sci.* **108** 11842–7

Chen W *et al* 2013 Human embryonic stem cell-derived mesenchymal stem cell seeding on calcium phosphate cement-chitosan-RGD scaffold for bone repair *Tissue Eng.* A **19** 915–27

Coburn J *et al* 2012 Bioinspired nanofibers support chondrogenesis for articular cartilage repair *Proc. Natl Acad. Sci. USA* **109** 10012–7

Corona B T *et al* 2012 Further development of a tissue engineered muscle repair construct *in vitro* for enhanced functional recovery following implantation *in vivo* in a murine model of volumetric muscle loss injury *Tissue Eng.* A **18** 1213–38

Cowan C M *et al* 2004 Adipose-derived adult stromal cells heal critical-size mouse calvarial defects *Nat. Biotechnol.* **22** 560–7

Dennis R, Kosnik P, Gilbert M and Faulkner J A 2001 Excitability and contractility of skeletal muscle engineered from primary cultures and cell lines *Am. J. Physiol. Cell Physiol.* **280** C288–95

Dupont K M *et al* 2010 Human stem cell delivery for treatment of large segmental bone defects *Proc. Natl Acad. Sci. USA* **107** 3305–10

Dvir T *et al* 2009 Prevascularization of cardiac patch on the omentum improves its therapeutic outcome *Proc. Natl. Acad. Sci.* **106** 14990–5

Ehrlich H 2004 Understanding experimental biology of skin equivalent from laboratory to clinical use in patients with burns and chronic wounds *Am. J. Surg.* **187S** 29S–33S

Erro E *et al* 2013 Bioengineering the liver: scale-up and cool chain delivery of the liver cell biomass for clinical targeting in a bioartificial liver support system *BioRes. Open Access* **2** 1–11

Falanga V *et al* 1998 Rapid healing of venous ulcers and lack of clinical rejection with an allogeneic cultured human skin equivalent *Arch. Dermatol.* **134** 293–9

Foroughi J *et al* 2011 Torsional carbon nanotube artificial muscles *Science* **334** 494–7

Forouzandeh F *et al* 2010 Local expression of indolamine 2,3-dioxygenase suppresses T-cell-mediated rejection of an engineered bilayer skin substitute *Wound Rep. Reg.* **18** 614–23

Fossum M, Nordenskjold A and Kratz G 2004 Engineering of multilayered urinary tissue *in vitro* *Tissue Eng.* **10** 175–80

Freed L E and Vunjak-Novakovic G 1995 Tissue engineering of cartilage *Biomedical Engineering Handbook* ed J Bronzino (Boca Raton, FL: CRC Press), pp 1788–806

Gabouev A I *et al* 2003 *In vitro* construction of urinary bladder wall using porcine primary cells reseeded on acellularized bladder matrix and small intestinal submucosa *Int. J. Artif. Organs* **26** 935–42

Glorioso J M *et al* 2015 Pivotal preclinical trial of the spheroid reservoir bioartificial liver *J. Hepatol.* **63** 388–98

Godier-Furnemont A F G *et al* 2011 Composite scaffold provides a cell delivery platform for cardiovascular repair *Proc. Natl Acad. Sci. USA* **108** 7974–9

Greaves N S *et al* 2013 The role of skin substitutes in the management of chronic cutaneous wounds *Wound Rep. Reg.* **21** 194–210

Haines C S *et al* 2014 Artificial muscles from fishing line and sewing thread *Science* **343** 868–72

Heliotis M *et al* 2006 Transformation of a prefabricated hydroxyapatite/osteogenic protin-1 implant into a vascularized pedicled bone flap in the human chest *Int. J. Oral Maxillofac. Surg.* **35** 265–9

Hibino N *et al* 2010 Late-term results of tissue-engineered vascular grafts in humans *J. Thorac. Cardiovasc. Surg.* **139** 431–6 e1–2.

Jain R K 2003 Molecular regulation of vessel maturation *Nat. Med.* **9** 685–93

Jank B J, Xiong L and Moser P T 2015 Engineered composite tissue as a bioartificial limb graft *Biomaterials* **61** 246–58

Jimenez P A and Jimenez S E 2004 Tissue and cellular approaches to wound repair *Am. J. Surg.* **187S** 56S–64S

Johnston K A and Humes H D 2011 Extracorporeal renal replacement *Principles of Regenerative Medicine* 2nd edn ed A Atala, R Lanza, J A Thompson and R Nerem (San Diego, CA: Elsevier/Academic), pp 943–53

Juhas M *et al* 2014 Biomimetic engineered muscle with capacity for vascular integration and functional maturation *in vivo Proc. Natl Acad. Sci. USA* **111** 5508–13

Jukes J *et al* 2008 Endochondral bone tissue engineering using embryonic stem cells *Proc. Natl Acad. Sci. USA* **105** 6840–5

Jungebluth P *et al* 2014 *Nat. Protoc.* **9** 2164–79

Kang S-B, Olson J L, Atala A and Yoo J J 2012 Functional recovery of completely denervated muscle: implications for innervation of tissue-engineered muscle *Tissue Eng.* A **18** 1912–20

Kaushal S *et al* 2001 Functional small-diameter neovessels created using endothelial progenitor cells expanded *ex vivo Nat. Med.* **7** 1035–40

Kim S, Fissell W H, Humes H D and Roy S 2015 Current strategies and challenges in engineering a bioartificial kidney *Front. Biosci.* **7** 215–28

Kim S *et al* 2016 Diffusive silicon nanopore membranes for hemodialysis applications *PLoS One* **11** e159526

Kinoshita Y *et al* 1996 Functional reconstruction of the jaw bones using poly(L-lactide) mesh and autogeneic particulate cancellous bone and marrow *Tissue Eng.* **2** 327–41

Kirsner R S *et al* 2015 Comparative effectiveness of a bioengineered living cellular construct vs a dehydrated human amniotic membrane allograft for the treatment of diabetic foot ulcers in a real-world setting *Wound Rep. Reg.* **23** 737–44

Kulkarni S S, Orth R, Ferrari M and Moldovan N I 2004 Micropatterning of endothelial cells by guided stimulation with angiogenic factors *Biosens. Bioelectron.* **19** 1401–7

Laflamme M A and Murry C E 2011 Heart regeneration *Nature* **473** 326–35

Landeen L K *et al* 1992 Characterization of a human dermal replacement *Wounds* **4** 167–75

Lanza R P *et al* 1999 Xenotransplantation of cells using biodegradable microcapsules *Transplantation* **67** 1105–11

Lawson J H *et al* 2016 Bioengineered human acellular vessels for dialysis access in patients with end-stage renal disease: two phase 2 single-arm trials *Lancet* **187** 2026–34

Levenberg S *et al* 2005 Engineering vascularized skeletal muscle tissue *Nat. Biotechnol.* **23** 879–84

L'Heureux N *et al* 1998 A completely biological tissue-engineered human blood vessel *FASEB J.* **12** 47–56

L'Heureux N *et al* 2006 Human tissue engineered blood vessel for adult arterial revascularizaton *Nat. Med.* **12** 361–5

Lima M D *et al* 2012 Electrically, chemically, and photonically powered torsional and tensile actuation of hybrid carbon nanotube yarn muscles *Science* **338** 928–32

Lum Z P *et al* 1992 Xenografts of microencapsulated rat islets into diabetic mice *Transplantation* **53** 1180–3

Ludwig B, Rotem A, Schmid J and Weir G C 2012 Improvement of islet function in a bioartificial pancreas by enhanced oxygen supply and growth hormone releasing hormone agonist *Proc. Natl Acad. Sci. USA* **109** 5022–7

Lv S *et al* 2010 Designed biomaterials to mimic the mechanical properties of muscles *Nature* **465** 69–73

MacNeil S 2007 Progress and opportunities for tissue-engineered skin *Nature* **445** 874–80

Madden L *et al* 2015 Bioengineered human myobundles mimic clinical responses of skeletal muscle to drugs *eLife* **4** e04885

Maki T, Monaco A P, Mullon C J P and Solomon B A 1996 Early treatment of diabetes with porcine islets in a bioartificial pancreas *Tissue Eng.* **2** 299–306

Mandal B B, Park S-H, Gil E S and Kaplan D L 2011 Multilayered silk scaffolds for meniscus tissue engineering *Biomaterials* **32** 639–51

Marcacci M *et al* 2006 Articular cartilage engineering with Hyalograft C: 3-year clinical results *Clin. Orthop. Rel. Res.* **435** 96–105

Marston W A, Sabolinski M L, Parsons N B and Kirsner R S 2014 Comparative effectiveness of a bilayered living cellular construct and a porcine wound dressing in the treatment of venous leg ulcers *Wound Rep. Reg.* **22** 334–40

Mullon C and Solomon B A 2000 HepatAssist liver support system *Principles of Tissue Engineering* 2nd edn ed R P Lanza, R Langer and J Vacanti (New York: Academic), pp 553–8

Naito H *et al* 2010 Engineering bioartificial tracheal tissue using hybrid fibroblast-mesenchymal stem cell cultures in collagen hydrogels *Interact. Cardiovasc. Thorac. Surg.* **12** 156–61

Ng J, Kara S, Jonathan B and Vunjak-Novakovic G 2017 Biometic approaches for bone tissue engineering *Tissue Eng.* B **23** 480–93

Niklason L *et al* 1999 Functional arteries grown *in vitro Science* **284** 489–93

Noishiki Y, Tomizawa Y, Yamane Y and Matsumoto A 1996 Autocrine angiogenic vascular prosthesis with bone marrow transplantation *Nat. Med.* **2** 90–3

Nomoto M *et al* 2013 Bioengineered trachea using aurologous chondrocytes for regeneration of tracheal cartilage in a rabbit model *Laryngoscope* **123** 2195–201

Oberpennig F, Meng J, Yoo J J and Atala A 1999 *De novo* reconstitution of a functional mammalian urinary bladder by tissue engineering *Nat. Biotechnol.* **17** 149–55

Oo Z Y *et al* 2011 The performance of primary human renal cells in hollow fiber bioreactors for bioartificial kidneys *Biomaterials* **32** 8806–15

Opitz F *et al* 2004 Tissue engineering of aortic tissue; dire consequence of suboptimal elastic fiber synthesis *in vivo Cardiovasc. Res.* **63** 719–30

Ott H C *et al* 2008 Perfusion-decellularized matrix:using nature's platform to engineer a bioartificial heart *Nat. Med.* **14** 213–21

Ott H C *et al* 2010 Regeneration and orthotopic transplantation of a bioartificial lung *Nat. Med.* **16** 927–33

Pashneh-Tala S, Eng B, MacNeil S and Claeyssens L 2016 The tissue-engineered vascular graft—past, present, and future *Tissue Eng.* B **22** 68–100

Perez R A *et al* 2014 Therapeutic bioactive microcarriers: co-delivery of growth factors and stem cells for bone tissue engineering *Acta Biomater.* **10** 520–30

Petersen T H *et al* 2010 Tissue-engineered lungs for *in vivo* implantation *Science* **329** 538–41

Pileggi A *et al* 2006 Reversal of diabetes by pancreatic islet transplantation into a subcutaneous, neovascularized device *Transplantation* **81** 1318–24

Qi M 2014 Transplantation of encapsulated pancreatic islets as a treatment for patients with type 1 diabetes mellitus *Adv. Med.* **2014** 429710

Quarto R, Kutepov S M and Kon E 2001 Repair of large bone defects with the use of autologous marrow stromal cells *New Eng. J. Med.* **344** 385–6

Raredon M S B *et al* 2016 Biomimetic culture reactor for whole-lung engineering *BioResearch* **5** 1

Rey-Rico A, Madry H and Cucchiarini M 2016 Hydrogel-based controlled delivery systems for articular cartilage repair *Biomed. Res. Int.* **2016** 1215263

Sheehy S J *et al* 2015 Tissue engineering whole bones through endochondral ossification: regenerating the distal phalanx *BioRes. Open Access* **4** 229–41

Shigemura N *et al* 2006 autologous transplantation of adiose tissue-derived stromal cells ameliorates pulmonary emphysema *Am. J. Transplant.* **6** 2592–29600

Song J J 2013 Regeneration and experimental orthotopic transplantation of a bioengineered kidney *Nat. Med.* **19** 646–51

Soon-Shiong P *et al* 1994 Insulin independence in a type 1 diabetic patient after encapsulated islet transplantation *Lancet* **343** 950–1

Steinke M *et al* 2015 Host-integration of a tissue-engineered airway patch: two-year follow-up in a single patient *Tissue Eng.* A **21** 573–9

Stevens M M *et al* 2005 *In vivo* engineering of organs: the bone bioreactor *Proc. Natl Acad. Sci. USA* **102** 11450–5

Stocum D L 2012 Regenerative medicine of epidermal structures Regenerative Biology and Medicine 2nd edn (San Diego, CA: Elsevier/Academic) chapter 11

Strain A and Neuberger J M 2002 A bioartificial liver-state of the art *Science* **295** 1005–9

Sullivan S J *et al* 1991 Biohybrid artificial pancreas: long-term implantation studies in diabetic, pancreatectomized dogs *Science* **252** 718–21

Sun V *et al* 1996 normalization of diabetes in spontaneously diabetis cynomologus monkeys by xenografts of microencapsulated porcine islets without immunosuppression *J. Clin. Invest.* **98** 1417–422

Tatara A M, Wong M E and Mikos A G 2014 *In vivo* bioreactors for mandibular reconstruction *J. Dent. Res.* **93** 1196–202

Tilles A W *et al* 2002 Bioengineering of liver assist devices *J Hepatobiliary Pancreat. Surg.* **9** 686–96

Toh W S *et al* 2010 Cartilage repair using hyaluronn hydrogel-encapsulated human embryonic stem cell-derived chondrogenic cells *Biomaterials* **31** 6968–80

Uygun B E *et al* 2010 Organ reengineering through development of a transplantable recellularized liver graft using decellularized liver matrix *Nat. Med.* **16** 814–21

Vegas A J *et al* 2016a Combinatorial hydrogel library enables identification of materials that mitigate the foreign body response in primates *Nat. Biotechnol.* **34** 345–52

Vegas A J *et al* 2016b Long-term glycemic control using polymer-encapsulated human stem cell-derived beta cells in immune-competent mice *Nat. Med.* **22** 306–11

Vigodarzeve G C and Mantero S 2014 Skeletal muscle tissue engineering: strategies for volumetric constructs *Front. Physiol.* **5** 362

Wakatani S *et al* 1998 Repair of large full-thickness articular cartilage defects with allograft articular chondrocytes embedded I a collagen gel *Tissue Eng.* **4** 429–44

Wang D A *et al* 2007 Multifunctional chondroitin sulphate for cartilage tissue-biomaterial integration *Nat. Mater.* **6** 385–92

Warnke P H *et al* 2004 Growth and transplantation of a custom vascularized bone graft in a man *Lancet* **364** 766–70

Warnke P H *et al* 2006 Man as living bioreactor: fate of an exogenously prepared customized tissue-engineered mandible *Biomats.* **31** 3163–7

Wolfe S P, Hsu E, Reid L M and Macdonald J M 2002 A novel multicoaxal hollow fiber bioreactor for adherent cell types. Part I. Hydrodynamic studies *Biotechnol. Bioeng.* **7** 783–90

Wu F J, Peshwa M V, Cerra F B and Hu W-S 1995 Entrapment of hepatocyte spheroids in a hollow fiber bioreactor as a potential bioartifical liver *Tissue Eng.* **1** 29–40

Yuan J and Poulin P 2014 Fibers do the twist *Science* **343** 845–6

Yoo J J, Ashkar S and Atala A 1996 Creation of functional kidney structures with excretion of kidney-like fluid *in vivo Pediatrics* **98S** 605

Yoo J J, Meng J, Oberpenning F and Atala A 1998 Bladder augmentation using allogeneic bladder submucosa seeded with cells *Urology* **51** 221–5

Zakharova L *et al* 2010 Transplantaton of cardiac progenitor cell shet onto infarcted heart promotes cardiogenesi and improves function *Cardiovasc. Res.* **87** 40–9

Zhang W *et al* 2004 Effects of mesenchymal stem cells on differentiation, maturation, and function of human monocyte-derived dendritic cells *Stem Cells Dev.* **13** 263–71

Zhang, Zhou J, Lu Q and Hu S 2008 A novel small-diameter vascular graft *in vivo* behavior of biodegradable three-layered tubular scaffolds *Biotechnol. Bioeng.* **99** 1007–10016

Zhang Y, Kropp B P, Lin H-K, Cowan R and Cheng E Y 2004 Bladder regeneration with cell-seeded small intestinal submucosa *Tissue Eng.* **10** 181–7

Zakharova M *et al* 2010 Transplantation of cardiac progenitor cell sheet onto infarcted heart promotes cardiogenesis and improves function *Cardiovasc. Res.* **87** 40–9

IOP Publishing

Foundations of Regenerative Biology and Medicine

David L Stocum

Chapter 10

Into the adjacent possible

Summary

What will be the next advances in our understanding of regenerative biology and how can they be translated into regenerative medicine? In his book *Reinventing the Sacred*, Stuart Kauffman (2008) writes of the 'adjacent possible', those fermenting but not yet fully formed concepts and technologies that will emerge in the not too distant future. What will emerge from the adjacent possible cannot be predicted with total accuracy because it is driven by an evolutionary process of unpredictable combinations and permutations of current ideas and technologies and occasionally revolutionary ideas that open whole new frontiers of basic and applied research. Matt Ridley, in his book *The Evolution of Everything* (2015), has shown how this process applies to virtually every aspect of human endeavor, from education to the economy. Even though we cannot make detailed predictions, current technical innovations and experimental advances on the edge of the adjacent possible give us some insight into the directions that regenerative biology and medicine might take over the next decade or two. In this chapter, I will summarize several of the conceptual and technical innovations that I believe are likely to impact future research, followed by the applications of these advances to regenerative biology and medicine.

10.1 Technical innovations

10.1.1 Gene editing with CRISPR/Cas9

High on the list of technical innovations is a revolutionary gene editing tool, CRISPR/Cas9. CRISPR/Cas9 is a molecular complex consisting of the endonuclease Cas9 (CRISPR-associated 9) and associated CRISPR guide RNAs that target specific base sequences in DNA. CRISPR stands for *clustered regularly interspaced short palindromic repeats* of base sequences. These repeated sequences were first discovered in bacteria, where they are part of a defense mechanism against invading viruses or bacteriophages. The repeats are complementary to invading viral or phage

doi:10.1088/978-0-7503-1626-2ch10

DNA sequences and bind the Cas9 nuclease to the invading DNA, which is then cut and degraded.

This system was adapted by Feng Zhang (Broad Institute), Jennifer Doudna (University of California Berkeley), Emmanuelle Charpentier (Max Planck Institute Berlin), Philippe Horvath (Du Pont Nutrition and Health) and Virginijus Siksnys (Vilnius University) to cut target eukaryotic DNA at specific sites by inserting synthetic guide mRNAs and mRNA for Cas9 into plasmids that were then transfected into cells to cut these sequences (Gasiunas *et al* 2012, Charpentier and Doudna 2013, Doudna and Charpentier 2014, Waddington *et al* 2016). CRISPR/Cas9 can be used to remove genes or to insert genes into the genome using the cell's DNA repair system (figure 10.1). Many synthetic guide RNAs have been developed that are complementary to specific DNA sequences of interest in cells (Kelley *et al* 2016), and mRNAs encoding endonucleases smaller than Cas9 have been developed that are easier to get into the plasmids used for transfection (Mali *et al* 2013,

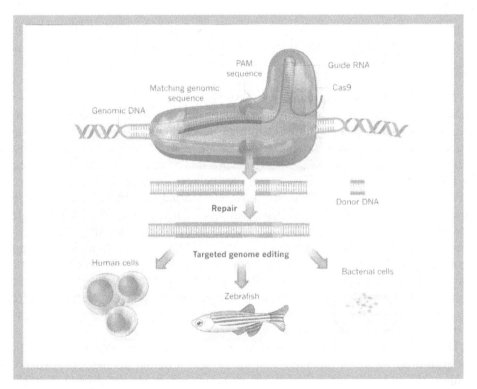

Figure 10.1. How CRISPR/Cas9 works to edit the genome. Messenger RNA for Cas9, along with guide mRNA and donor DNA is introduced in plasmids into cells. The base sequence of the guide RNA is complementary to the base sequence of the DNA that is to be edited, and positions Cas9 at the cut site. The PAM (Protospacer Adjacent Motif) is a sequence of 2–6 bases immediately following the DNA sequence targeted by Cas9. The PAM sequence is necessary to destabilize the target DNA enough to separate its strands and enable formation of the target DNA/guide RNA hybrid. When the host DNA is cut, repair mechanisms introduce the donor DNA into the genome. Reproduced with permission from (Charpentier and Doudna 2013). Copyright 2013 Nature.

Kiani *et al* 2015). CRISPR/Cas9 has applications in replacing faulty genes with good copies, revealing what genes are essential to cell survival (Wang *et al* 2015), introducing mutant DNA sequences associated with development and disease (Cohen 2016, Paquet *et al* 2016), humanizing animal models (Birling *et al* 2017), and targeting gene loci that regulate switch points for complex transcriptional circuits (Zalatan *et al* 2015, Chin 2017). The potential uses for this technology in medicine are legion. Genetically faulty stem cells can be corrected, immune cells engineered to attack cancer cells, genes that inhibit regeneration altered or eliminated, scarring minimized, and genes conferring regenerative power possibly added to regeneration-deficient or-incompetent tissues. It is highly probable that CRISPR technology will become progressively more sophisticated, greatly expanding its use in biology and medicine, and leading to other novel techniques of genetic manipulation (Kitada *et al* 2018).

10.1.2 Cell reprogramming

Cellular reprogramming is used to derive the types and numbers of cells required to replace damaged tissue, to study the onset of disease, and to screen agents for their efficacy to combat disease.

10.1.2.1 Indirect reprogramming via iPSCs

Refined reprogramming protocols to generate iPSCs and to differentiate them to specific cell phenotypes are under continual development for potential transplant. Induced PSCs can provide the human models necessary to study the mechanisms of many monogenic and complex diseases *in vitro* that cannot be completely mimicked by animal models, and serve to screen for therapeutic drug candidates and toxic substances (Passier *et al* 2016, for a review). What are the pivotal molecular junctions where cells involved in degenerative diseases are turned to the dark side? Having found these points, will it be possible to devise individualized therapies to prevent or halt the disease process and induce tissue regeneration? For example, CRISPR/Cas9 has been used to edit mutations in the amyloid precursor protein associated with Alzheimer's disease into mouse fibroblasts derived from iPSCs. Comparison of these edited fibroblasts and unedited fibroblasts differentiated from the iPSCs indicated an increase in the amyloid β-3 protein in the edited derivatives that is characteristic of the disease (Paquet *et al* 2016).

The studies of many degenerative diseases rely on mouse models that may or may not accurately represent the disease on the cellular level in humans. Comparison of human cells with cells from widely used mouse mutant models of diseases can tell us just how similar or dissimilar human and mouse models are.

10.1.2.2 Direct reprogramming

Direct reprogramming is achieved by inducing the transdifferentiation of one cell type into another without rewinding to an undifferentiated state. This type of reprogramming thus requires fewer steps than indirect reprogramming to achieve the desired phenotype. If it can be adapted to *in vivo* use, cells are more likely to adopt the mature new phenotype because they are in a natural three-dimensional environment surrounded by other cells from which they receive differentiation signals.

New methods are being sought to directly transdifferentiate cells. For example, fibroblasts have been transdifferentiated *in vitro* into neural stem cells and cardiomyocytes using cocktails of small molecules that can modify gene expression (Cao *et al* 2016, Zhang *et al* 2016). A modified CRISPR/Cas9 has been used to directly activate the three genes constituting the master neuronal induction network in mouse fibroblasts and produce neurons (Black *et al* 2016), and 76 pairs of transcription factors have been identified that convert fibroblasts to cells with core neuronal features and diverse transcriptional patterns predictive of their responses to pharmacological drugs (Tsunemoto *et al* 2018).

10.1.3 Imaging

Light microscopy of fixed and sectioned tissues stained with dyes or fluorescent antibodies to proteins, and electron microscopy have been the most widely used imaging techniques in the biological sciences. New imaging techniques combine optics, chemistry and transparent tissues to image live tissues, organs and embryos in real time, eliminating the need for laborious reconstruction of 3D images from stained sections (Peplow 2017, for a review). Collectively, these approaches have the potential to reveal a fine-grained picture of many different kinds of regenerative processes in response to trauma, or the effects of agents that enhance or reduce regenerative capacity. For example, adult neural stem cells of the injured zebrafish telencephalon, which lies just under the thin transparent skull, have been tracked individually after labeling with fluorescent green and red proteins to determine patterns of cell division, revealing that cell divisions are symmetric and deplete the pool of stem cells (Barbosa *et al* 2015).

Super resolution fluorescent microscopes that break the diffraction limit are able to image 3D dynamic molecular activity and subcellular structure by computationally combining many images of fluorescent probes introduced into cells (Bourzac 2013, Fessenden 2016). Light sheet microscopy uses a sheet of laser light to illuminate and repetitively capture single thin planes within tissues, after which a 3D rendition of the material can be obtained by integrating the planes. This technology has been used to examine the development of living embryos of *Drosophila*, zebrafish and the mouse (Sternad *et al* 2015, Keller *et al* 2008, Tomer *et al* 2012, Udan *et al* 2014). Gastrulation of the optically opaque *Xenopus* embryo has been live-imaged using time-lapse x-ray microtomography to reveal morphogenetic movements of the germ layers (Moosmann *et al* 2013).

Retinoic acid (RA), an important developmental molecule, has long been postulated to exist in gradients that specify differentiation patterns, but detecting such gradients relies on biochemical samples from tissue at different positions that are difficult to obtain and measure. A new method of visualizing RA has revealed an endogenous two-tailed source-sink retinoic acid gradient in live zebrafish embryos, with high point in the center of the embryonic rostral–caudal axis

(Shimozono *et al* 2013). The gradient was detected by genetically engineering blue and yellow fluorescent proteins into the retinoic acid receptor (RAR). In the absence of RA, the receptor fluoresces blue. RA binding to the ligand-binding domain of the receptor causes a conformational change that transfers energy from the blue to the yellow fluorescent protein, thus tracking the presence of free RA.

Another new imaging tool is a 'see-through' method of rendering fixed tissues transparent in 3D. CLARITY is one such method, in which fixed brain tissue is infiltrated with a hydrogel that binds proteins, neurochemicals, DNA and RNA in place (Chung *et al* 2013). Lipids, which are responsible for the opacity of the brain tissue, are then washed out with detergent (SDS) to make the structure transparent for studies with dyes and antibodies. Another clearing agent that does not require lipid removal and preserves fluorescent proteins and lipophilic dyes is a simple saturated fructose solution in water with 0.5% α-thioglycerol, called SeeDB (the DB stands for Deep Brain). SeeDB enabled visualization of the neuronal wiring diagram in the mouse olfactory bulb (Ke *et al* 2013). These methods can potentially be combined with various types of microscopy to describe the fine details of tissue architecture at the cellular and subcellular level.

'Cell painting' is another imaging approach designed to pick up subtle changes in a cell's profile in response to agents such as drugs and toxins that affect cell shape, structure and function (Camp and Cicerone 2017). The 'paints' are dyes that bind to cell components such as the cell membrane, Golgi body, actin filaments, mitochondria and DNA. The painted cells are imaged using advanced imaging technology before and after exposure to an agent of interest.

10.1.4 Biomimetic ECM

10.1.4.1 Functionalized biomaterial scaffolds
The development of biomimetic scaffolds to better support the proliferation, spatial patterning, and differentiation of cells into a functional tissue or organ is of prime importance (Ratner *et al* 2013, for a review). New biomaterials are continually being designed to mimic the structural, mechanical and chemical properties and thus the function of natural tissue-specific ECMs. Soft biomaterials such as collagen, elastin and other biopolymers are a particular focus of this effort, because soft tissue ECMs are the most difficult to mimic (Bauer *et al* 2014, for a review). To design these scaffolds, important input will come from polymer physical chemists and biochemists. The objective is to design 'smartness' into soft biomaterials that allows them to (1) conform precisely to the shape of a tissue gap, (2) transition from a liquid state at room temperature to a solid state at body temperature, (3) be 'tuned' to assume different degrees of hardness during solidification, (4) degrade and release growth factors in response to MMPs and (5) to respond homeostatically to fluctuations in body chemistry and mechanical stimuli. Successful natural or synthetic biomaterials will mimic natural ECM in their mechanical characteristics, surface chemistry, geometry, and micro and nanostructure.

The primary material of choice for soft biomaterial scaffolds is hydrogel (Bauer *et al* 2014). Hydrogels can be made from many different types of natural biomaterials such as collagen, and synthetic materials such as polyglycolic acid. A liquid azidobenzoic hydroxypropyl chitosan hydrogel has been designed that conforms precisely to wound contours and can be photo-crosslinked into a flexible and transparent membrane (Lu *et al* 2010). Other such hydrogels require no photo cross-linking and automatically undergo phase transition with change from room temperature to body temperature. Hydrogels are hydrophilic but they suffer the drawback of lacking the order to orient them into specific shapes. Liquid crystals, on the other hand, can be oriented to have structure, but are hydrophobic. These two properties have been combined in a polymer of a liquid crystalline elastomer that forms into a hydrogel having both a hydrophilic and crystalline structure that mimics the same mechanical properties as soft tissues (Torbati and Mather 2016).

Hydrogels containing stalks of catalyst-bearing microstructures separated from a reactant-containing layer have provided proof of principle that self-regulating homeostatic scaffolds can be fabricated. The gel can be made to swell or contract by a chemical stimulus. When swollen, the stalks protrude into the reactant layer and initiate a chemical reaction. When the gel contracts, the stalks bend down out of the reactant layer, turning off the chemical reaction. The design is highly customizable due to a broad choice of chemistries, tunable mechanics, and physical simplicity (He *et al* 2012).

10.1.4.2 Nano- and microtechnology

Nanotechnology involves the creation and use of materials and devices at the size scale of intracellular structures and molecules, typically on the order of less than 100 nm; for microtechnology the scale is on the order of $1-200$ μm (Zarbin *et al* 2013, Christ *et al* 2013, for reviews). Nano- and micro-particles are used for imaging, delivery of molecules to target sites, or to encapsulate cells, either as independent particles or incorporated into scaffolds to deliver therapeutic or regeneration-promoting agents. These particles take a variety of forms: quantum dots and gold nanoparticles for tracking and imaging cells; liposomes, micelles and cation/anion complexes encapsulating small molecule drugs, peptides, proteins or nucleic acids; functionalized particles that localize to cell receptors and release their payload in response to enzymes or internal and external conditions such as pH and temperature change or light; and micro-particles for encapsulation of therapeutic cells such as β-cells for the treatment of diabetes.

A simple, inexpensive and effective technique that can produce a variety of polymeric nanoparticles or fibers is electrospinning (Bauer *et al* 2014, for a review). Electrospinning applies an electrostatic force between a polymer solution ejected from a syringe and a metal collector plate at some distance from the syringe. The ejected polymer solution can be spun into particles and fibers of different sizes and architectures, depending on parameters such as molecular weight, solution properties (viscosity, conductivity), electric potential, distance between syringe and collector, temperature, and motion of the collector. An emerging natural biomaterial for use in making electrospun scaffolds is the keratin of hair, which can form self-assembled structures that regulate cellular recognition and behavior (Rouse and van Dyke 2010).

10.1.5 Organoids

Populations of embryonic and adult cells have an incredible ability to sort out and self-organize into miniature tissue complexes, as shown many decades ago by pioneering *in vitro* studies by Johannes Holtfreter (Townes and Holtfreter 1955) and Malcolm Steinberg (Steinberg 1978). These miniatures today are called 'organoids'. The progeny of adult stem cells, or derivatives of ESCs and iPSCs can also self-organize into organoids (figure 10.2). Organoid formation offers a new opportunity for understanding developmental processes such as cell interaction and morpho-genesis, and for modeling of disease (Lancaster and Knoblich 2014, Willyard 2015, Clevers 2016, Harrison *et al* 2017, for reviews). Organoids also have great potential for building biomimetic tissues and organs, either in the presence or absence of a biodegradable scaffold (Bhatia and Ingber 2014). Because their 3D structure is much closer to the natural state, they are more useful than 2D cultures for analyzing disease states *in vitro* and for screening drugs for both negative and therapeutic effects. To study these interactions, different tissue organoids can be linked together to form a 'body-on-a chip' (Skardal *et al* 2016).

10.1.6 Bioprinting

One of the most exciting possibilities for the future of regenerative medicine is the replication of human tissues and organs by bioprinting from a digital blueprint derived from medical images or molds of organs (Ventola 2014, Savage 2016). Bioprinting is derived from the industrial world of 'additive manufacturing', where a process called laser sintering uses a high-powered laser to fuse small particles of plastic, metal, ceramic or glass powders into desired three-dimensional shapes according to a

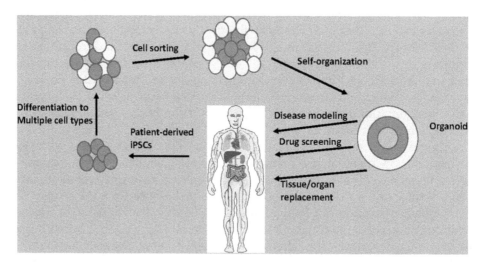

Figure 10.2. Creation of an organoid. iPSCs are induced to differentiate into the multiple cell types comprising an organ. These cells sort out from one another in their proper spatial relationships and self-organize to form the organoid, a miniature version of the organ. Organoids can be used to model disease, screen drugs, or in tissue and organ replacement.

digitized virtual pattern. Bioprinting does the same with natural or synthetic biomaterials and cells to build 3D structures. A bioprinter (of which there are a number of commercial models) adds cells, cell clusters or organoids ('ink') from printing nozzles to scaffold material ('paper') layer by layer. The nozzles can be moved in three axes to produce 3D tissue (Mironov *et al* 2009, Murphy and Atala 2014).

A major focus of current research is scaffold-free bioprinting, in which the printer adds cell clusters or organoids together that sort out and self-organize into larger structures. The basic cell clusters used are spheroids, toroids, or honeycombs of cells (Mironov *et al* 2009, Norotte *et al* 2009, Skardal *et al* 2010, Livoti and Morgan 2009) (figure 10.3). An instrument called the BioP3 (for pick, place and perfuse) has been designed to hand-construct stacks of these building units, and an automated unit is in the offing (Blakely *et al* 2014).

The major limitation to bioengineering tissues and organs to scale is the difficulty of providing the construct with blood vessels that can join with ingrowing host vessels after implantation. All cells in the body are within only a few microns of capillaries that, in addition to providing nutrients, oxygen, and waste removal, provide tissue-specific endothelial angiocrine factors to surrounding cells essential for their development and maintenance (Rafii *et al* 2016). Three-dimensional bioprinting addresses the problem of vascularization by either creating micro-channels in scaffold material that can be endothelialized, or incorporating endothelial cells into the cellular 'ink'.

Bioprinting technology will eventually be coupled to computer assisted design and manufacture (CAD/CAM) technology to fabricate synthetic scaffolds of complex shapes and tissue patterns using digitized blueprints made from three-dimensional medical images, followed by cellularization of the scaffolds to make biomimetic tissues or organs. Proof of principle has come from experiments in which a mandibular condyle template was machined from decellularized trabecular bone and seeded with human MSCs (Grayson *et al* 2010), and a bioprinted composite polycaprolactone–mineral scaffold was shown to be osteoinductive when seeded with adipose stem cells (Nyberg *et al* 2017). Anatomically shaped bone scaffolds

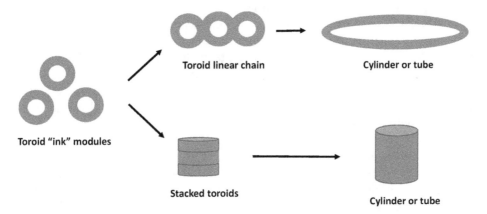

Figure 10.3. Toroids as bioprinter 'ink' to make scaffold-free cylindrical/tubular structures.

could be made from other scaffold materials as well by printing from a 3D scan of the bone.

The ultimate in medical bioprinting would be to develop a 'replicator' that can digitally scan an injury or burn on a patient and print new tissue *in situ* to repair the damage (see Skardal *et al* 2012). The replicator would have attachments that conform precisely to the morphology of bones, muscles, skin and other tissues that have suffered tissue loss, and deliver whatever types of cellular units and/or scaffolding into the injury required to reconstitute vascularized tissue. Such a device, called the 'Autodoc', has featured prominently in some of the science fiction stories of Larry Niven, and the concept was a feature of the 1940 story 'Farewell to the Master' by Harry Bates, which was made into the film 'The Day the Earth Stood Still'.

10.2 Applications of edge technologies

10.2.1 Pharmaceutical induction of regeneration

The ability to pharmaceutically induce repair and regeneration *in situ* would be transformative for regenerative medicine, and intense efforts will continue to be expended in this direction. Pharmaceutical approaches, where feasible, are the least expensive and least invasive approach to tissue regeneration, an advantage that cannot be overestimated. They take advantage of the natural regenerative ability of a tissue or confer regenerative capacity on tissues that do not regenerate naturally.

10.2.1.1 Regeneration-inducing soluble molecules
The search for more effective soluble wound repair and regeneration-promoting agents will continue in multiple directions. Combinatorial chemical libraries of molecular derivatives are being screened for their ability to enhance the properties of materials, drug discovery and optimization, initiating and enhancing regenerative pathways, and enhancing the reprogramming of cells. An untapped source of soluble regenerative agents is the traditional plant medicines that have been used for millennia in India and China (Singh 2007, Corson and Crews 2007). For example, curcumin, the yellow pigment of the curry spice turmeric, has been used for centuries as an antioxidant and anti-inflammatory wound-healing agent. The Indian government has compiled the Traditional Knowledge Digital Library, which lists more than 120 000 formulations from 100 ancient texts that could be mined for potential uses in regenerative medicine. Several of these formulations have been and are in clinical trials, but many others will be slower to reach that stage because they are available only in small quantities from their biological source and cannot yet be chemically synthesized, and/or their molecular mechanism of action has not been determined.

Soluble molecules mediate the paracrine and autocrine effects of cells on regeneration. For example, conditioned medium of hypoxic cardiomyocyte cultures transfected with the *Akt* survival gene reverses the loss of cardiomyocytes in infarcted mouse hearts (chapter 7). It might be possible to identify the factors in the conditioned medium responsible for this effect and assess their clinical efficacy in preventing cardiomyocyte apoptosis and scarring, and to regenerate cardiomyocytes. Furthermore, an important line of research would be to identify the full range of immunomodulatory and anti-inflammatory paracrine factors produced by adult stem cells such as MSCs for possible chemical synthesis and clinical use.

Regeneration-competent tissues activate complex cascades of signaling molecules that activate the proliferation of stem cells to repair or restore the injured tissue. These cascades are difficult to reproduce for therapy because we do not have a complete inventory of their elements, the timing of the activities of each element, and the concentrations at which the elements are effective. However, it might not be necessary to know everything about the cascades, but only to identify the signals that initiate them so they would then proceed autonomously. Some success has been had with this approach in regenerating cartilage and bone across critical size segment defect gaps using scaffolds that deliver growth factors (chapter 7).

The use of controlled release particles or microcapsules for delivering drugs or regeneration-promoting factors to injured tissues is a rapidly advancing part of regenerative medicine. For example, a tunable material has been described that is composed of the twin base linkers guanine and cytosine functionalized with lysine (Song *et al* 2011). In solution, groups of six twin linkers self-organize into rosettes that stack to form a nanotube structure. Drugs such as tamoxifen or growth factors can be mixed in to bind to the rosettes. The solution transforms to a solid upon injection, degrades and releases the factors bound to it. Alternatively, this material could be loaded with lentiviral growth factor gene constructs that transfect in-migrating cells and support their proliferation and differentiation into new tissue (Pelled *et al* 2010).

Affinity-controlled release (ACR) is a strategy for the sustained and tunable release of protein therapeutic molecules in a neutral aqueous environment. ACR works by adding therapeutic protein (TP) to binding ligands that are covalently linked to a polymer matrix such as a hydrogel. A dynamic equilibrium is established between ligand-bound and free diffusible TP that changes in response to local conditions. The challenge has been to find binding ligands that provide the desired release profiles. The search for such ligands has resulted in the discovery of a variety of novel peptide-, protein- and oligonucleotide-based ligands for the sustained release of several growth factors (Pakulska *et al* 2016).

A major problem in the use of neurotrophic and other molecules to stimulate spinal cord regeneration is the diffusion away from the lesion of the injected molecules. In China, Chen *et al* (2018) have described a novel bioconduit (called the NeuroRegen scaffold) that deals with this problem. The conduit consisted of ordered collagen fibers functionalized with hybrid agonist molecules consisting of a collagen binding domain (CBD) and a neurotrophin or growth factor molecule (CDB-BDNF, CDB-NT3, CDB-SDF1α), hybrid antagonists to myelin breakdown proteins (CDB-EphA4-LBD, CDB-PlexinB1-LBD), and antibodies to NSC and MSC cell surface recognition molecules. This environment captures and retains endogenous or transplanted NSCs and MSCs at the lesion site, neutralizes myelin breakdown proteins, and promotes the proliferation and differentiation of the NSCs into neurons and glia, as well as axon regeneration. The MSCs (from bone marrow or umbilical cord) in the conduit offer additional paracrine factors to promote regeneration. After tests on rats and dogs showed that the agonists, antagonists and cells remained in place, the conduit was evaluated on both acute and chronic SCI patients, after surgical resection of scar tissue. Over 12 months post-operation, no adverse effects of the treatment were noted, voluntary movements of the lower limbs and urine sensation were regained, and there was recovery of interrupted neural conduction.

10.2.1.2 Direct reprogramming

Direct reprogramming uses defined transcription factors or signaling molecules *in vitro* or *in vivo* to transdifferentiate somatic cells directly to other cell types or their progenitors as an approach to provide cells for transplant or create them *in situ*. Several groups of investigators have reported the direct reprogramming *in vitro* of mouse and human skin fibroblasts to neurons by induction with different combinations of lineage-specific transcription factor genes in viral vectors (Pang *et al* 2011, Yoo *et al* 2011, Ambasudhan *et al* 2011). Fibroblasts from healthy donors and Parkinson's disease patients were converted to dopaminergic neurons (Caiazzo *et al* 2011). Fibroblast marker genes were down regulated and genes characteristic of the dopaminergic phenotype were up regulated. There was no difference in the ability of fibroblasts from healthy and Parkinson's donors to undergo transdifferentiation, and reprogramming took place without cell division and dedifferentiation. Five lentiviral gene constructs, *Ascl1*, *Brn2/4*, *Myt1l*, *Zic1* and *Olig2* that regulate neuron differentiation from progenitors have been used to transdifferentiate mouse fibroblasts into neurons (Vierbuchen *et al* 2010).

Three retroviral gene constructs, *Gata4*, *Mef2c* and *Tbx5*, transdifferentiated mouse fibroblasts to cardiomyocytes *in vitro* (Ieda *et al* 2010). Fibroblasts have been converted to cardiomyocytes inside the mouse heart, and studies to reprogram fibroblasts directly to cardiomyocytes *in situ* are underway in pig hearts (Srivastava and DeWitt 2016, Chen *et al* 2017).

Transfection of fibroblasts with *Gata4*, *Hnf1α* and *Foxa3*, combined with inactivation of the p19[Arf] gene induced hepatocyte formation (Huang *et al* 2014), as did three specific combinations of two transcription factors composed *Hnf4α* plus either *Foxa1*, *Foxa2*, or *Foxa3* (Sekiya and Suzuki 2011). These transdifferentiations were induced *in vitro*, but acinar cells of the mouse pancreas have been transdifferentiated *in vivo* to β-cells by transfection with genes for three transcription factors central to pancreas differentiation, *Px1*, *Neurog3* and *MafA* (Zhou and Melton 2008). These results are proof of principle that transdifferentiation *in vivo* may be a viable therapeutic approach.

Transdifferentiation of cells by cell extracts has also been reported. The 293T line of fibroblasts were reprogrammed to T-cell-like cell when exposed *in vitro* to nuclear and cytoplasmic extracts derived from stimulated human T-lymphocytes (Hakalien *et al* 2002). The treated fibroblasts exhibited nuclear uptake and assembly of lymphocyte transcription factors, activation of a chromatin remodeling complex, histone acetylation, and activation of genes specific for T-cells. They also responded to antigen stimulation by up-regulation of a T-cell-specific pathway. In addition, 293T fibroblasts exposed to a neuronal cell precursor extract expressed a neurofilament protein in polarized outgrowths resembling neurites (Hakalien *et al* 2002). Changes in gene expression were stable over several weeks in culture. The reprogrammed fibroblasts did not express all the morphological features of T-cells or neurons, suggesting that their molecular profile was not completely reprogrammed. To what degree the T-cells are able to function as lymphocytes remains unknown. Assuming that further research on the use of soluble factors for reprogramming could identify the active factors in such extracts, they could potentially be used for *in vivo* reprogramming *in situ.*

A most spectacular case of transdifferentiation is the reprogramming of *Xenopus* embryonic gut tissue to an eye by misexpressing a variety of ion channels (Levin 2013). Manipulation of ion channels may be a mechanism applicable to changing cell populations locally in therapeutic ways.

10.2.1.3 Gene editing by CRISPR/Cas9

CRISPR/Cas9 has already been used in several potential therapeutic applications. The defective gene causing retinitis pigmentosis has been replaced in hiPSCs-derived from the skin cells of a patient with the disease, paving the way for the derivation of cells for transplant to cure the disease (Bassuk *et al* 2016). Human embryonic kidney cells have been engineered to sense proinflammatory cytokines produced by psoriatic epidermis and counter these by production of anti-inflammatory cytokines (Schukur *et al* 2015). Age-related hearing loss in mice due to mutations in the $Cdh23^{ahl}$ allele has been reversed by using CRISPR/Cas9-mediated homology directed replacement of the faulty allele at the zygote stage, as demonstrated by auditory-evoked brainstem response of rescued adults and by electron microscopy of regenerated hair cells in their cochleae (Mianne *et al* 2016). No off-target effects of the CRISPR/Cas9 were detected by genome sequencing. CRISPR/Cas9 was also used to ameliorate the genetic liver disease, tyrosinemia, in mice by a protocol that first delivered guide RNA and the normal gene DNA to the liver, followed by delivery of Cas9 mRNA in lipid nanoparticles (Yin *et al* 2016). CRISPR/Cas9 or Cpf1(a nuclease smaller than CRISPR)-mediated germline gene editing has been used to successfully correct the dystrophin gene and partially restore dystrophin expression and muscle function *in vivo* in the skeletal and cardiac muscle of post-natal *mdx* mice by transplantation of corrected satellite cells (Nelson *et al* 2016, Tabebordbar *et al* 2016).

Glaucoma is a common eye disease characterized by high intraocular pressure that left untreated can lead to retinal damage. The eye contains a cellular trabecular network. The protein myocilin is found in trabecular network cells. The function of the protein is unclear, but mutations in the gene for myocilin cause glaucoma. CRISPR/Cas9 editing of the myocilin gene *in vivo* was able to lower intraocular pressure in glaucomatous mouse eyes and prevent further damage (Jain *et al* 2017).

Many genetic diseases are caused by single base changes (point mutations), one of which is β-thalassemia, a rare blood disorder caused by a recessive point mutation in the hemoglobin gene of the hematopoietic stem cells. Hematopoietic cells in such people cannot carry enough oxygen for normal activities. Liang *et al* (2017) used CRISPR in conjunction with a disabled Cas9 enzyme tied to another enzyme that can swap base pairs, in this case a G (the mutant base) to an A, to correct the mutation. To get enough embryos to test the outcome of this switch, they found one person with the disease and converted this person's skin fibroblasts to iPSCs. The iPSCs were used to create embryonic clones with the β-thalassemia mutation. When they were subjected to the base switch, 40% of the clones converted the G back to an A. While this is not clinically acceptable, it is proof of principle that single base editing has promise as cure for this disease. Another blood disease caused by a point mutation in the hemoglobin molecule is sickle-cell anemia. A sickle-cell gene mutation has been corrected in patient-derived stem and progenitor hematopoietic cells (Dever *et al* 2016). In both β-thalassemia and sickle-cell disease, corrected autogeneic hematopoietic stem cells could be re-introduced into patients to repopulate the hematopoietic system.

To provide every cell in the adult body with normal copies of mutated genes that cause inherited disease, it is necessary to correct the genes at very early stages of development. CRISPR/Cas9 or Cpf1-mediated germline gene editing has been used successfully to correct the dystrophin gene and prevent muscular dystrophy in *mdx* mice (Long *et al* 2014, Zhang *et al* 2016). The dystrophin gene has been corrected *in vitro* in iPSCs of Duchenne muscular dystrophy patients by Cpf1-mediated correction of a nonsense mutation or exon skipping. Cardiomyocytes differentiated from these cells have functional dystrophin expression and exhibit enhanced contraction over control cardiomyocytes (Zheng *et al* 2017).

Mutations in a gene called MYBPC3, which encodes the cardiac myosin-binding protein, cause late cardiac hypertrophy in human populations. Using CRISPR/Cas9 to replace and repair the mutant sequence in preimplant human embryos achieved a high percentage of embryos with a normal MYBC3 gene (Ma *et al* 2017). Still another group has attempted to introduce a mutation into human embryos that confers resistance to HIV infection (Kang *et al* 2016). The embryos in these experiments were not allowed to develop further than a few cells.

Conceivably, gene editing of the immune system could be extended to pigs to enable xenotransplantation of pig organs to humans, thus obviating the need to evoke regeneration or construct a bioartificial tissue (Reardon 2015, Feng *et al* 2015, for reviews). If pig organ DNA can be edited to suppress expression of the vascular cell surface molecules that lead to the immune destruction of pig xenotransplants, pigs may become the transplant surgeon's new best friends. Above and beyond immunorejection, one of the concerns about xenotransplanting pig organs to humans is the risk of transmitting porcine endogenous retroviruses (PERVS) to the human recipient. CRISPR/Cas9 has recently been used to inactivate the PERVs in a porcine primary cell line and create PERV-free pigs from this line by SCNT (Niu *et al* 2017, Denner 2017)

New and better methods of gene editing are sure to be developed and applied to a wide variety of questions about developmental processes, correct genetic diseases, and augment the function of cells in animals and plants.

10.2.2 Cell transplants: derivatives of iPSCs

Derivatives of human iPSCs will be the primary future source of cells for transplant. The use of genetically edited hiPSCs by CRISPR/Cas9 will be beneficial in cases where a genetic defect requires correction.

Cells of the cornea, lens and retinal pigmented epithelium are among the first to be clinically targeted for replacement by derivatives of hiPSCs. hiPSCs have been differentiated to epithelial stem cells, which were then differentiated to corneal cells (Hayashi *et al* 2016). Sheets of these corneal cells were transplanted to the eyes of rabbits from which corneal epithelial stem cells had been depleted, where they restored healthy epithelium. The lens can be removed in a minimally invasive way through a small hole in the lens capsule while preserving the anterior lens epithelial stem cells intact. This procedure activates the stem cells to form a new lens. The operation has been performed on rabbits, macaque monkeys and on 12 human infants, with no complications (Lin *et al* 2016). However, activation of lens-forming stem cells in adults with cataracts is an undesirable feature that occurs in a substantial number of cases after artificial lens replacement for cataracts (Servick 2017).

Wet macular degeneration is associated with physical disruption and impairment of the support function of the RPE to overlying photoreceptors and underlying choroidal vasculature. Human autologous iPSCs have been derived from skin fibroblasts of a 77 year old patient suffering from age-related wet macular degeneration. The iPSCs were subsequently differentiated to a sheet of RPE cells that was transplanted under the neural retina (Mandai *et al* 2017). A year after transplantation, the RPE sheet had remained intact and although the patient's vision had not improved, it was stabilized.

Neural cells are of particular interest for the study of neurodegenerative disease and replacement of damaged neurons. Human iPSCs have been made from skin fibroblasts of patients with a number of both early and late-onset neurodegenerative diseases. Early onset neurodegenerative diseases for which hiPSCs have been created are Hutchinson–Gilford progeria (Liu *et al* 2011, Zhang *et al* 2011), Rett syndrome (an X-linked neurodevelopmental disorder) (Marchetto *et al* 2010), familial dysautonomia (FD, a rare fatal peripheral neuropathy caused by aberrant splicing of the *IKBKAP* gene) (Lee *et al* 2009), spinal muscular atrophy (Ebert *et al* 2009), and schizophrenia, a disease that has very strong heritability (Brennand *et al* 2011). Neurons derived by directed differentiation of these hiPSCs recapitulated cellular, biochemical and genetic features of the diseases. Rett syndrome could be partially reversed by IGF-1 and gentamycin, and exposure of FD cells to the plant hormone kinetin significantly reduced the mutant *IKBKAP* splice form. Neurons derived from the iPSCs of schizophrenia patients displayed elevated expression of the schizophrenia risk gene *TCF4*, as well as decreases in neuronal connectivity that were reversed by the antipsychotic drug loxapine. Late-onset neurodegenerative diseases for which iPSCs have been made are Parkinson's (Wernig *et al* 2008), Alzheimer's (Qiang *et al* 2011, Israel *et al* 2012), ALS (Dimos *et al* 2008, Richard and Maragakis 2015, Hedges *et al* 2016) and Huntington's (Carter *et al* 2014). The ALS patient was an 82 year old female patient with a rare *SOD1* mutation (Leu 144 to Phe) associated with a slowly progressing form of ALS. Both motor neurons and glia have been differentiated from these iPSCs. These results demonstrate the potential of using hiPSC derivatives to investigate the molecular aberrations associated with neurodegenerative diseases and to screen for therapeutic drugs.

Experiments transplanting neurons derived from human PSCs *in vitro* have begun to explore their therapeutic effects on late-onset neurodegenerative disease. Induced PSCs derived from a Huntington's patient were differentiated to GABAergic striatal neurons, the neurons most susceptible to degeneration in this disease. When transplanted into the striatum of a rat model of HD, significant behavioral recovery was observed (Jeon *et al* 2012). This initial effect, however, was followed at 33 weeks by clear signs of re-emerging HD pathology. Autologous neural progenitors derived from fibroblasts of MPTP-induced Parkinsonian monkeys showed long-term survival after transplantation into the brains of these monkeys, with little sign of inflammatory cells or reactive glia, suggesting their potential use for treatment of Parkinson's disease (Emborg *et al* 2013, Hallet *et al* 2015). Functional dopaminergic neuron precursors made from human ESCs and transplanted into the striatum of Parkinsonian 6-OHDA mice and rats, or MPTP monkeys survived well and differentiated into dopaminergic neurons. Behavioral deficits were fully corrected in the rodent models; functional outcomes were not reported for the monkeys (Kriks *et al* 2011).

An interesting and ambitious goal of regenerative technology is the ReAnima project, a phase I trial approved by an Institutional Review Board of the NIH and India, to revive 20 brain-dead patients on life support by infusing a cocktail of MSCs and peptides directly into the brain. The project is a joint venture between the Philadelphia-based company Bioquark and Revita Life Sciences in India. These companies hope to gain insights into a wide variety of consciousness disorders due to trauma, such as coma and vegetative states, and diseases such as Alzheimer's and Parkinson's. There are reports in the medical literature that persons have come out of a brain-dead state via the augmentation of residual brain activity, which might include stem cell activation to repair damage.

Hematopoietic stem cells have been derived from hiPSCs differentiated to hematogenic endothelium and then to HSCs (Sugimura *et al* 2017), or by conversion of mouse adult endothelial cells to HSCs (Lis *et al* 2017). The HSCs derived by either of these routes engraft into the bone marrow in experimental animal studies.

Ten years of finding ways to induce pluripotent stem cells from somatic cells has not yet produced definitive clinical applications, but the availability of these cells, coupled with gene editing technology, is transforming the biological research that will lead to those applications (Scudellari 2016, Cyranowski 2018). Genetic diseases can be investigated by gene editing to produce disease-associated mutations in healthy iPSCs for comparison to the originals, or in organoids to study developmental and disease processes.

10.2.3 Biomimetic tissues

Several examples indicate that advances in biomimetic tissue construction are edging closer to clinical reality.

10.2.3.1 Intervertebral discs

Degeneration of intervertebral discs is a major cause of back pain and incapacitation in the human population. Intervertebral discs engineered with ovine adult fibroblasts and collagen fibrils have been shown to maintain disc space height, produce new ECM, and show dynamic mechanical properties similar to native intervertebral discs when implanted into the rat spine (Bowles *et al* 2011).

10.2.3.2 Vascular core scaffold

Surgeons will need bioartificial tissues complete with blood vessels that can be stitched into host vasculature for immediate perfusion. A recent advance in this direction is the AngioChip, a biodegradable scaffold of poly(octamethylene maleate (anhydride) citrate) (POMaC) with vascular channels (Zhang *et al* 2016). The scaffold is built by stamping thin layers of POMaC with channels 50–100 μm wide and pierced by 20 μm diameter microholes. The layers are stacked into 3D rectangular chips to form a mesh of synthetic channels. Chips are then placed in a bioreactor and perfused with a suspension of human umbilical vein endothelial cells to form blood vessels. In response to thymosin $\beta4$, endothelial cells were shown to migrate through the microholes of the system and undergo angiogenesis to form a vascular network.

The Angiochip has been used as a vascular core around which to build functional liver and heart tissue by adding either rat or hESC-derived hepatocytes or cardiomyocytes to the parenchymal spaces surrounding the vascular channels. Anastomosis of AngioChips into rat femoral vessels was followed by immediate perfusion of the tissue with blood. The chips have immediate application for both toxicity and therapeutic drug screening, but will need to be scaled up for replacement of damaged tissues.

10.2.3.3 Bioprinted tissues

Bioprinting is making rapid progress in tissue construction. Several different types of tissues have been bioprinted with an eye toward clinical use (Sears *et al* 2016, for a review). Aortic valves have been printed using alginate/gelatin hydrogels and smooth muscle and valve leaflet interstitial cells (Duan *et al* 2012) or adipose-derived stem cells (Hockaday *et al* 2014). A polycaprolactone splint for a collapsed bronchial airway in a two-month-old infant was bioprinted from a CT-based design and sutured to the collapsed region (Zopf *et al* 2013) (figure 10.4). The splint kept the bronchus open and no unforeseen problems were reported a year later. The splint was expected to degrade over a three-year period.

Polymer-derived ceramics have been manufactured that can be cured with UV light in a bioprinter and then pyrolized to form 3D structures of complex shape (Eckel *et al* 2016). This process has the potential to manufacture bones of various shapes. Knee and dental replacements are now being custom bioprinted to the precise dimensions required for proper fit. A prototype of a bioartificial ear has been bioprinted using a hydrogel material and chondrocytes. The ear includes a sound-capturing antenna made of silver nanoparticles that can signal to the auditory nerve (Manoor *et al* 2013).

An integrated tissue–organ printer (ITOP) is now available that can fabricate stable, human-scale tissue constructs of any shape (Kang *et al* 2016). Cells of different types are mixed in printing reservoirs with hydrogel consisting of gelatin, fibrinogen, hyaluronic acid, glycerol and high glucose DMEM culture medium. The

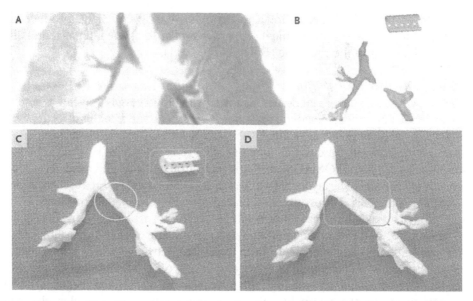

Figure 10.4. Customized, bioresorbable broncheal splint fabricated from polycaprolactone by 3D printing from a computed tomographic image and implanted into a two-month-old infant. (A) Image of collapsed bronchus. (B) Computed tomography-based design of bronchial splint (red). (C) 3D printed cast of patient's airways without the splint in place (yellow circle) and the splint (orange rectangle). (D) Cast of airway with splint in place. The repair was successful. Reproduced with permission from (Zopf *et al* 2013). Copyright 2013 Massachusetts Medical Society.

hydrogel is extruded onto supporting polycaprolactone in a pattern that leaves microchannels for diffusion. The fibrinogen is crosslinked using thrombin, and the remaining unlinked components washed out. To achieve the correct shape of the construct, CT scan or MRI clinical image data is processed by CAD software. The ITOP has been used to print several types of tissues. Human amniotic fluid stem cells were printed to differentiate bone in the precise shape of defects in human mandibular and rat calvarial bones. Constructs in the shape of human ears were printed from rabbit ear chondrocytes and then matured by implanting them in the dorsal subcutaneous space of athymic mice. Small muscle constructs were printed using mouse myoblasts that underwent maturation after subcutaneous implantation and connection with the peroneal nerve. These results show that, with further development, the ITOP device will be capable of bioprinting both simple and complex tissue shapes that conform to defect shapes and mature after implantation.

A printer to make replacement tissue *in situ* will not be far behind and prototypes are already being tested for repair of skin wounds and burns. Skin-like structures of fibroblasts and keratinocytes have been successfully bioprinted *in vitro* and have been found to integrate into host skin after implantation in mice (Koch *et al* 2012, Michael *et al* 2013). A bioprinter has been developed at the Wake Forest Institute for Regenerative Medicine to print skin cells onto burn wounds, using a scanner to determine wound size and depth. This device has also been used to print amniotic stem cells onto a mouse excisional wound model. Although the amniotic cells did not form skin cells, they secreted a number of paracrine factors that accelerated the healing of the wounds (Skardal *et al* 2012). We can now visualize a future therapy in which the cell types comprising the dermis and epidermis of the skin, including hair follicle stem cells, can be printed directly onto a wound in the correct spatial configuration. In addition, such bioprinted skin has the potential to eliminate the use of animals in testing cosmetic products for unwanted side effects.

Complete elimination of the need for scaffold material in bioprinting will be the next wave of bioprinting technology. A new type of bioprinter (Regenova) developed in Japan uses 'Kenzan' (translation: pinholder for flowers) methodology to assemble cell spheroids without scaffolds (Moldovan *et al* 2017, for a review). Spheroids are spitted contiguously on microneedles (kenzans), where the cells can sort out and fuse into larger constructs that can then be removed from the needle array. The needles can be patterned into any shape. Taniguchi *et al* (2018) prepared spheroids from mixed cell suspensions of rat chondrocytes, MSCs, and endothelial cells to generate tubular biomimetic tracheas using the Regenova bioprinter (figure 10.5). The constructs were matured in a bioreactor and transplanted into nine rats with silicone stents to prevent collapse of the transplant and to support the graft until it had obtained a sufficient blood supply. The grafts maintained their shape and stiffness over the 23 day period of

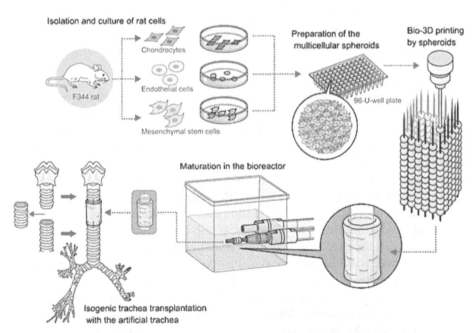

Figure 10.5. Construction of a biomimetic trachea using the Regenova bio-3D printer. The printer uses kenzan methodology to arrange multicellular spheroids, consisting of chondrocytes from rib cartilage, endothelial cells, and MSCs, on a tubular needle array. Once the spheroids were fused into the tubular structure of the trachea, the construct was matured in a bioreactor perfused with medium supporting chondrocyte and endothelial cell proliferation, before being transplanted into rats to replace a short segment of resected trachea. Reproduced with permission from (Taniguchi *et al* 2018). Copyright 2018 Oxford University Press.

observation. The rats had some breathing difficulties, however, that need to be addressed before clinical applications are considered.

The Indiana-based company Techshot and its collaborators are looking to bioprint scaffold-free organs under zero-gravity conditions. They have bioprinted the lower half of a human infant's heart in less than 30 s aboard an aircraft flying arcs to produce weightlessness, using a proprietary ink of adult stem cells and growth factors. The ultimate goal is to print human organs on the International Space Station, allow the organs to grow and mature for a month, and return the completed organ to earth. Space science and medicine are thus joined in this venture.

10.2.4 Organoids

Both scaffold-free bioprinting and organoid formation are dependent on the self-organization of cells into tissues. Rapid advances in the creation and use of organoids are underway. The hope is to use organoids to understand normal development, the development of cancer, screen drugs for toxicity and therapeutic value, and to replace damaged or missing tissues. Because organoids are 3D structures with the tissue phenotypes, functions and interactions of normal organs, they are even more useful than simple 3D cell cultures for screening the effects of therapeutic drugs, or for the efficacy of CRISPR/Cas9 technology to genetically modify form and function. Most research to date has focused on making organoids of gastrointestinal tissues (stomach, intestine, liver and pancreas), but this technology is exploding to encompass virtually all organ systems.

10.2.4.1 Liver and pancreas

A progenitor cell that can form both pancreatic and hepatic organoids has been isolated from the ducts of both liver and pancreas of mice (Dorrell *et al* 2014). This cell has also been found in human pancreatic ductal cells, and can be expanded and converted into β-cells (Lee *et al* 2013). The cell has a unique surface signature and both the pancreatic and hepatic versions show extensively overlapping transcriptomes. Upon injury to the liver or pancreas the progenitor cells activate the Lgr5 R-spondin receptor, which is also a marker for intestinal crypt stem cells (Barker *et al* 2010), and acquire clonogenic expansion capacity. A defined differentiation medium induced mouse pancreatic and hepatic organoids from clones of single Lgr5+ cells in Matrigel (Huch *et al* 2013). The hepatic organoids expressed markers for mature hepatocytes and were able to engraft in the liver and partially rescue tyrosinemic mice after intrasplenic injection. Interestingly, pancreatic organoids were also able to generate hepatocyte-like cells after transplantation to tyrosinemic mice, indicating functionality similar to hepatic organoids. Liver bud organoids have also been derived from a single isogenic hiPSC population genetically engineered to express GATA binding protein (GATA6) induced by doxycycline (Dox) (Guye *et al* 2016). This system has the advantage of generating the many cell types associated with a liver bud from a single cell source rather than combining hiPSC-derived hepatocyte progenitors with other cell types.

A transgenic mouse line has been created that expresses the Doxycycline-inducible TetO promoter driving the pancreatic transcription factors Ngn3, Pdx1 and Mafa (Ariyachet *et al* 2016). After STZ ablation of β-cells, Dox treatment of these animals reprogrammed a significant percentage of antral stomach cells, which share transcriptional similarity to β-cells, to cells with close molecular and functional resemblance to β-cells. While induction of β-cells from antral stomach tissue *in vivo* might seem at first glance to be a therapy for diabetes, this would diminish the number of native stomach endocrine cells that regulate many physiological processes. To bypass this problem, organoids were established from transgenic antral tissue on a polyglycolic acid (PGA) scaffold, and transplanted onto the omental flap of immunodeficient animals. Dox-induced β-cell formation in the antral organoids maintained normoglycemia in the mice after STZ depletion of β-cells.

The establishment of organoids derived from mouse and human ductal pancreatic tumor cells has highlighted their value in understanding how this cancer arises (Boj *et al* 2015). Organoids derived from cells at different stages of the disease exhibited the cellular and molecular characteristics specific to each stage. When transplanted to immunodeficient mice the organoids recapitulated all of the stages of tumor development from early grade neoplasms to locally invasive and metastatic carcinomas. Transcriptional and proteomic analyses revealed the genes and biochemical pathways altered during disease progression, many of which take place in human tissue, thus providing a system to further understand pancreatic cancer, suggest potential points of intervention, and to screen for drugs that target those points. How pancreatic cancer cells metastasize is a mystery that is being investigated using organoids of pancreatic cancer cells (Roe *et al* 2017). Organoids from other types of tumors could serve similar purposes.

10.2.4.2 Stomach and intestine

The stomach is derived from foregut endoderm of the embryo. Stomach organoids have been derived from mouse ESCs by induction of the mesenchymal gene *Barx1*, which is essential for specifying stomach development (Noguchi *et al* 2015). Both the body and the antrum of the stomach were formed in three-dimensional culture and mature stomach cells were differentiated with a gene expression profile similar to that of adult stomach. The stomach cells secreted both pepsinogen and gastric acid. McCracken *et al* (2014) have derived gastric tissue organoids from human pluripotent cells.

The intestine is derived from mid and hindgut endoderm. Intestinal organoids have been created from fragments of mouse small and large intestine (Kaihara *et al* 2000; Grikscheit *et al* 2002). When transplanted, the organoids developed with typical intestinal structure, including a mucosa with crypt-villous structures, and smooth muscle layers, and were able to rescue mice after massive small bowel resection (Sala *et al* 2011). Lgr5+ progenitor cells are probably the origin of the organoids, as indicated by the fact that colon organoids can be derived from single clonally expanded GFP+/Lgr5+ intestinal crypt stem cells (Sato *et al* 2009, Yui *et al* 2012, Sato and Clevers 2013). These cells were able to repair superficially injured colonic epithelium, forming self-renewing crypts that were histologically and functionally normal. Organoids of small intestine have been produced from hESCs and hiPSCs differentiated first to endoderm (Spence *et al* 2011, Watson *et al* 2014). *In vitro*, these organoids differentiated into tissue resembling native human intestine with crypt–villus architecture, submucosal and smooth muscle layers, which matured further after engraftment *in vivo* under the kidney capsule of mice and demonstrated digestive functions. By growing intestinal stem cells in hydrogels of different stiffness, Gjorevski *et al* (2016) showed that separate phases of organoid formation require different ECM components and mechanical environments. The expansion phase of ISCs required a stiffer matrix with fibronectin adhesion, whereas differentiation required a softer matrix with laminin adhesion.

10.2.4.3 Lung

Lung organoids have been derived from human ESCs and iPSCs by first inducing them with Activin A to form endoderm. The endoderm was induced to a foregut lineage by exposure to WNT and FGF signals while simultaneously inhibiting BMP and TGF-β signaling, and then into the lung lineage by manipulating FGF and SHH signaling (Wells and Spence 2014, Mondrinos *et al* 2014, Dye *et al* 2015, Aurora and Spence 2016). The organoids consisted of epithelial and mesenchymal compartments of the lung organized into the normal structural features. They possessed proximal airway epithelium with basal cells and ciliated cells surrounded with smooth muscle, and distal alveolar-like epithelium. Both histological examination and transcriptional profiles indicated that the organoids were immature and similar to human fetal lung tissue.

10.2.4.4 Kidney

Progress is being made on constructing organic neokidney tissue. Mixtures of dissociated mouse renal cells can reorganize on synthetic tubular scaffolds into functional glomeruli and tubule tissue (chapter 9). Kidney organoids have been generated from hiPSCs (Morizane *et al* 2015, Takasato *et al* 2015). These organoids contain nephrons with glomeruli containing podocytes, distal and proximal tubules, and early loops of Henle. They were comparable in structure to human first trimester kidney tissue and their transcriptional profiles matched human first trimester kidney tissue (Roost *et al* 2015). CRISPR/Cas9 has been used to introduce genes (*PDK1* or *PDK2*) for polycystic kidney disease into hiPSCs-derived kidney organoids (Freedman *et al* 2015). Subsequent deletion of the inserted genes disrupted the formation of cysts in the organoids, suggesting a potential therapy for this disease.

10.2.4.5 Brain

Following the pioneering work of Yoshiki Sasai and Motosugu Eiraku, who produced organoids of cerebral cortex and eye from aggregates of mouse ESCs (Eiraku *et al* 2011, Sasai 2013), a simple and reproducible 3D culture system was described for generating cortical organoids from human PSCs that contain both deep and superficial neurons and astrocytes and are transcriptionally equivalent to fetal cortex (Pasca *et al* 2015). The neurons formed functional synapses that were electrophysiologically active, and participated in network activity. Brain organoids derived from human and non-human primate pluripotent stem cells are now a focus to '…answer questions about brain development and evolutionary innovation, and to gain insights into human disease' (Pasca 2018, for a review).

Cortical organoids with long survival times *in vitro* have been made from rat embryo primary cortical neurons grown on scaffolds (Tang-Schomer *et al* 2014). These organoids are electrophysiologically functional and respond to injury like their counterparts *in vivo*. Brain organoids consisting of forebrain, midbrain, choroid plexus, hippocampus, choroid plexus and retina, were self-organized from human neurectoderm differentiated from ESC embryoid bodies and grown in drops of Matrigel (Lancaster *et al* 2013). One can imagine spinal cord organoids that could be used to bridge gaps in transected or crushed cord. Further work has shown that brain organoids can be developed over extended periods of time, generate a broad diversity of neural cell types and robust neural networks, with electrical responses to light in organoids containing retina (Quadrato *et al* 2017).

Cerebral organoids derived from hiPSCs and infected with the Zika virus develop features resembling microcephaly (Qian *et al* 2017, for a review). Since the formation of these organoids takes place on a time scale much shorter than in an actual human embryo, they are useful in deciphering the causes of microencephalopathy, which is a major feature of Zika virus infection. The same would be true as well of CNS organoids derived from the iPSCs of patients with other neural disorders.

10.2.4.6 Skin

Embryoid bodies have been generated from iPSCs that when stimulated by Wnt10b and implanted under the kidney capsule of SCID mice differentiated into an integumentary organ system (IOS) consisting of epidermis with hair follicles and sebaceous glands. The IOS tissues were then transplanted onto the backs of nude mice, where they were integrated with surrounding skin tissue with proper connections to epidermis, dermis, arrector pili muscles, fat and nerve fibers (Takagi *et al* 2016). This technology, when humanized, will have application to regenerative therapies for burns and baldness, as well as being an alternative to testing cosmetics and drugs on animals.

10.2.5 Organs generated by interspecies chimerism

Interspecies chimeras have promise for growing organs for transplant (Wu *et al* 2016, for a review). The general method is to inject PSCs of one species into blastocysts of another species where the master developmental regulatory genes for the desired organ have been deleted or suppressed. The technique has been used to differentiate a mouse rat-sized pancreas by disabling the rat *PDX-1* master pancreatic gene and injecting mouse ESCs or iPSCs into the rat blastocyst (Yamaguchi *et al* 2017). The mouse PSCs contributed to all the tissues of the rat hosts, but the pancreas of these chimeras was composed totally of mouse pancreatic cells. Islets harvested from these pancreata normalized blood sugar level for over 370 days when transplanted into STZ diabetic mice. The technique could potentially be used to grow human organs in pigs.

10.2.6 Biomimetic whole organisms

The concept of bioengineering whole adult organisms has been tested in experiments making a miniature jellyfish (Nawroth *et al* 2012). The jellyfish is basically a floating bell with tentacles that swims by repetitive contractions of muscle in the bell. Millimeter scale jellyfish organoids were produced by making an eight-lobed (arms) bell of polydimethylsiloxane (PDMS) combined with a layer of rat cardiomyocytes. These 'medusoids' could replicate jellyfish swimming behavior by cardiomyocyte-powered contraction of the lobes.

10.2.7 Appendage regeneration

The grand challenge of appendage regeneration is to achieve the regeneration of a human limb. Studies of urodele limb regeneration will continue to be a source of insights to apply to regeneration-deficient anuran limbs. Why amputated urodele appendages regenerate whereas the amputated appendages of other vertebrate taxa do not is a long-standing evolutionary question. It has been proposed that the selective factor involved is the kind of functional demands placed on the appendages, these being higher in mammals and reflected in more complex structure (Jazwinska and Sallin 2016).

A significant breakthrough for developmental and evolutionary research on regeneration is the recent sequencing of the axolotl genome (Nowoshilow *et al* 2018). Combined with CRISPR/Cas9 technology (Fei *et al* 2016), the axolotl gene sequence will more easily enable the identification and manipulation of genes essential for regeneration, and comparison of the axolotl genome with the genomes of regeneration-deficient tetrapods might give insights into why the axolotl can regenerate appendages, while others cannot.

In chapter 5, mention was made that genes for the regeneration of adult limbs may be lacking to one degree or another in non-urodelean tetrapod vertebrates. These genes could hypothetically be provided using CRISPR/Cas9 gene editing. Such an idea is already being applied by George Church of Harvard University to de-extinct the woolly mammoth by editing the genome of Asian elephant fibroblasts to include thermoregulatory mammoth genes that are adaptive for cold climate such as short ears, fat production, long hair, and high-octane hemoglobin (see Shapiro 2015). Using CRISPR/Cas9, we might use gene editing to first regenerate anuran limbs as a proof of principle. Genes key to urodele limb regeneration such as Prod1 plus others identified from the axolotl genome could first be edited into the genome of anuran limb fibroblasts, which could then be grafted in a fibrin clot to the wound surface of recipient limbs amputated at the same level from which the fibroblasts were derived. Alternatively, the transfection might be achieved *in situ* by introducing plasmids via a Biodome (see ahead). If these cells successfully support regeneration, the same could then be done with mouse digit/limb fibroblasts and if successful there, with human limb fibroblasts.

Research will increasingly focus on mammalian appendage regeneration, especially since progress has been made on understanding how mice regenerate digit tips and a regenerative response has been elicited from the mouse non-regenerating second phalange by BMP (Simkin *et al* 2015, Quijano *et al* 2016). Research on mouse digits will be extended to understanding the roles of oxygen concentration, manipulation of ECM degradation, BMP and Wnt signaling, the nail bed, and the source of the cells that form the blastema in the amputated mouse digit tip. CRISPR/Cas9 technology will be developed to test the effects of gene deletion to regenerating digit tissue and gene addition to non-regenerating tissue.

Several translational ideas have been proposed for the regeneration of a mouse digit, or possibly limb. First is the idea that in-depth understanding of the soluble factors involved in digit tip regeneration and their regulation can lead to the formulation of a molecular cocktail that initiates a regenerative cascade. This mixture could be delivered by a bioreactor such as the Biodome, a device that can be fitted over an amputated appendage and deliver growth and signaling factors, control pH, hydration, oxygenation, and provide electrical stimulation (Hechavarria *et al* 2010, Quijano *et al* 2016, Golding *et al* 2016). One can also imagine a Biodome delivering reprogramming factors such as 5-azacytidine and PDGF-AB to convert the skin fibroblasts of a mammalian appendage to multipotential mesenchymal cells that produce new limb tissue (Chandrakanthan *et al* 2016) or the making of a blastema by providing hiPSC mesenchymal derivatives that self-organize into bone, cartilage and muscle at the amputation plane and are guided in the formation of these tissues by vascularized scaffolds and soluble factors. In fact, aged human bone marrow mesenchymal stem cells subjected to specific 3D culture environments have been reported to dedifferentiate and autonomously form aggregates of cells resembling blastemas (Pennock *et al* 2015). Dedifferentiation was associated with autonomously controlled autophagy that promoted cytoplasmic remodeling, mitochondrial regression, and a bioenergetic shift from oxidative phosphorylation to anaerobic metabolism. This bioenergetic shift has been previously observed in regenerating urodele limbs (Schmidt 1968, Rao *et al* 2009, 2014).

A second idea is to implant iPSC-derived limb progenitor-like cells, plus a mixture of soluble factors (BMP2, Fgf8, Wnt3a, thymosin β4) that stimulate their proliferation and differentiation, to the non-regenerating stumps of amputated P2 digits in fibrin clots. This has been done in mice using GFP-labeled cells, with the interesting and important result that a tapering extension of bone with a marrow cavity was regenerated consisting of both donor and host cells (Chen *et al* 2017) (figure 10.6). Since amputated P2 stumps regenerate nothing when the soluble factor mix alone is implanted, it would appear that the progenitor-like donor cells recruit and organize host cells to participate in regeneration of the new bone.

A third idea being explored by the laboratory of Michael Levin at Tufts University is that during development the morphogenetic pattern of organs and appendages is laid out by a bioelectric code of resting potential gradients produced across tissues by ion channels/pumps and linked into networks by gap junctions. This pattern maps to the epigenetic codes and patterns of gene activity associated with development. For tail and limb regeneration the bioelectric code is a memory system that reproduces the original code and thus the original transcription program and anatomical structure. In a series of articles, Levin and colleagues have described the bioelectric code and provided extensive evidence for its existence and function.

In addition, they have provided many examples of manipulation of the code that rescue birth defects, reprogram tumors, and result in the rearrangement of large-scale pattern (Levin 2011, 2013, Tseng and Levin 2012, 2013, Mustard and Levin 2014, Pezzulo and Levin 2015). For example, a number of cell membrane channels associated with eye formation in *Xenopus* embryos induced eye formation in the gut, tail, or lateral plate mesoderm when misexpressed in these regions. Furthermore, the induction of H^+ efflux/Na^+ influx by the sodium ionophore monensin initiated the whole cascade of events leading to tail regeneration during a regeneration-refractory period of the tail bud during *Xenopus* development, and also to digit regeneration in stage 57 limbs (chapter 5).

A fourth idea is to bioengineer a limb by providing a scaffold with compartments for each tissue type of the limb arranged in the exact shape of the limb tissues and provided with vascular channels. Proof of concept for this idea has been provided by experiments in which skinned rat and primate limbs were decellularized leaving the skeleton and the ECM of muscles, tendons, ligaments, nerves and blood vessels. Repopulation of the scaffold was accomplished by injecting the muscle compartment with myoblasts and fibroblasts and the vascular compartment with endothelial cells (Jank *et al* 2015). The composite was then cultured in a bioreactor for 10 days and skin added, followed by further culture to 21 days. Histological analysis showed

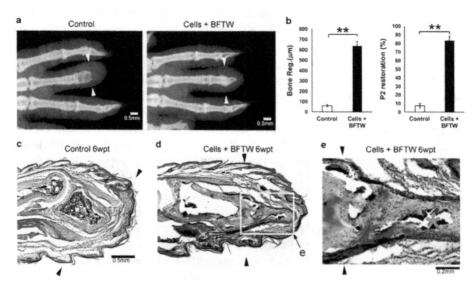

Figure 10.6. Effect of implanting limb bud-like progenitor cells in a fibrin clot with a mixture of BMP/Fgf/Wnt/thymosin β4 under the wound epithelium of a mouse digit amputated midway through the second (normally non-regenerating) phalange. (A) Left: control, either no treatment or growth factors alone. Right: cells plus growth factor combination, showing regeneration of a tapered bone extension. (B) Quantitative difference in length of regenerating bone between control and implanted phalange. (C) Histological structure of control phalange at six weeks post-clot transplant showing lack of any regeneration, compared to (D) cell-implanted phalange showing regenerated bone (box) with marrow cavity. (E) Boxed area in (D) at higher magnification with the marrow cavity marked with a red asterisk. Arrowheads indicate plane of amputation. Reproduced from (Chen *et al* 2017). Nature Publishing Group, Open Access.

that muscle and endothelium were present and electrical stimulation of the muscle indicated a force generation of about 80% that of neonatal muscle. The construct was grafted to a host animal and demonstrated blood flow from host to graft, and contraction of muscles after electrical stimulation. To use this concept clinically would require the printing of a three-dimensional artificial limb scaffold from digitized medical images and populating it with the various cell types that make up the limb. Alternatively, a prosthetic appendage with neural interface could be combined with organic muscle and vascular system to nourish the muscle, creating a cyborg limb.

10.3 Bioethics in regenerative medicine

There has always been opposition to scientific and medical advances, often based on ethical and/or religious grounds. For example, the theory of evolution has been opposed since its inception, even though there is now over 150 years of paleonto-logical, anatomical and molecular evidence for its correctness. In the 1950s, there initially were deep divisions within the medical profession and the public over the feasibility and ethics of organ transplantation. Likewise, there were opposing views as to the ethics of inoculating children with the killed polio vaccine developed by Jonas Salk. In the 1970s, a major issue was the birth of children by *in vitro* fertilization. Today, in addition to anti-evolutionists, there are anti-vaxers and resistance to genetically modified crops based on false or misinterpreted informa-tion, even though the evidence is overwhelming that all of these things are safe, have benefitted society, and are accepted by the overwhelming majority. Furthermore, they are governed by strict rules and protocols that are part of the formal field of bioethics.

However, the evidence is not yet overwhelming that the powerful technologies discussed in this chapter are safe to apply clinically. We faced a similar situation in the 1970s over the use of recombinant DNA to alter the genome. That was resolved at a 1976 conference in Asilomar, California, which resulted in the drafting of guidelines for use of the technology. Today the uses of powerful biomedical technologies such as CRISPR are routinely debated in the light of potential ethical and moral problems associated with their outcomes, some of which are outlined ahead. For an excellent review of the bioethical issues faced in regenerative medicine, see King *et al* (2009).

10.3.1 Pluripotent stem cell research

To make hESCs, the ability of the blastocyst to develop into a complete organism is destroyed, even though its cells live on. The fact that the embryo as an integrated whole is destroyed to create hESCs has engendered a debate about the moral status of the embryo that has made research on hESCs part of the politics of abortion (Stocum 2012). On the one extreme are those who want all research on hESCs banned. No research on existing cell lines should be permitted, and no new hESC lines should be derived from leftover blastocysts produced by *in vitro* fertilization. The argument against all hESC research is based on the premise that the embryo has the moral right to existence, because from the moment of conception the fertilized egg is a genetically distinct potential human life, a person. The making of designer blastocysts by SCNT would be considered particularly immoral, because it is abortion for a selfish end. Another fear that has been raised is that SCNT will inevitably lead to human reproductive cloning, although there is virtually total agreement in the scientific and political world that such cloning should be banned, if it is even possible.

The counter argument is that, although it represents a potential human, the blastocyst is not a sentient person. Its destruction is not abortion because that term is medically defined as termination of a pregnancy after implantation, and the blastocyst has not yet implanted. Furthermore, the blastocyst has not attained personhood because it still has the potential to form duplicate embryos and is not sentient, i.e. does not have the capacity to register sensory input from the environment, a capacity that is not attained until at least eight weeks post-fertilization. Thus, it would be immoral not to conduct research on hESCs that has the potential to alleviate suffering. The cost to society in terms of lost economic productivity and to patients and their families in terms of physical and emotional suffering, is so enormous that if there is a moral conflict between the right of the embryo to exist and the needs of patients, the moral rights of sentient humanity should prevail.

The ability to make hiPSCs circumvents most of the ethical objections associated with ESCs, but we are not yet in a position where we can use derivatives of these cells for clinical purposes. Making autogeneic hiPSCs is an expensive proposition, whether it is done directly or indirectly, requiring expensive facilities, technology and expertise, and there are other biological issues to be resolved that will add to the cost. The derived cells for transplant must be shown to be as functionally robust as the originals, and over a long period of time. Prior to transplant, they must be shown to be a 100% differentiated population; inclusion of any undifferentiated hiPSCs could lead to teratoma formation within the host tissue. The establishment of biobanks of allogeneic hiPSCs that have passed these tests and represent a wide range of immune-compatibility could theoretically avoid the need to repeat the procedure each and every time cells are needed, but the costs of maintaining such banks would be considerable.

10.3.2 Interspecies chimeras

In 1896, H G Wells published *The Island of Dr Moreau*, the story of a vivisectionist who surgically created chimeras of men and beasts. This was not all that far-fetched, for the embryologists soon showed that it is possible to make such species chimeras by embryonic grafting techniques. Experimental embryologists have long used interspecies grafts of cells that differ in some marker to follow the developmental fate of the grafted cells, giving valuable information about the origin of the different parts of the embryo and the mechanisms of their development. Human embryonic cells can also be grafted to the embryos of other species, and interspecies chimeras created by SCNT.

Earlier in this chapter the development of a rat-sized mouse pancreas in a rat host was described, but there have also been examples of interspecies grafts involving human neural stem cells (Stocum 2012). Making interspecies chimeras, either by grafting or SCNT, is a serious bioethical issue, particularly in non-human primates. Human NSCs have been grafted to the brain ventricles of fetal bonnet monkeys (*Macaca radiata*). The cells migrated into the cortex and subventricular zone, where they differentiated into cortical neurons and glia, and contributed to the NSC population (Ourednik *et al* 2001). No chimeric animals have been brought to term, however. Such experiments have potential benefits for discovering how neural progenitors make a brain, but the downside is the potential alteration (presumably an increase) of cognitive capacity if large numbers of human neural progenitors are grafted. One can even imagine an experiment in which the whole brain of a spontaneously aborted human embryo is grafted in place of the brain of an embryonic non-human primate. The same would apply to interspecies chimeras derived by SCNT, or embryos formed by mixing the blastomeres of two species. For these reasons, the National Academy of Sciences has recommended that such experiments be subject to special review (Greene *et al* 2005).

10.3.3 CRISPR concerns

The tremendous power of gene editing by CRISPR and its applications to medicine and human reproduction has generated debate on whether and how this technology should be used in people (Ledford 2016). For example, Bredenoord (2017) have written that '…ethical approaches are needed for the first clinical transplantations of organoids, particularly when organoid technologies are combined with gene editing.' Even though the applications may be years away, the US Academies of Sciences, Engineering and Medicine have compiled a report (The National Academies Press 2017) on human applications of genetic engineering. The Nuffield Council on Bioethics completed a report on ethical questions related to human reproduction in 2017, and an independent group of European ethicists wants a steering committee formed to ensure the safety and reliability of CRISPR technology before being used for medical purposes. In the meantime, the technology to produce gene-edited livestock is ready to be implemented, which will further inflame the controversy over GMO foods.

CRISPR technology is part of the toolbox of synthetic biology, a science that seeks to modify existing genomes in ways that augment cell functions, regenerate appendages, de-extinct extinct creatures, or even to create new organisms. In addition to medical applications, CRISPR technology could ultimately advance to the point where it might be possible to make designer babies with augmented mental and physical capabilities that confer immense advantages, a theme that is a staple of the science fiction genre, but which could become science reality. How should such power, that could lead to class distinctions of a quality and magnitude never before seen, be handled morally and ethically? The same can be said of the potential convergence of augmentation biology with artificial intelligence (AI), the creation of which is being intensely pursued. What kind of beings might emerge from this research? Once evolution has generated a biological organism with analytical and creative intelligence (*Homo sapiens*), is an inevitable next step the creation by that intelligence of life forms that have the potential to exceed—or succeed—it?

10.3.4 Cutting Gordian knots

I have suggested (Stocum 2012) that in the final analysis, the different moral views on biomedical research are Gordian knots not fully resolvable *a priori* through bioethical argument and analysis, particularly in pluralistic societies, though such arguments and analyses are absolutely necessary. Such knots can be cut only by an assessment of the utility of an idea or technology that weighs its potential benefits against its potential costs, including those costs that might have a demoralizing or dehumanizing effect. Historically, the only measure that determines whether a controversial technology or idea survives and is adopted by society is whether or not it provides a high benefit to cost ratio across a population. That utility must always be measured within a framework of moral principles that includes a thorough understanding and assessment of the potential outcomes of the technology, even though we cannot predict with certainty what the long-term effects of our decisions would be. But if new technologies prove beneficial to enough people and do not harm the rest, they will eventually be accepted and become part of our social and moral fabric, even if at first ridiculed or opposed, as were organ transplants and *in vitro* fertilization. Today's Frankenstein becomes tomorrow's benefactor. Alternatively, the unacceptable will generate an alternative route to the goal. A good example is the development of hiPSC technology to generate pluripotent stem cells without destroying embryos. Even so, a very real danger is that the ultimate acceptance of a technology may in the end benefit only those wealthy enough to take advantage of it, leaving the vast majority of the population without access. This was the theme of the science fiction film *Elysium*, and is a reality today in the dubious pricing strategies of our pharmaceutical companies, which deny access even to life-saving drugs that are cheap to manufacture, by raising the prices of these drugs by double and triple digit percentages to unaffordable levels (Quigley 2017).

10.3.5 Cashing in on hype

Regenerative medicine is a field with a high 'wow' factor, but its current capabilities are over-hyped by almost everyone associated with it. While its promise is great, most of regenerative medicine has yet to reach a clinically acceptable level. Unfortunately, the promise of regenerative medicine is falsely advertised by some medical practitioners as clinical fact, with safe and proven treatments, particularly with regard to stem cell therapies (Sugarman 2008).

There are nearly 600 unregulated regenerative medicine centers across the US and many more worldwide. A study of US stem cell clinic websites has been conducted to determine what kinds of adult stem cell therapies are being advertised, types of stem cells being marketed, and conditions treated (Turner and Knoepfler 2016). The top five types of conditions treated were orthopedic, pain, sports injuries, neurological and immune. The top five stem cell sources for treatment were adipose cells, bone marrow cells, amnion stem cells, blood cells and placental stem cells. This study and a survey conducted by Lau *et al* (2008) concluded that most of the marketing claims for stem cell treatments have little or no peer-reviewed evidence to support them. Treatments can cost $5000–$25 000, and most are not covered by health insurance. These clinics often take advantage of people who are desperate, particularly those with neurological diseases such as Parkinson's and ALS.

An example of the problem is the case of three women, aged 72–78, who each paid $5000 to US Stem Cell, a Florida treatment center operating under the previous name of Bioheart, Inc (Begley 2017). Each woman received injections into both eyes of a mixture of enzymatically dissociated autogeneic adipose cells (whether these were purified is unknown) and platelet-enriched plasma to improve their vision. There was no preclinical evidence that injection of fat cells into eyes is safe or effective, nor was it clear what the injections would accomplish. Injection into both eyes at once is always considered risky, even in the case of lens replacement for cataracts, which is the most established and successful eye surgery done today. Whereas previously the women only had problems reading small print, they now became essentially blind, with detached retinas and hemorrhages. This company initially had a clinical trial listed in a government database, but the trial was withdrawn before enrolling anyone.

Patients who are looking for *bona fide* experimental treatments should go to the clinicaltrials.gov database. Legitimate trials from whatever source can be recognized by the fact that they do not generally charge patients for treatment, whereas suspect trials do.

10.4 Concluding statement

We are entering a new age of research that will see the convergence of new technologies from biology, chemistry, engineering and informatics science to further understand how tissues, organs and appendages regenerate naturally and to develop new and more effective therapies for regenerative medicine. For students looking to enter a field that is deeply interesting for its biology while having clear potential applications to alleviating the ills and injuries of human life, as well as for resolution of ethical, moral and legal issues, look no further than regenerative biology and medicine!

References

Ambasudhan R *et al* 2011 Direct reprogramming of adult human fibroblasts to functional neurons under defined conditions *Cell Stem Cell* **9** 113–8

Ariyachet C *et al* 2016 Reprogrammed stomach tissue as a renewable source of functional β cells for blood glucose regulation *Cell Stem Cell* **18** 410–21

Aurora M and Spence J R 2016 hPSC-derived lung and intestinal organoids as models of human fetal tissue *Dev. Biol.* **42** 2–238

Barbosa J S *et al* 2015 Live imaging of adult neural stem cell behavior in the intact and injured zebrafish brain *Science* **348** 789–93

Barker N, Bartfield S and Clevers H 2010 Tissue-resident adult stem cell populations of rapidly self-renewing organs *Cell Stem Cell* **7** 656–70

Bassuk A G *et al* 2016 Precision medicine: genetic repair of retinitis pigmentosa in patient-derived stem cells *Sci. Rep.* **6** 19969

Begley S 2017 Three patients blinded by stem cell procedure, physicians say *STAT* March 15

Bauer A J P, Song F, Windsor L J and Li B 2014 Current development of collagen-based biomaterials for tissue repair and regeneration *Soft Mater.* **12** 359–70

Bhatia S N and Ingber D E 2014 Microfluidc organs-on-chips *Nat. Biotechnol.* **32** 760–72

Birling M-C, Herault Y and Pavlovic G 2017 Modeling human disease in rodents CRISPR/Cas9 genome editing *Mamm. Genome* **28** 291–301

Black J B *et al* 2016 Targeted epigenetic remodeling of endogenous loci by CRISPR/Cas9-based transcriptional activators directly converts fibroblasts to neuronal cells *Cell Stem Cell* **19** 406–14

Blakely A M, Manning K L, Tripathi A and Morgan J R 2014 Bio-Pick, Place and Perfuse: a new instrument for 3D tissue engineering *Tissue Eng.* C **21** 737–46

Boj S F *et al* 2015 Organoid models of human and mouse ductal pancreatic cancer *Cell* **160** 324–38

Bourzac K 2013 Beyond the limits *Nature* **525** S50–4

Bowles R D, Gebhard H H, Harti R and Bonassar L J 2011 Tissue-engineered intervertebral discs produce new matrix, maintain disc height and restore biomechanical function to the rodent spine *Proc. Natl Acad. Sci. USA* **108** 13106–11

Bredenoord A L 2017 Human tissues in a dish: the research and ethical implcations of organoid technology *Science* **355** 260

Brennand K J *et al* 2011 Modelling schizophrenia using human induced pluripotent stem cells *Nature* **473** 221–5

Caiazzo M *et al* 2011 Direct generation of functional dopaminergic neurons from mouse and human fibroblasts *Nature* **476** 224–7

Camp C and Cicerone M T 2017 A larger palette for biological imaging *Nature* **544** 423–4

Cao N, Huang Y and Zheng J 2016 Conversion of human fibroblasts into functional cardiomyocytes by small molecules *Science* **352** 1216–20

Carter R L *et al* 2014 Reversal of cellular phenotypes in neural cells derived from Huntington's disease monkey-induced pluripotent stem cells *Stem Cell Rep.* **3** 585–93

Chandrakanthan V *et al* 2016 PDGF-AB and 5-azacytidne induce conversion of somatic cells into tissue-regenerative multipotent stem cells *Proc. Natl Acad. Sci. USA* **113** E2306–15

Charpentier E and Doudna J A 2013 Rewriting a genome *Nature* **495** 50–1

Chen Y, Xu H and Lin G 2017 Generation of iPSC-derived limb progenitor-like cells for stimulating phalange regeneration in the adult mouse *Cell Discov.* **3** 17046

Chen Y, Yang Z, Zhao Z-A and Shen Z 2017 Direct reprogramming of fibroblasts into cardiomyocytes *Stem Cell Res. Therap.* **8** 118

Chen B, Xiao Z, Zhao Y and Dai J 2018 Functional biomaterial-based regenerative microenvironment for spinal cord injury repair *Natl Sci. Rev.* **4** 530–2

Chin J W 2017 Expanding and reprogramming the genetic code *Nature* **550** 53–8

Christ G J, Saul J M, Furth M E and Anderson K-E 2013 The pharmacology of regenerative medicine *Pharmacol. Rev.* **65** 1091–133

Chung K *et al* 2013 Structural and molecular interrogation of intact biological systems *Nature* **497** 332–7

Clevers H 2016 Modeling development and disease with organoids *Cell* **165** 1586–97

Cohen J 2016 Mice made easy *Science* **354** 539–42

Corson T W and Crews C M 2007 Molecular understanding and modern application of traditional medicines: triumphs and trials *Cell* **130** 769–74

Cyranowski D 2018 The cells that sparked a revolution *Nature* **555** 428–30

Denner J 2017 Advances in organ transplant from pigs *Science* **357** 1238–9

Dever D P *et al* 2016 CRISPR/Cas9 globin gene targeting in human haematopoietic stem cells *Nature* **539** 384–9

Dimos J T *et al* 2008 Induced pluripotent stem cells generated from patients with ALS can be differentiated into motor neurons *Science* **321** 1218–21

Dorrell C *et al* 2014 The organoid-initiating cells in mouse pancreas and liver are phenotypically and functionally similar *Stem Cell Res.* **13** 275–83

Doudna J A and Charpentier E 2014 The new frontier of genome engineering with CRISPR-CAS9 *Science* **346** 1258096

Duan B, Hockaday L A, Kang K H and Butcher J T 2012 3D bioprinting of heterogeneous aortic valve conduits with alginate/gelatin hydrogels *J. Biomed. Mater. Res.* A **101** 1255–64

Dye B R *et al* 2015 *In vitro* generation of human pluripotent stem cell derived lung organoids *eLife* **2015** 4e05098

Ebert A D *et al* 2009 Induced pluripotent stem cells from a spinal muscular atrophy patient *Nature* **457** 277–80

Eckel Z C *et al* 2016 Additive manufacturing of polymer-derived ceramics *Science* **351** 58–62

Eiraku M *et al* 2011 Self-organizing optic-cup morphogenesis in three-dimensional culture *Nature* **472** 5156

Emborg M E *et al* 2013 Induced pluripotent stem cell-derived neural cells survive and mature in the non-human primate brain *Cell Rep.* **3** 646–50

Fei J-F *et al* 2016 Tissue-and time-directed electroporation of Cas9 protein-gRNA complexes *in vivo* yields efficient multigene knockout for studying gene function in regeneration *Npj Reg. Med.* **1** 16002

Feng W *et al* 2015 The potential of the combination of CRISPR/Cas9 and pluripotent stem cells to provide human organs from chimaeric pigs *Int. J. Mol. Sci.* **16** 6545–56

Fessenden M 2016 Metabolomics: small molecules, single cells *Nature* **540** 153–5

Freedman B S *et al* 2015 Modelling kidney disease with CRISPR-mutant kidney organoids derived from human pluripotent epiblast spheroids *Nat. Commun.* **6** 8715

Gasiunas G, Barrangou R, Horvath P and Siksnys V 2012 Cas-crRNA ribonucleoprotein complex mediates specific DNA cleavage for adaptive immunity in bacteria *Proc. Natl Acad. Sci. USA* **109** E2579–86

Gjorevski N *et al* 2016 Designer matrices for intestinal stem cell and organoid culture *Nature* **519** 560–4

Golding A *et al* 2016 A tunable silk hydrogel device for studyng limb regeneration in adult *Xenopus laevis PLoS One* **11** e0155618

Grayson W L *et al* 2010 Enginering anatomically shaped human bone grafts *Proc. Natl Acad. Sci. USA* **107** 3299–304

Greene M *et al* 2005 Moral issues of human—non-human primate neural grafting *Science* **309** 385–6

Grikscheit T C *et al* 2002 Tissue engineered colon exhibits function *in vivo Surgery* **132** 200–4

Guye P *et al* 2016 Genetically engineering self-organization of human pluripotent stem cells into a liver bud-like tissue using Gata6 *Nat. Comm.* **7** 10243

Hakalien A M *et al* 2002 Reprogramming fibroblasts to express T-cell functions using cell extracts *Nat. Biotechnol.* **20** 460–6

Hallet P J *et al* 2015 Successful function of autologous iPSC-derived dopamine neurons following transplantation in a non-human primate model of Parkinson's disease *Cell Stem Cell* **16** 269–74

Harrison S E *et al* 2017 Assembly of embryonic and extraembryonic stem cells to mimic embryogenesis *in vitro Science* **356** 153

Hayashi R *et al* 2016 Co-ordinated ocular development from human iPS cells and recovery of corneal function *Nature* **531** 376–80

He X *et al* 2012 Synthetic homeostatic materials with chemo-mechanical-chemical self-regulation *Nature* **487** 214–8

Hechavarria D *et al* 2010 BioDme regenerative sleeve for biochemical and biophysical stimulation of tissue regeneration *Med. Eng. Phys.* **32** 1065–73

Hedges E C 2016 The use of stem cells to model amyotropic lateral sclerosis and frontotemporal dementia from basic research to regenerative medicine *Stem Cells Int.* **2016** 9279516

Hockaday L A, Duan B, Lang K H and Butcher J T 2014 3D-printed hydrogel technologies for tissue-engineered heart valves *3D Printing* **1** 122–36

Huang P *et al* 2014 Direct reprogramming of hman fibroblasts to functional and expandable hepatocytes *Cell Stem Cell* **14** 370–84

Huch M *et al* 2013 *In vitro* expansion of single Lgr5$^+$ liver stem cells induce by Wnt-driven regeneration *Nature* **494** 247–50

Ieda M *et al* 2010 Direct reprogramming of fibroblasts into functional cardiomyocytes by defined factors *Cell* **142** 375–86

Israel M A *et al* 2012 Probing sporadic and familial Alzheimer's disease using induced pluripotent stem cells *Nature* **482** 216–20

Jain A *et al* 2017 CRISPR-Cas9-based treatment of myocilin-associated glaucoma *Proc. Natl Acad. Sci. USA* **114** 11199–204

Jank B J *et al* 2015 Engineered composite tissue as a bioartificial limb graft *Biomaterials* **61** 246–56

Jazwinska A and Sallin P 2016 Regeneration vs scarring in vertebrate appendages and heart *J. Pathol.* **238** 233–46

Jeon I *et al* 2012 Neuronal properties, *in vivo* effects, and pathology of a Huntington's disease patient-derived induced pluripotent stem cells *Stem Cells* **30** 2054–62

Kaihara S *et al* 2000 Long-term follow-up of tissue engineered intestine after anastomosis to native small bowel *Transplantation* **69** 1927–32

Kang H-W *et al* 2016 A 3D bioprinting system to produce human-scale tissue constructs with structural integrity *Nat. Biotechnol.* **34** 312–9

Kauffman S A 2008 *Reinventing the Sacred* (New York: The Persons Book Group)

Ke M-T, Fujimoto S and Imai T 2013 SeeDB a simple and morphology-preserving optical clearing agent for neuronal circuit reconstruction *Nat. Neurosci.* **16** 1154–61

Keller P J *et al* 2008 Reconstruction of zebrafish early embryonic development by light-sheet microscopy *Science* **322** 1065–9

Kelley M L *et al* 2016 Versatility of chemically synthesized guide RNAs for CRISPR-Cas9 genome editing *J. Biotechnol.* **233** 74–83

Kiani S *et al* 2015 Cas9 gRNA engineering for genome editing, activation and repression *Nat. Methods* **12** 1051–4

King N M P, Coughlin C N and Furth M E 2009 Ethical issues in regenerative medicine *Wake Forest Intellectual Property Journal* **9** 1380162

Kitada T, DiAndreth B, Teague B and Weiss R 2018 Programming gene and engineered-cell therapies with synthetic biology *Science* **359** 651

Koch L *et al* 2012 Skin tissue generation by laser cell printing *Biotechnol. Bioeng.* **109** 1855–63

Kriks S *et al* 2011 Dopamine neurons derived from human ES cells efficiently engraft in animal models of Parkinson's disease *Nature* **480** 547–51

Lancaster M A *et al* 2013 Cerebral organoids model human brain development and microcephay *Nature* **501** 373–9

Lancaster M A and Knoblich J A 2014 Organogenesis in a dish modeling development and disease using organoid technologies *Science* **345** 283

Lau D *et al* 2008 Stem cell clinics online: the direct-to-consumer portrayal of stem cell medicine *Cell Stem Cell* **3** 591–4

Ledford H 2016 CRISPR concerns *Nature* **538** 17

Lee G *et al* 2009 Modeling pathogenesis and treatment of familial dysautonomia using patient-specific iPSCs *Nature* **461** 402–808

Lee J *et al* 2013 Expansion and conversion of human pancreatic ductal cells into insulin-secreting endocrine cells *Elife* **2** e00940

Levin M 2011 The wisdom of the body: future techniques and approaches to morphogenetic fields in regenerative medicine, developmental biology and cancer *Future Med.* **6** 667–73

Levin M 2013 Reprogramming cells and tissue patterning via bioelectrical pathways: molecular mechanisms and biomedical opportunities *WIREs Syst. Biol. Med.* **5** 657–76

Liang P *et al* 2017 Correction of β-thalassemia mutant by base editor in human embryos *Protein Cell* **8** 811–22

Lin H *et al* 2016 Lens regeneration using endogenous stem cells with gain of visual function *Nature* **531** 323–8

Lis R *et al* 2017 Conversion of adult endothelium to immunocompetent haematopoietic stem cells *Nature* **545** 439–45

Liu G H *et al* 2011 Recapitulation of premature aging with iPSCs from Hutchinson–Gilford progeria syndrome *Nature* **472** 221–5

Livoti C M and Morgan J R 2009 Self-assembly and tissue fusion of toroid-shaped minimal building units *Tissue Eng.* A **16** 2051–61

Long C *et al* 2014 Prevention of muscular dystrophy in mice by CRISPR/Cas9-mediated editing of germline DNA *Science* **345** 1184–8

Lu G, Ling K, hao P, Xu Z, Deng C, Zheng H, Huang J and Chen J 2010 A novel *in situ*-formed hydrogel wound dressing by the photocross-linking of a chitosan derivative *Wound Rep. Reg.* **18** 70–9

Ma H *et al* 2017 Correction of a pathogenic gene mutation in human embryos *Nature* **548** 413–9

Mali P *et al* 2013 RNA-guided human genome engineering via Cas9 *Science* **339** 823–6

Mandai M *et al* 2017 Autologous induced stem-cell-derived retinal cells for macular degeneration *New Eng. J. Med.* **376** 1038–46

Manoor M S *et al* 2013 3D printed bionic ears *Nano Lett.* **13** 2634–9

Marchetto M C N *et al* 2010 A model for neural development and treatment of Rett syndrome using human induced pluripotent stem cells *Cell* **143** 527–39

McCracken K W *et al* 2014 Modelling human development and disease in pluripotent stem-cell-derived gastric organoids *Nature* **516** 400–9

Mianne J *et al* 2016 Correction of the auditory phenotype in C57BL/6N mice via CRIAPR/Cas9-mediated homology directed repair *Genome Med.* **8** 16

Michael S *et al* 2013 Tissue engineered skin substitutes created by laser-assisted bioprinting form skin-like structures in the dorsal skin fold chamber in mice *PLoS One* **8** e57741

Mironov V *et al* 2009 Organprinting: tissue spheroids as building blocks *Biomaterials* **30** 2164–74

Moldovan N I, Hibino N and Nakayama K 2017 Principles of the Kenzan method for robotic cell spheroid-based three-dimensional bioprinting *Tissue Eng.* B **23** 237–44

Mondrinos M J, Jones P L, Finck C M and Leikes 2014 Engineering *de novo* assembly of fetal pulmonary organoids *Tissue Eng.* A **20** 2892–907

Moosmann J *et al* 2013 X-ray phase-contrast *in vivo* microtomography probes new aspects of *Xenopus* gastrulation *Nature* **494** 374–7

Morizane R *et al* 2015 Nephron organoids derived from human pluripotent stem cells model kidney development and injury *Nat. Biotechnol.* **33** 1193–200

Murphy S and Atala A 2014 3D printing of tissues and organs *Nat. Biotechnol.* **8** 773–85

Mustard J and Levin M 2014 Bioelectrical mechanisms for programming growth and form: taming physiological networks for soft body robotics *Soft Robot.* **1** 169–91

Nawroth J C *et al* 2012 A tissue engineered jellyfish with biomimetic propulsion *Nat. Biotechnol.* **10** 792–7

Nelson C E *et al* 2016 *In vivo* genome editing improves muscle function in a mouse model of Duchenne muscular dystrophy *Science* **351** 403–7

Niu D *et al* 2017 Inactivation of porcine endogenous retrovirus in pigs using CRISPR-Cas9 *Science* **357** 1303–7

Noguchi T K *et al* 2015 Generation of stomach tissue from mouse embryonic stem cells *Nat. Cell Biol.* **17** 984–93

Norotte C, Marga F S, Niklason L E and Forgacs G 2009 Scaffold-free vascular tissue engineering using bioprinting *Biomaterials* **30** 5910–7

Nowoshilow S *et al* 2018 The axolotl genome and the evolution of key tissue formation regulators *Nature* **554** 50–5

Nyberg E *et al* 2017 Coparison of 3D printed poly-ε-caprolactone scaffolds functionalized with tricalcium phosphate, hydroxyapatite, Bio-Oss, or decellularized bone matrix *Tissue Eng.* A **23** 503–14

Ourednik V *et al* 2001 Segregation of human neural stem cells in the developing primate forebrain *Science* **293** 1820–4

Paquet D *et al* 2016 Efficient introduction of specific homozygous and heterozygous mutatons using CRISPR/Cas9 *Nature* **533** 125–9

Pasca A M *et al* 2015 Functional cortical neurons and strocytes from human pluripotent stem cells in 3D culture *Nat. Methods* **12** 671–8

Pasca S P 2018 The rise of three-dimensional human brain cultures *Nature* **553** 437–45

Passier R, Orlova V and Mummery C 2016 Complex tissue and disease modeling using hiPSCs *Cell Stem Cell* **18** 309–21

Peplow M 2017 The next big hit in Molecule Hollywood *Nature* **544** 408–10

Pezzulo G and Levin M 2015 Re-membering the body: applications of computational neuroscience to the top-down control of regeneration of limbs and other complex organs *Integr. Biol.* **7** 1487

Pakulska M M, Miersch S and Shoichet M S 2016 Designer protein delivery: from natural to engineered affinity-controlled release systems *Science* **351** 1279

Pang Z *et al* 2011 Induction of human neuronal cells by defined transcription factors *Nature* **476** 220–3

Paquet D *et al* 2016 Efficient introduction of specific homozygous and heterozygous mutations using CRISPR/Cas9 *Nature* **533** 125–9

Pasca A M *et al* 2015 Functional cortical neurons and strocytes from human pluripotent stem cells in 3D culture *Nat. Methods* **12** 671–78

Passier R, Orlova V and Mummery C 2016 Complex tissue and disease modeling using hiPSCs *Cell Stem Cell* **18** 309–21

Pelled G *et al* 2010 Direct gene therapy for bone regeneration: gene delivery, animal models, and outcome measures *Tissue Eng. Part B* **16** 13–20

Pennock R *et al* 2015 Human cell differentiation in mesenchymal condensates through controlled autophagy *Sci. Rep.* **5** 13113

Qian X *et al* 2017 Using brain organoids to understand Zika virus-induced microcephaly *Development* **144** 952–7

Qiang L *et al* 2011 Directed conversion of Alzheimer's disease patient skin fibroblasts into functional neurons *Cell* **146** 359–71

Quadrato G *et al* 2017 Cell diversity and network dynamics in photosensitive human brain organoids *Nature* **545** 48–53

Quijano L M *et al* 2016 Looking ahead to engineering epimorphic regeneration of a human digit or limb *Tissue Eng. B* **22** 1–12, 251–62

Quigley F 2017 *Prescription for the People* (Ithaca, NY: Cornell University Press)

Rafii S, Butler J M and Ding B-S 2016 Angiocrine functions of organ-specific endothelial cells *Nature* **529** 316–25

Rao N *et al* 2009 Proteomic analysis of blastema formation in regenerating axolotl limbs *BioMed Central Biol.* **7** 83

Rao N *et al* 2014 Proteomic analysis of fibroblastema formation in regenerating hind limbs of *Xenopus laevis* froglets and comparison to axolotl *BioMed Central Dev. Biol.* **14** 32

Ratner B D, Hoffman A S, Schoen F and Lemons J E 2013 Biomaterials science: an evolving, multidisciplinary endeavor *Biomaterials Science: An Introduction To Materials in Medicine* (New York: Academic)

Reardon S 2015 New life for pig-to-human transplants *Nature* **527** 152–4

Richard J-P and Maragakis N J 2015 Induced pluripotent stem cells from ALS patients for disease modeling *Brain Res.* **1607** 15–25

Ridley M 2015 *The Evolution of Everything* (London: Fourth Estate)

Roe J-S *et al* 2017 Enhancer reprogramming promotes pancreatic cancer metastasis *Cell* **170** 875–88

Roost M S *et al* 2015 KeyGenes, a tool to probe tissue differentiation usng a huan fetal transcriptional atlas *Stem Cell Rep.* **4** 1112–24

Sala F G *et al* 2011 A multicellular approach forms a significant amount of tissue-engineered small intestine in the mouse *Tissue Eng. A* **17** 1841–50

Sasai Y 2013 Next-generation regenerative medicine:organogenesis from stem cells in 3D culture *Cell Stem Cell* **12** 520–30

Sato T *et al* 2009 Single Lgr stem cells build crypt-villus structures *in vitro* without a mesenchymal niche *Nature* **459** 262–5

Sato T and Clevers H 2013 Growing self-organizing mini-guts from a single intestinal stem cell: mechanism and applications *Science* **340** 1190–4

Savage N 2016 The promise of printing *Nature* **540** S56–7

Schmidt A J 1968 *Cellular Biology of Vertebrate Regeneration and Repair* (Chicago: University of Chicago Press)

Schukur L, Geering B, Charpin-El Hamri G and Fussenegger M 2015 Implantable synthetic cytokine converter cells with AND-gate logic treat psoriasis *Sci. Transl. Med.* **7** 318ra201

Scudellari M 2016 A decade of iPS cells *Nature* **534** 310–2

Sears N A, Seshadri Rhruv R, Dhavalikar Prachi S and Cosgriff-Hernandez E 2016 A review of three-dimensional printing in tissue engineering *Tissue Eng.* B **22** 298–310

Sekiya S and Suzuki A 2011 Direct conversion of mouse fibroblasts to hepatocyte-like cells by defied factors *Nature* **475** 390–3

Servick K 2017 Stem cell approach for cataracts challenged *Science* **356** 1318–9

Shapiro B 2015 *How to Clone a Mammoth* (Princeton, NJ: Princeton University Press)

Shimozono S *et al* 2013 Visualization of an endogenous retinoic acid gradient across embryonic development *Nature* **496** 363–6

Simkin J *et al* 2015 The mammalian blastema: regeneration at our fingertips *Regeneration* **2** 93–105

Singh S 2007 From exotic spice to modern drug? *Cell* **130** 765–8

Skardal A, Zhang J and Prestwich G D 2010 Bioprinting vessel-like constructs using hyaluronan hydrogels crosslinked with tetrahedral polyethylene glycol tetracyclates *Biomats* **31** 6173–81

Skardal A *et al* 2012 Bioprinted amniotic fluid-driven stem cells accelerate healing of large skin wounds *Stem Cells Transl. Med.* **11** 792–802

Skardal A, Shupe T and Atala A 2016 Organoid-on-a-chip and body-on a chip systems for drug screening and disease modeling *Drug Discov. Today* **21** 1399–411

Song S, Chen Y, Yan Z, Fenniri H and Webster T J 2011 Self-assembled rosette nanotubes for incorporating hydrophobic drugs in physiological environments *Int. J. Nanomed.* **6** 101–7

Spence J R *et al* 2011 Directed differentiation of human pluripotent stem cells into intestinal tissue *in vitro Nature* **470** 105–9

Srivastava D and DeWitt N 2016 *In vivo* cellular reprogramming: The next generation *Cell* **166** 1386–96

Steinberg M S 1978 Cell–cell recognition in multicellular assembly: levels of specificity *Symp. Soc. Exp. Biol.* **32** 25–49

Sternad P *et al* 2015 Inverted light-sheet microscope for imaging mouse pre-implantation development *Nat. Methods* **13** 139–42

Stocum D L 2012 *Regenerative Biology and Medicine* 2nd edn (NewYork: Elsevier/Academic Press) ch 15

Sugarman J 2008 Human stem cell ethics: beyond the embryo *Cell Stem Cell* **2** 529–33

Sugimura R *et al* 2017 Haematopoietic stem and progenitor cells from human pluripotent stem cells *Nature* **545** 432–8

Tabebordbar M *et al* 2016 *In vivo* gene editing in dystrophic mouse muscle and muscle stem cells *Science* **35** 407–11

Takagi R *et al* 2016 Bioengineering a 3D integumentary organ system from iPS cells using an *in vivo* transplantation model *Sci. Adv.* **2** e1500887

Takasato M *et al* 2015 Kidney organoids from human iPS cells contain multiple lineages and model human nephrogenesis *Nature* **526** 564–8

Tang-Schomer M D *et al* 2014 Bioengineered functional brain-like cortical tissue *Proc. Natl Acad. Sci. USA* **111** 13811–6

Taniguchi D *et al* 2018 Scaffold-free trachea regeneration by tissue engineering with bio-3D printing *Interact. Cardiovasc. Thorac. Surg.* **26** 745–52

Tomer R, Khairy K, Amat F and Keller P J 2012 Quantitative high-speed imaging of entire developing embryos with simultaneous multiview light-sheet microscopy *Nat. Methods* **9** 755–63

Torbati A H and Mather P T 2016 A hydrogel-forming liquid crystalline elastomer exhibiting soft shape memory *J. Polym. Sci.* B **54** 38–52

Townes P L and Holtfreter J 1955 Directed movements and selective adhesion of embryonic amphibian cells *J. Exp. Zool.* **128** 53–120

Tseng A-S and Levin M 2012 Transducing bioelectric signals into epigenetic pathways during tadpole tail regeneration *Anat. Rec.* **295** 1541–51

Tseng A and Levin M 2013 Cracking the bioelectric code *Commun. Integr. Biol.* **6** 1–8

Tsunemoto R *et al* 2018 Diverse reprogramming codes for neuronal identity *Nature* **557** 375–80

Turner L and Knoepfler P 2016 Selling stem cells in the USA: assessing the direct-to-consumer industry *Cell Stem Cell* **19** 154–7

Udan R S *et al* 2014 Quantitative imaging of cell dynamics in mouse embryos using light-sheet microscopy *Development* **141** 4406–14

Ventola C L 2014 Medical applications for 3D printing:current and projected uses *P&T* **39** 704–11

Vierbuchen T *et al* 2010 Direct conversion of fibroblasts to functional neurons by defined factors *Nature* **463** 1035–41

Waddington S N, Privolizzi, Karda R and O'Neill H C 2016 A broad overview and review of CRISPR-Cas9 technology and stem cells *Curr. Stem Cell Rep.* **2** 9–20

Wang T *et al* 2015 Identification and characterization of essential genes in the human genome *Science* **350** 1096–101

Watson C L *et al* 2014 An *in vivo* model of human small intestine using pluripotent stem cells *Nat. Med.* **20** 1310–4

Wells J M and Spence J R 2014 How to make an intestine *Development* **141** 752–60

Wernig M *et al* 2008 *In vitro* reprogramming of fibroblasts into a pluripotent ES-cell-like state *Nature* **448** 318–24

Willyard C 2015 Rise of the organoids *Nature* **523** 520–2

Wu J *et al* 2016 Stem cells and interspecies chimeras *Nature* **540** 51–9

Yamaguchi T *et al* 2017 Interspecies organogenesis generates autologous functional islets *Nature* **542** 191–6

Yin H *et al* 2016 Therapeutic genome editing by combined viral and non-viral delivery of CRISPR system components *in vivo Nat. Biotechnol.* **34** 328–33

Yoo A S *et al* 2011 MicroRNA-mediated conversion of human fibroblasts to neurons *Nature* **476** 228–31

Yui S *et al* 2012 Functional engraftment of colon epithelium expanded *in vitro* from a single adult Lgr5$^+$ stem cell *Nat. Med.* **18** 618–23

Zalatan J G *et al* 2015 Engineering complex synthetic transcriptional programs with CRISPR RNA scaffolds *Cell* **160** 339–50

Zarbin M A, Arlow T and Ritch R 2013 Regenerative nanomedicine for vision restoration *Mayo Clin. Proc.* **88** 1480–90

Zhang J *et al* 2011 A human iPSC model of Hutchinson Gilford progeria reveals vascular smooth muscle and mesenchymal stem cell defects *Cell Stem Cell* **8** 31–45

Zhang M *et al* 2016 Pharmacological reprogramming of fibroblasts into neural stem cells by signaling-directed transcriptional activation *Cell Stem Cell* **18** 653–67

Zhang B *et al* 2016 Biodegradable scaffold with built-in vasculature for organ-on-a-chip engineering and direct surgical anastomosis *Nat. Mater.* **15** 669–78

Zheng Y *et al* 2017 CRISPR-Cpf1 correction of muscular dystrophy mutations in human cardiomyocytes and mice *Sci. Adv.* **3** e1602814

Zhou Q and Melton D A 2008 *In vivo* reprogramming of adult pancreatic exocrine cells to β-cells *Nature* **455** 627–32

Zopf D A *et al* 2013 Bioresorbable airway splint created with a three-dimensional printer *New Eng. J. Med.* **368** 2043–6

CPSIA information can be obtained
at www.ICGtesting.com
Printed in the USA
BVHW011728210922
647657BV00010B/363

9 780750 319454